Animal Cell Culture

Animal Cell Culture

Essential Methods

Editor

John M. Davis

School of Life Sciences, University of Hertfordshire, Hatfield, Hertfordshire, UK

WILEY-BLACKWELL

A John Wiley & Sons, Ltd., Publication

This edition first published 2011, © 2011 by John Wiley & Sons Ltd

Wiley-Blackwell is an imprint of John Wiley & Sons, formed by the merger of Wiley's global Scientific, Technical and Medical business with Blackwell Publishing.

Registered Office
John Wiley & Sons Ltd, The Atrium, Southern Gate, Chichester, West Sussex, PO19 8SQ, UK

Other Editorial Offices:
9600 Garsington Road, Oxford, OX4 2DQ, UK
111 River Street, Hoboken, NJ 07030-5774, USA
The Atrium, Southern Gate, Chichester, West Sussex, PO19 8SQ, UK

For details of our global editorial offices, for customer services and for information about how to apply for permission to reuse the copyright material in this book please see our website at www.wiley.com/wiley-blackwell

The right of the author to be identified as the author of this work has been asserted in accordance with the UK Copyright, Designs and Patents Act 1988.

Library of Congress Cataloging-in-Publication Data

Animal cell culture : essential methods / editor John M. Davis.
     p. cm.
  Includes index.
  ISBN 978-0-470-66658-6 (pbk.)
1. Cell culture–Technique. 2. Tissue culture–Technique. I. Davis, John, 1952-
  QH585.2.A565 2011
  571.6′381–dc22   2010037008

A catalogue record for this book is available from the British Library.
This book is published in the following formats: ePDF: 978-0-470-66982-2; Wiley Online Library: 978-0-470-66981-5; ePub: 978-0-470-97563-3

Typeset in 10.5/12.5pt Times by Aptara Inc., New Delhi, India
Printed and bound in Singapore by Fabulous Printers Pte Ltd

Second Impression   2011

For Carole, Tim and Helen, with love, as always
and
In memory of Ian Pearce, who often enquired about the progress of this book, but never saw it completed – thank you for the music

Contents

4 Basic Techniques and Media, the Maintenance of Cell Lines, and Safety   **91**

John M. Davis

6 Cryopreservation and Banking of Cell Lines 185
Glyn N. Stacey, Ross Hawkins and Roland A. Fleck

7 Primary Culture of Specific Cell Types and the Establishment of Cell Lines 205
Kee Woei Ng, Mohan Chothirakottu Abraham, David Tai Wei Leong, Chris Morris and Jan-Thorsten Schantz

8 Cloning 231
John Clarke, Alison Porter and John M. Davis

Colour plates fall between pages 164 and 165.

Contributors

Mohan Chothirakottu Abraham   John Stroger Hospital of Cook County, Chicago, IL 60612, USA

John Clarke   Haemophilia Centre, St Thomas' Hospital, Westminster Bridge Road, London SE1 7EH, UK.

Sue Clarke   ImmunoBiology Limited, Babraham Research Campus, Babraham, Cambridge, CB22 3AT, UK

John M. Davis   School of Life Sciences, University of Hertfordshire, College Lane, Hatfield, Hertfordshire, AL10 9AB, UK

Janette Dillon   MedImmune, Granta Park, Cambridge, CB21 6GH, UK

Roland A. Fleck   National Institute for Biological Standards and Control, Blanche Lane, South Mimms, Hertfordshire, EN6 3QG, UK

Stephen F. Gorfien   Cell Systems Division, Life Technologies Corporation, 3175 Staley Road, Grand Island, NY 14072, USA

Colin Gray   School of Medicine and Biomedical Sciences, University of Sheffield, Beech Hill Road, Sheffield, S10 2RX, UK

Jennifer Halsall   Eden Biodesign, Speke Road, Liverpool, L24 8RB, UK

Ross Hawkins   National Institute for Biological Standards and Control, Blanche Lane, South Mimms, Hertfordshire, EN6 3QG, UK

David W. Jayme   Department of Biochemistry and Physical Sciences, Brigham Young University – Hawaii, 55-220 Kulanui Street #1967, Laie, HI 96762–1294, USA

David Tai Wei Leong   Cancer Science Institute, National University of Singapore, CeLS, 28 Medical Drive, Singapore 117456

Chris Morris   Health Protection Agency Culture Collections, Centre for Emergency Preparedness and Response, Porton Down, Wiltshire, SP4 0JG, UK

Kee Woei Ng   Division of Materials Technology, School of Materials Science & Engineering, Nanyang Technological University, 50 Nanyang Avenue, Singapore 639798

Barbara Orton Quality Assurance, Bio-Products Laboratory, Dagger Lane, Elstree, Hertfordshire, WD6 3BX, UK

Alison Porter Lonza Biologics plc, Bath Road, Slough, Berkshire, SL1 4DX, UK

Peter L. Roberts Research & Development Department, Bio-Products Laboratory, Dagger Lane, Elstree, Hertfordshire, WD6 3BX, UK

Cathy Rowe European Collection of Cell Cultures, Porton Down, Wiltshire, SP4 0JG, UK

Jan-Thorsten Schantz Department of Plastic and Hand Surgery, Klinikum rechts der Isar der Technischen Universität München, Ismaningerstrasse 22, 81675 Munich, Germany

Glyn N. Stacey National Institute for Biological Standards and Control, Blanche Lane, South Mimms, Hertfordshire, EN6 3QG, UK

Peter Thraves European Collection of Cell Cultures, Porton Down, Wiltshire, SP4 0JG, UK

Daniel Zicha Cancer Research UK, 44 Lincoln's Inn Fields, London, WC2A 3PX, UK

Preface

This book brings together what I have found – from 37 years using mammalian cells *in vitro* in both academia and industry – to be the core concepts and techniques for success in the performance of cell culture, whether it be used to perform research, to generate and use test or model systems, or to produce biopharmaceuticals or other products.

Animal Cell Culture: Essential Methods is a successor to the second edition of the highly successful *Basic Cell Culture: A Practical Approach* [1]. However, the change of both title and publisher has provided the scope to do far more than simply revise and update an existing text. While all the basics are still here, coverage has been widened by the addition of dedicated chapters on cryopreservation, on the design and use of serum- and protein-free media, and on systems for culture scale-up.

Animal (including human) cell culture is becoming ever more important, not only in academic research (e.g. stem cells, regenerative medicine) but also in industry (e.g. for the production of monoclonal antibodies, vaccines and other products, and for toxicological testing). Thus an increasing number of scientists, technicians and students will be requiring the grounding of knowledge and skills encapsulated in this book. Similarly, the growing use of cell culture within a regulatory environment, such as in the toxicity or safety testing of pharmaceuticals, cosmetics and general chemicals, means that more and more individuals will need to work within the framework of Good Laboratory Practice, which is addressed in the final chapter of this book.

Thus it is my hope that not only will all those new to cell culture find this book a useful and concise introduction to the field, but that experienced workers will also find it of continuing use as a handy source of reference to all those tips, techniques and approaches that are at the heart of cell culture and keep the laboratory running smoothly.

Finally, my unreserved thanks go to all the authors who contributed chapters to this book. Without their unstinting efforts in sharing their considerable expertise, this book would never have moved from concept to reality.

John M. Davis
Hatfield, October 2010

Reference

1. Davis, J. (ed.) (2002) *Basic Cell Culture: A Practical Approach*, 2nd edn, Oxford University Press, Oxford, UK.

Abbreviations

2D	two-dimensional
3D	three-dimensional
ACDP	Advisory Committee on Dangerous Pathogens (UK HSE)
ATCC	American Type Culture Collection
AMV	avian myeloblastosis virus
B19	parvovirus B19 (also known as erythrovirus B19)
BSA	bovine serum albumin
BSE	bovine spongiform encephalopathy
BSI	British Standards Institution
BSS	balanced salt solution
BUN	blood urea nitrogen
BVDV	bovine viral diarrhoea virus
CCD	charge-coupled device (camera) *or* central composite design (statistical technique)
CDM	chemically defined media
CFE	colony-forming efficiency
cfu	colony-forming units
CFR	Code of Federal Regulations (United States)
CHO	Chinese hamster ovary (cell line)
CJD	Creutzfeld–Jakob disease
CMF-PBS	Ca^{2+}- and Mg^{2+}-free phosphate-buffered saline
CMF-DPBS	Ca^{2+}- and Mg^{2+}-free Dulbecco's phosphate-buffered saline
CM+PS	complete medium plus penicillin and streptomycin
C of A	certificate of analysis
CPA	cryoprotective agent
cpm	counts per minute (radioactivity)
DAPI	4′,6-diamidino-2-phenylindole
DIC	differential interference contrast
DMEM	Dulbecco's modified Eagle's medium
DMF	dimethyl formamide
DMSO	dimethylsulphoxide
DNA	deoxyribonucleic acid
DNAse	deoxyribonuclease
DO	dissolved oxygen

DOE	design of experiments (statistical analysis)
DPBS	Dulbecco's PBS
DPBSA	Dulbecco's PBS A
DSMZ	Deutsche Sammlung von Mikroorganismen und Zellkulturen (German culture collection)
EBSS	Earle's balanced salt solution
ECACC	European Collection of Cell Cultures (Porton Down, UK)
ECM	extracellular matrix
EDTA	ethylenediaminetetraacetic acid
EGF	epidermal growth factor
EHS	Englebreth–Holm–Swarm (tumour)
ELISA	enzyme-linked immunosorbent assay
EMA	European Medicines Agency
EMC	encephalomyocarditis virus
EMEA	European Medicines Evaluation Agency, now renamed as the European Medicines Agency
EMEM	Eagle's minimal essential medium
ESSS	Earle's spinner salt solution
FACS	fluorescence-activated cell sorting
FBS	fetal bovine serum (= fetal calf serum)
FDA	United States Food and Drug Administration
FEP	fluorinated ethylene propylene (for most purposes, synonymous with PTFE)
FGF	fibroblast growth factor
FLAP	fluorescence localization after photobleaching
FRAP	fluorescence recovery after photobleaching
FRET	fluorescence resonance energy transfer
GCCP	good cell culture practice
GGT	gamma-glutamyl transferase
GLP	Good Laboratory Practice
GMEM	Glasgow minimal essential medium
GMO	genetically modified organism
GMP	Good Manufacturing Practice
HAV	hepatitis A virus
HBSS	Hanks' balanced salt solution
HCV	hepatitis C virus
HEPA	high efficiency particulate air (filters)
HEPES	N-(2-hydroxyethyl) piperazine-N'-(2-ethanesulphonic acid)
hES	human embryonic stem (cells)
HIV	human immunodeficiency virus
HMC	Hoffman modulation contrast
HPA	UK Health Protection Agency
HPLC	high-performance (or high-pressure) liquid chromatography
HSE	UK Health and Safety Executive
HTLV	human T-lymphotropic virus(es)

HUVECs	human umbilical vein endothelial cells
HVAC	heating, ventilation and air conditioning
IATA	International Air Transport Association
IBR	see IBRV
IBRV	infectious bovine rhinotracheitis virus
ICH	International Conference on Harmonization of Technical Requirements for Registration of Pharmaceuticals for Human Use
IGF	insulin-like growth factor
IMDM	Iscove's modified Dulbecco's medium
IPA	isopropyl alcohol (= propan-2-ol)
iPS	induced pluripotent stem cells
IRM	interference reflection microscopy
IT	information technology
IVCD	integral viable cell density
JMEM	Joklik's minimal essential medium
JPEG	Joint Photographic Experts Group (format)
LAL	*Limulus* amoebocyte lysate
LDH	lactate dehydrogenase
LFC	laminar flow cabinet
LN	liquid nitrogen
LZW	Lempel-Ziv-Welch (compression)
MAD	Mutual Acceptance of Data (agreement)
MBA	microbioreactor array
MCB	master cell bank
MEM	minimal essential medium
MHRA	Medicines and Healthcare products Regulatory Agency (UK)
MLV	murine leukaemia virus
MoMLV	Moloney murine leukaemia virus
mRNA	messenger RNA
MSC	microbiological safety cabinet (= biological safety cabinet in the USA)
MTS	3-(4,5-dimethylthiazol-2-yl)-5-(3-carboxymethoxyphenyl)-2-(4-sulfophenyl)-2H-tetrazolium, inner salt
MTT	3-[4,5-dimethylthiazol-2-yl]-2,5-diphenyl-tetrazolium bromide; this chemical is also known as thiazolyl blue tetrazolium bromide
MVM	minute virus of mice
NA	numerical aperture
NCTC	National Collection of Type Cultures (Porton Down, UK)
NIBSC	National Institute for Biological Standards and Control (South Mimms, UK)
OECD	Organization for Economic Cooperation and Development
PBS	phosphate-buffered saline
PCR	polymerase chain reaction
PDL	population doubling level
PDT	population doubling time
PEG	polyethylene glycol

PES	polyethersulphone
PET	poly(ethylene) terephthalate
PETG	poly(ethylene) terephthalate glycol
PFM	protein-free media
PI	propidium iodide
PI-3	parainfluenza-3 (virus)
PMS	phenazine methosulphate
PPLO	pleuropneumonia-like organisms (= mycoplasma)
PTFE	poly(tetrafluoroethylene)
PVDV	polyvinyl deoxyfluoride
QA	quality assurance (department)
QC	quality control
RFLP	restriction fragment length polymorphism
RNA	ribonucleic acid
RO	reverse osmosis
rpm	revolutions per minute
RPMI	Roswell Park Memorial Institute
RT	reverse transcriptase
SDA	Sabouraud agar
SFM	serum-free media
SGOT/AST	aspartate aminotransferase
SGPT/ALT	alanine aminotransferase
SIV	simian immunodeficiency virus
SOP	standard operating procedure
SPE	solid phase extraction
STED	stimulated emission depletion
STR	short tandem repeat (profiling)
SV40	simian virus 40
TCA	trichloroacetic acid
TIFF	tagged image file format
TIRF	total internal reflection fluorescence
TSA	tryptone soy agar
TSE	transmissible spongiform encephalopathy
TTP	thymidine triphosphate
U	units (usually of enzyme or antibiotic activity)
UDAF	unidirectional airflow (cabinet)
UKSCB	United Kingdom Stem Cell Bank
UV	ultraviolet (light)
vCJD	variant Creutzfeld–Jakob disease
WCB	working cell bank
WFI	water for injection
WHO	World Health Organization
WTS-1	2-(4-iodophenyl)-3-(4-nitrophenyl)-5-(2,4-disulfophenyl)-2H-tetrazolium, monosodium salt

WTS-3	2-(4-iodophenyl)-3-(2,4-dinitrophenyl)-5-(2,4-disulfophenyl)-2H-tetrazolium, sodium salt
WTS-4	2-benzothiazolyl-3-(4-carboxy-2-methoxyphenyl)-5-[4-(2-sulfoethyl carbamoyl) phenyl]-2H-tetrazolium
WTS-5	2,2′-dibenzothiazolyl-5,5′-bis[4-di(2-sulfoethyl)carbamoylphenyl]-3,3′-(3,3′-dimethoxy-4,4′-biphenylene)ditetrazolium, disodium salt
XTT	2,3-bis-(2-methoxy-4-nitro-5-sulfophenyl)-2H-tetrazolium-5-carboxanilide

Protocols

1 The Cell Culture Laboratory

2 Sterilization

3 Microscopy of Living Cells

4 Basic Techniques and Media, the Maintenance of Cell Lines and Safety

5 Development and Optimization of Serum- and Protein-free Culture Media

6 Cryopreservation and Banking of Cell Lines

7 Primary Culture of Specific Cell Types and the Establishment of Cell Lines

8 Cloning

9 The Quality Control of Animal Cell Lines and the Prevention, Detection and Cure of Contamination

10 Systems for Cell Culture Scale-up

1

The Cell Culture Laboratory

Sue Clarke[1] and Janette Dillon[2]

[1] *ImmunoBiology Limited, Babraham Research Campus, Babraham, Cambridge, UK*
[2] *MedImmune, Granta Park, Cambridge, UK*

1.1 Introduction

Cell culture dates back to the early twentieth century (Table 1.1) by which time some progress had already been made in cryopreservation, the long-term storage of mammalian cells in a viable state.

The laboratory process of cell culture allows cells to be manipulated and investigated for a number of applications, including:

- studies of cell function, for example metabolism;

- testing of the effects of chemical compounds on specific cell types;

- cell engineering to generate artificial tissues;

- large-scale synthesis of biologicals such as therapeutic proteins and viruses.

The pioneering work of Ross Harrison in 1907 [1] demonstrated that culturing tissue *in vitro* (in glass) not only kept cells alive, but enabled them to grow as they would *in vivo* (in life). However, the early development of cell culture technology was hindered by issues of microbial contamination. The growth rate of animal cells is relatively slow compared with that of bacteria. Whereas bacteria can double every 30 minutes or so, animal cells require around 24 hours. This makes animal cell cultures vulnerable to contamination, as a small number of bacteria soon outgrow a larger population of animal cells. However, tissue culture became established as a routine laboratory method by the 1950s with the advent of defined culture media devised by Eagle and others. The discovery of antibiotics

Animal Cell Culture: Essential Methods, First Edition. Edited by John M. Davis.
© 2011 John Wiley & Sons, Ltd. Published 2011 by John Wiley & Sons, Ltd.

Table 1.1 The early years of cell and tissue culture.

Late nineteenth century – Methods established for the cryopreservation of semen for the selective breeding of livestock for the farming industry

1907 – Ross Harrison [1] published experiments showing frog embryo nerve fibre growth *in vitro*

1912 – Alexis Carrel [2] cultured connective tissue cells for extended periods and showed heart muscle tissue contractility over 2–3 months

1948 – Katherine Sanford *et al.* [3] were the first to clone – from L-cells

1952 – George Gey *et al.* [4] established HeLa from a cervical carcinoma – the first human cell line

1954 – Abercrombie and Heaysman [5] observed contact inhibition between fibroblasts – the beginnings of quantitative cell culture experimentation

1955 – Harry Eagle [6] and, later, others developed defined media and described attachment factors and feeder layers

1961 – Hayflick and Moorhead [7] described the finite lifespan of normal human diploid cells

1962 – Buonassisi *et al.* [8] published methods for maintaining differentiated cells (of tumour origin)

1968 – David Yaffe [9] studied the differentiation of normal myoblasts

---

by Fleming was of course another major milestone that facilitated prolonged cell culture by reducing contamination issues.

In the 1940s and 1950s major epidemics of (among others) polio, malaria, typhus, dengue and yellow fever stimulated efforts to develop effective vaccines. It was shown in 1949 that poliovirus could be grown in cultures of human cells, and this became one of the first commercial 'large-scale' vaccine products of cultured mammalian cells. By the 1970s methods were being developed for the growth of specialized cell types in chemically defined media. Gordon Sato and his colleagues [10] published a series of papers on the requirements of different cell types for attachment factors such as high molecular weight glycoproteins, and hormones such as the insulin-like growth factors. These early formulations and mixtures of supplements still form the basis of many basal and serum-free media used today (see Chapters 4 and 5).

Recombinant DNA technology (also known as genetic engineering) was developed in the 1970s and it soon became apparent that large complex proteins of therapeutic value could be produced from animal cells. Another milestone came in 1975 with the production of hybridomas by Köhler and Milstein [11]. These cell lines, formed by the fusion of a normal antibody-producing cell with an immortal cancer (myeloma) cell, are each capable of the continuous production of antibody molecules with (in the modern embodiment of the technology) a single, unique amino acid sequence. By 2007, the centenary year of tissue culture, such monoclonal antibodies were being commercially produced in multi-kilogram quantities.

Large-scale culture applications have led to the manufacture of automated equipment, and today's high-end cell culture robots are able to harvest, determine cell viability and perform all liquid handling. The Cellmate™, for example, is a fully automated system for T-flasks and roller bottles [12] that was first produced for Celltech's manufacturing operations, and which has since been used in vaccine production. The latest version includes software to support validation if it is used in processes requiring compliance with regulations such as 21CFR Part 11 (see Chapter 11, Section 11.3.5). Also on the market are automatic cell culture devices that handle the smaller volumes used by high-throughput laboratories. This recognizes the growing importance of cell-based assays, particularly in the pharmaceutical industry. The Cello™ is an automated system for the culture of adherent and non-adherent mammalian cells in 6-, 24-, 96- and 384-well plates for the selection of optimal clones and cell lines. It automates operations from seeding through expansion and subculturing, and thereby decreases the time required for cell line development.

In the biomedical field cultured cells are already used routinely for a variety of applications, for example Genzyme's Epicel® (cultured epidermal keratinocyte autografts) for burns patients and Carticel® (cultured autologous chondrocytes) for cartilage repair, as well as at *in vitro* fertilization (IVF) clinics where the zygote is cultured – usually for a few days – prior to implantation in the mother's uterus. Stem cell research is another cell culture application that holds huge promise for the future, especially now specific cell programming is possible. Although much stem cell research used to depend on the use of embryonic stem cells (obtained from early-stage embryos,) scientists can, at least in some cases, now change differentiated somatic cells into stem cells (iPS – induced pluripotent stem cells) using genetically engineered viruses, mRNA or purified proteins, thus avoiding the ethical issues surrounding the use of embryos as a source of cells. These iPS cells are similar to classic embryonic stem cells in many of their molecular and functional features, and are capable of differentiating into various cell types, such as beating cardiac muscle cells, neurons and pancreatic cells [13]. Stem cells can potentially be used to replace or repair damaged cells, and promise to drastically change the treatment of conditions such as cancer, Alzheimer's and Parkinson's diseases, and even paralysis.

1.2 Methods and approaches

1.2.1 Cell culture laboratory design

When planning a new cell culture facility, it is important to clearly identify the type of work anticipated within the laboratory, as much will depend on the nature and scale of the culture to be performed. For any design, the access doors need to be large enough for the passage of any major equipment (an obvious point, but sadly one often overlooked). Even routine small-scale work, such as much of the tissue culture undertaken in healthcare, biotechnology and academia, has varied needs that require careful consideration in the planning stage.

Certain types of laboratory involved in highly specific work – such as IVF laboratories, environments dedicated to the production of biopharmaceuticals under Good Manufacturing Practice (GMP) conditions, or work with Hazard Group 4 pathogens (biological agents that are likely to cause severe human disease and pose a serious hazard to laboratory workers, are likely to spread to the community and for which there is usually no effective prophylaxis or treatment available) [14] – are not dealt with here. They require expert help for laboratory design because of the need to comply with stringent legislation and/or the highly significant associated health and safety risks.

Some of the questions that need to be answered before commencing the design of a laboratory are set out below. This list of questions is by no means exhaustive.

1.2.1.1 Health and safety implications

- What is the highest Hazard Group of material to be handled [15]?

- Will all work need to be carried out at the related containment level [14]? If not, what facilities are required at what level? (The lower the containment level, the less onerous and expensive it is to build, equip, run and work in the laboratory.)

- Will genetically modified organisms (GMOs) be used? (Note that, in the UK, any work with (or storage of) GMOs falls under the Genetically Modified Organisms (Contained Use) Regulations 2000. These require that the laboratory is registered with the Health and Safety Executive for the performance of GMO work, and that various other safeguards are put in place, *before* any GMOs enter the laboratory. Many other countries also have regulations/legal requirements covering work with GMOs.)

These are extremely important considerations as the relevant health and safety legislation may require (or recommend) features that need to be incorporated into the laboratory design from the start (e.g. specific air handling, the need for a changing lobby, accessibility of autoclaves), as well as constraining the specifications of the equipment required within the laboratory.

1.2.1.2 The scale of the work

- How many and of what size are the largest vessels that will be used within the facility?

- Will they require support vessels? (For example, a large fermenter would also require media preparation and storage vessels as well as smaller fermenters in which to grow up the cell inoculum.)

- For both of the above, will they be fixed or mobile? Is special handling equipment required? How will they be cleaned and/or sterilized?

- How are any spillages to be dealt with?

Thought needs to be given to the safe preparation, handling, inactivation and disposal of cells and media, as well as the cleaning and sterilization (or disposal) of *all* the equipment used. Disposal and discard areas need to be clearly identified, and of sufficient size to cope with the amount of waste generated by the laboratory. Good practice dictates that they are kept tidy and cleaned regularly.

Medium- to large-scale manufacturing facilities may have very specialized requirements, for example systems for handling large volumes of media, such as lifting devices or under-floor kill tanks. Even laboratories using smaller-scale benchtop cultures, for example 10-l glass fermenters or wave-type bioreactors (see Chapter 10, Section 10.2.2.5), might find that trolleys and lipped benches help with risk management, especially if multiple units are to be in use at any one time.

1.2.1.3 Segregation requirements

Consideration needs to be given to the physical segregation of work:

- Are separate rooms needed for (for example) primary cell culture or the quarantine of incoming cells?

- Do several smaller rooms give more flexibility than one large one?

Often full segregation is not possible and in this situation the full implications of this need to be understood. Ideally, work place practices can then be implemented to reduce the risk of cross-contamination, for example the use of dedicated cabinets and incubators for specific cell types, with associated records of what was handled when, and by whom. Valuable stock cultures should be duplicated in incubators with independent services (electricity, CO_2) to avoid their complete loss. Use of an uninterruptible power supply is worth consideration in this respect. Thought must also be given to the flow of work within the room – try to keep dirty areas, such as those used for waste disposal, near the door, with critical clean areas containing items such as the microbiological safety cabinet(s) and incubator(s) as far from the entrance as possible (see below).

1.2.1.4 General considerations

One feature that is common to all cell culture laboratories is the need to prevent the contamination of the cell cultures with adventitious agents from external sources, that is operators and the environment. For this reason, the microbiological safety cabinet (MSC) is a central component to all tissue culture laboratories. It provides protection to the operator and, in the case of Class II cabinets, also offers the culture some protection against any microbes that might be present in the environment. As discussed in Section 1.2.3.1, the positioning of MSCs is extremely important – not only to ensure that the correct airflow is maintained, but also to provide an ergonomic environment for the user – and will have a significant impact on the design of the culture laboratory. For more demanding applications, such as in the production of biological medicines, MSCs will be situated in a dedicated culture suite supplied with sterile-filtered air. The air in such rooms is generally kept under slight positive pressure with regard to neighbouring areas, to ensure that non-sterile air is not drawn in [16].

In order to minimize further the risk of contamination to cultures, the passage of people through the cell culture laboratory should always be minimized, especially past the cleanest areas – the MSCs and incubators. Therefore, as mentioned above, these critical work areas should be positioned away from the entrance. Sticky mats can be placed on the floor immediately inside the laboratory to remove loose dust and grime from shoes. These mats should be changed regularly to prevent them becoming a source of contamination themselves. Disposable overshoes or shoes dedicated for use in the laboratory provide another option.

To reduce dust build-up within the laboratory, there should be as few horizontal surfaces as possible, commensurate with the work to be undertaken and along with any allowance for future developments (see Section 1.2.1.5). All surfaces must be cleaned regularly. In order to facilitate cleaning, all plumbing, cabling etc. on entering the laboratory should be boxed in, with any access points through walls, floors or ceilings being well sealed. Flooring should be flat and even, as seamless as possible and joined smoothly to the walls. Sufficient storage needs to be at hand for work surfaces to be kept clutter-free. Ideally, under-bench storage cupboards and drawers would be moveable to facilitate cleaning, rearrangement and removal. Even if the room is not designed to be fully fumigable (using formaldehyde or vaporized hydrogen peroxide – see Chapter 2, Section 2.2.4.1), it should at least lend itself to thorough cleaning with a biocidal agent (see Chapter 2, Section 2.2.4.2).

Once workers have entered, there should be sufficient space to change into clean laboratory coats, with adequate provision for storage of these coats and safety glasses when not in use. Hand wash sinks with soap and alcohol rub should be nearby to allow thorough hand cleaning on entry and exit. Eye wash stations are best positioned by the doors where they are easily accessible. Users require sufficient room for drawers or moveable trolleys of consumables to be at hand, and to have easy access to basic equipment such as the incubator, microscope and centrifuge.

Having entered, the need to exit/re-enter should be reduced by having all necessary small items of equipment within the laboratory. Ideally any equipment that does not need to be operated under sterile conditions (e.g. analytical flow cytometer, fluorescence microscope) should be housed in a separate but nearby room. Adequate stocks of unopened media and frozen reagents for use in the short term should be stored within a laboratory refrigerator or freezer, but larger quantities of unopened supplies should be housed elsewhere, preferably in dedicated clean storage.

1.2.1.5 Future needs

Often, requirements change within the lifetime of the laboratory due to fast-moving technology and changes in scientific focus. Therefore, it is worth considering what the requirements may be in the future, as designing in flexibility for upcoming work may be cost-effective in the long run. What seems routine now may well be superseded with time, and innovative technology and instrumentation may need to be brought in. Although one cannot predict exactly what these changes may be, it is worth ensuring that sufficient power, utility and computer network connection points are installed at the outset. Leaving plenty of room for workers, and free bench space, not only allows safe and full access to equipment when needed but can give scope for some rearrangement at a later date.

1.2.2 Services

The service requirements of tissue culture laboratories are very similar to those of other laboratories, with the additional need for gas supplies, for example carbon dioxide for incubators, oxygen and nitrogen for fermenters, and mixed carbon dioxide/air for the manual gassing of culture vessels.

The laboratory should have some form of continuous environmental control (heating/air conditioning) to ensure that there is little variation in room temperature during different times of the day or year. This is important as many pieces of equipment will not be able to operate as required (e.g. an incubator may not be able to maintain constant temperature, and could potentially malfunction) if the ambient temperature varies significantly. Many pieces of instrumentation also require a steady ambient temperature in order to give reproducible results.

Numerous power sockets are essential in the laboratory, both above and below bench level. Although most laboratory equipment will require only normal single-phase electricity (230 v, 15 A in the UK), it should be noted that certain items such as large centrifuges may require a single-phase electricity supply that will deliver a higher amperage (e.g. 30 A), and some items such as large fermenters and autoclaves may require a three-phase supply.

IT network points and communication ports may be necessary to network many of the pieces of computer-driven instrumentation that may be used in the laboratory. This can help to reduce the amount of paperwork moving in and out of the laboratory by giving workers access to the computer network from within the laboratory.

1.2.2.1 Water

High-purity water is essential in any cell culture laboratory, not only for preparing media and solutions but also for glassware washing. A source of ultra-pure or RO (reverse osmosis) water is therefore required. RO is a process that typically removes 98% of water contaminants. Tap-water fed to an RO unit should first be passed through a conventional water softener cartridge to protect the RO membrane. RO water can then be fed directly to a second-stage ultra-purification system, comprising a series of cartridges through which the water is filtered for ion exchange and the removal of organic contaminants, and finally through a microporous filter to exclude any particulate matter, including microorganisms. Water purity is monitored by measuring resistivity, which should reach about 18 megohm/cm. Most water purification units, supplied by companies such as Millipore (Milli-Q system), have semi-automatic cleansing cycles requiring minimum effort to maintain. Water should be collected and autoclaved or filtered immediately prior to use for sterile applications.

If in doubt about the quality of the water from a purification system, a simple *Limulus* amoebocyte lysate (LAL) assay should be performed to check endotoxin levels. If the water is found to be a source of endotoxin and the problem cannot be solved, then non-pyrogenic water for injection (WFI) can be purchased and used for preparing media and other critical solutions. Most laboratories now buy ready-made media and supplements for cell culture use.

Further details on the requirements for, and purification of, water for laboratory use can be found in reference 17.

1.2.2.2 Pressurized gases

Ideally, gas cylinders should be kept outside the main cell culture laboratory and the gas piped through, to maintain cleanliness in the sterile handling area. CO_2 cylinders should be secured to a rack and the gas fed via a pressure-reduction valve on the cylinder head, through pressure-rated tubing to the incubator intake ports. It is critical to maintain an uninterrupted gas supply, and it is advisable to have two cylinders connected to the CO_2 supply system via an automatic changeover unit that will switch to the second cylinder when pressure in the first drops below a certain level. (Note that some automatic changeover units require an electricity supply, and in the event of an interruption to the electricity these may malfunction or even cut off the gas supply. Thus a purely mechanical changeover unit may be preferable.) An in-line 0.2-μm filter is a useful precaution against dirt or microorganisms in the supply system, and some incubators will require the use of a further pressure-reduction valve to bring the CO_2 pressure down to that for which the incubator was designed.

In order to check the integrity of pressurized gas connections, a wash-bottle filled with soap solution (household washing-up liquid is ideal) for squirting around connections makes a cheap and effective leak detector. In particular, newly connected cylinders should be tested in this way; dirt on either face of the connection can cause significant leakage.

It is the responsibility of the laboratory manager, not the gas supplier, to ensure that gas systems, storage and operating practice on site comply with any relevant laws and regulations.

1.2.2.3 Liquid nitrogen

Liquid nitrogen (LN) is required for the long-term storage of cell stocks (see Chapter 6). All vessels containing LN must be kept and used in well-ventilated areas, and certainly *not* coldrooms or basements. LN refrigerators containing cell stocks (see Section 1.2.3.6) need to be kept in a secure location fairly close to the culture laboratory, but preferably not in the laboratory itself. If keeping such vessels in the laboratory is unavoidable, a risk analysis must be carried out first, taking into account the maximum amount of LN that could be released in the event of a spillage or vessel failure (and the volume of nitrogen gas this would become), the volume of the laboratory, and any other relevant factors such as air exchange rates of the ventilation system and the presence in the room of an oxygen depletion monitor (a good precaution in any event). Larger-volume stocks of LN for keeping the cell stores topped up are best stored in a secure enclosure on the outside of the building, and must be easily accessible for filling by a delivery vehicle. As these vessels will usually be pressurized, they must be designed specifically for the storage of LN, and maintained on a regular basis. For further details, and general advice on the handling of LN, see Chapter 4, Section 4.2.8.1.

1.2.3 Equipment

1.2.3.1 Microbiological safety cabinet

The most important piece of equipment within a cell culture laboratory is the MSC. Although cell culture can be performed on the open bench with the help of a Bunsen burner [16], laboratory techniques may produce aerosols that can contain hazards such as infectious agents which can be inhaled by laboratory workers. MSCs are used as primary barriers during the handling of materials that may contain or generate hazardous particles or aerosols, in order to prevent exposure of laboratory personnel and contamination of the general environment. Some MSCs are designed to also provide a clean work environment to protect cell cultures or sterile apparatus/solutions

All MSCs purchased in the UK must comply with British Standard BS EN 12469:2000, and the USA [18] and other countries have similar specifications. There are three classes

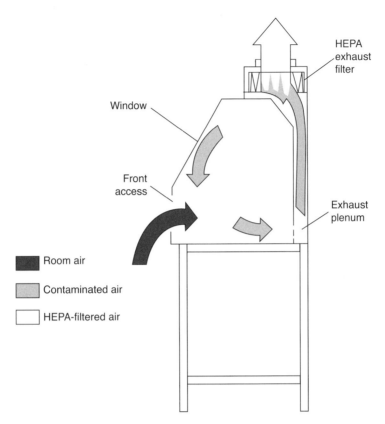

Figure 1.1 A diagrammatic cross-section of a Class I microbiological safety cabinet, showing airflow and HEPA filtration.

of microbiological safety cabinet available. They are used in different contexts, but all are suitable for cell culture work.

• **Class I** (see Figure 1.1) – *gives operator protection only*. These cabinets work by draw-ing air into the cabinet away from the worker and then the exhaust air passes through a HEPA (high efficiency particulate air) filter to remove particulate matter before it is discharged outside the building, thus protecting the environment.

• **Class II** (see Figure 1.2) – *both the work and the operator are protected*. This cabi-net functions as above, but air drawn in from the laboratory via the front access area immediately flows downwards through holes or slots in the front of the work surface into the base of the cabinet before being HEPA filtered and recirculated, so providing an in-flowing curtain of air that offers protection for the operator. Similarly, some of the recirculated filtered air flowing downwards within the hood is also draw through the same holes or slots in the front of the work surface, thus providing a curtain of air that protects the work from particles entering through the front access. Generally this is the cabinet of choice in a cell culture laboratory, allowing handling of all but the most hazardous cell lines.

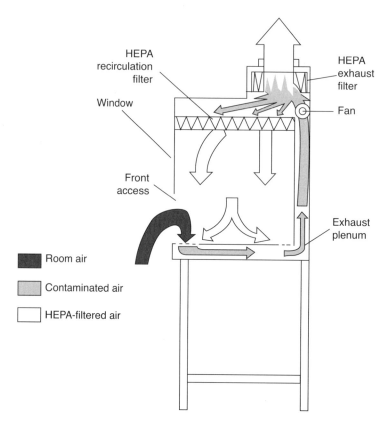

Figure 1.2 A diagrammatic cross-section of a Class II microbiological safety cabinet, showing airflow and HEPA filtration.

• **Class III** (see Figure 1.3) – *These hoods provide protection to the operator and the work*, for those working with Hazard Group 4 organisms. The hoods are totally enclosed, with access to the interior via glove ports, or an air lock. They are onerous to use, and only employed for the handling of extremely dangerous pathogens.

NOTE THAT THERE ARE OTHER TYPES OF AIRFLOW CABINETS THAT ARE NOT MICROBIOLOGICAL SAFETY CABINETS. Unidirectional airflow cabinets (UDAFs) – formerly known as laminar flow cabinets – come in horizontal and vertical versions. They protect the work area, **but not the operator**, by blowing HEPA-filtered air over the work surface. Thus THESE CABINETS MUST NOT BE USED FOR HANDLING CELL CULTURES. Horizontal UDAFs (where airflow is directed at the operator) can be used for the assembly of sterile apparatus, and vertical UDAFs may be used for the filtration of non-hazardous solutions, and the filling of such solutions into containers.

Extreme care needs to be given to the siting of MSCs. Not only is the space available important, but the presence of any obstructing features such as walls, windows and pillars that may disturb air flow will influence the positioning of the cabinet. MSCs need to be

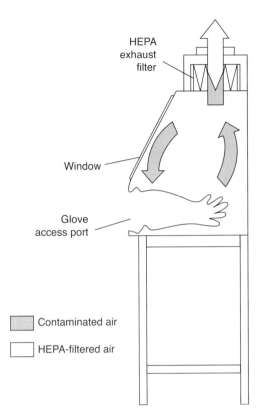

HEPA
exhaust
filter

Window

Glove
access port

■ Contaminated air

□ HEPA-filtered air

Figure 1.3 A diagrammatic cross-section of a Class III microbiological safety cabinet, showing airflow and HEPA filtration. Fresh air is drawn into the cabinet through a HEPA filter (not shown).

positioned to minimize traffic past them, and away from features and equipment that may disturb airflow in the vicinity. Examples can be seen in Figures 1.4 and 1.5.

There are many options available for configuration of Class II MSCs – the exhaust can be ducted to the exterior of the building or recirculated to the laboratory depending on the organisms to be used, the type of work to be undertaken and the procedures to be employed within the laboratory. (Note that it is essential that if cabinets are to be fumigated independently of the laboratory (see Chapter 2, *Protocol 2.6*) then they MUST exhaust outside the building.) Different manufacturers can supply varying widths and heights of cabinet; some even have adjustable base heights. Reputable suppliers will be happy to make site visits and advise on the best locations and hood types for your specific laboratory and applications.

Some of the considerations when choosing an MSC include the following.

- Are you likely to want to use volatile hazardous substances in the MSC (e.g. toxic chemicals, radioisotopes) where the hazard is non-particulate? If so, consult reference 18 for the different possible subtypes of Class II cabinet. Such cabinets must always be securely ducted to exhaust outside the building.

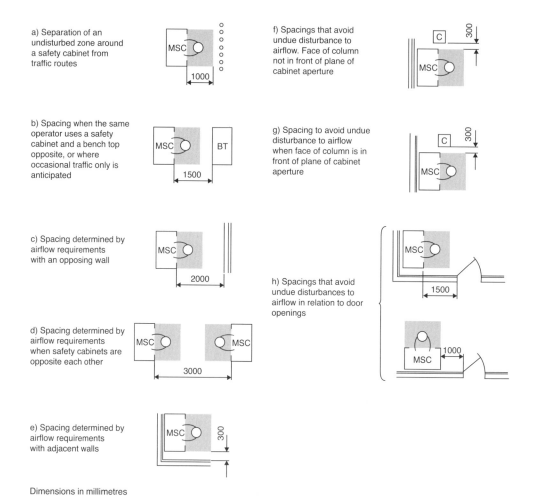

a) Separation of an undisturbed zone around a safety cabinet from traffic routes

b) Spacing when the same operator uses a safety cabinet and a bench top opposite, or where occasional traffic only is anticipated

c) Spacing determined by airflow requirements with an opposing wall

d) Spacing determined by airflow requirements when safety cabinets are opposite each other

e) Spacing determined by airflow requirements with adjacent walls

f) Spacings that avoid undue disturbance to airflow. Face of column not in front of plane of cabinet aperture

g) Spacing to avoid undue disturbance to airflow when face of column is in front of plane of cabinet aperture

h) Spacings that avoid undue disturbances to airflow in relation to door openings

Dimensions in millimetres

Figure 1.4 Recommendations for minimum distances for avoiding disturbance to the safety cabinet and its operator. For key, see page 14. Reproduced with permission from British Standard 5726:2005.

- Does the cabinet you are thinking of purchasing have airflow alarms? For the safety of the user, audible and visible alarms should be present to indicate when the airflow rates fall below specification.

- Do you require electrical sockets within the hood?

- Is the type of work surface in the cabinet suitable for your purposes? Thought should be given to any issues that may affect your work. The work surface can be made in several (lighter) sections or one large piece, but ease of cleaning is important.

- Most cabinets have the optional extra of ultraviolet (UV) lighting for decontamination. Would you find this useful? UV radiation is directional, and thus for it to be effective the cabinet must be totally empty. UV lamps are active microbicidally for a relatively short part of their working life. Efficacy must therefore be monitored regularly.

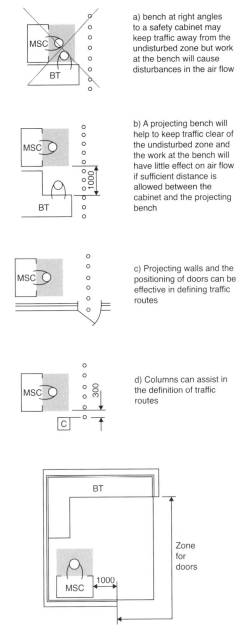

a) bench at right angles to a safety cabinet may keep traffic away from the undisturbed zone but work at the bench will cause disturbances in the air flow

b) A projecting bench will help to keep traffic clear of the undisturbed zone and the work at the bench will have little effect on air flow if sufficient distance is allowed between the cabinet and the projecting bench

c) Projecting walls and the positioning of doors can be effective in defining traffic routes

d) Columns can assist in the definition of traffic routes

e) In a small laboratory, the safety cabinet should be clear of personnel entering through the doors

f) Danger of too much air movement in front of safety cabinets should be alleviated by allowing more space between the apertures of the safety cabinets and the bench tops

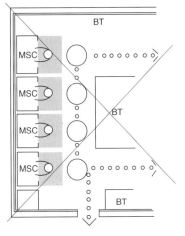

g) Danger of too much movement in front of safety cabinets should be avoided by allowing more space between the apertures of the safety cabinets and the bench tops

Dimensions in millimetres

Figure 1.5 Recommendations for minimum distances for avoiding disturbances to other personnel. For key, see page 14. Reproduced with permission from British Standard 5726:2005.

Key to figures 1.4 and 1.5

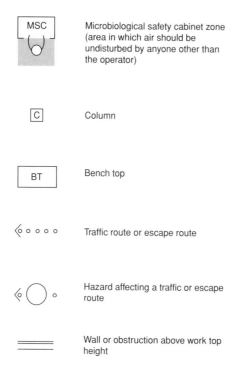

MSC	Microbiological safety cabinet zone (area in which air should be undisturbed by anyone other than the operator)
C	Column
BT	Bench top
◁ ∘ ∘ ∘ ∘	Traffic route or escape route
◁ ◯ ∘	Hazard affecting a traffic or escape route
═══	Wall or obstruction above work top height

NOTE In Figures 1.4 and 1.5, siting arrangements which should be avoided are overlaid with a cross.

- Are gas taps required within the hood?

- Are all internal surfaces accessible to enable thorough cleaning?

- Do you have additional requirements? For specialist laboratories, various pieces of equipment including microscopes and heated bases can be built in.

All MSCs must be regularly serviced and tested to ensure their continued safe performance. Twelve-monthly would be the minimum frequency for cabinets used in low-risk work, but if handling Hazard Group 2 or higher organisms 6-monthly or even more frequent testing may be appropriate. This must be decided on a case-by-case basis, with regard to the relevant risk assessment.

1.2.3.2 Incubators

Incubators for cell culture must maintain an environment which is optimal for cell growth. This requires that three parameters be measured, controlled and reliably maintained: temperature, humidity and carbon dioxide concentration. Generally for mammalian cell culture the incubator is set to 37 °C (or to be safe, 36.8 °C or 36.9 °C, as cells are more tolerant to a low incubation temperature than a high one) with a relative humidity level of 95% and with CO_2 concentration matched to the media in use (usually 5% – see Chapter 4, Section 4.1.2.1 *i Basal Medium*).

i. Temperature Different types of incubator maintain their internal temperature in different ways. Careful thought should be given to your particular situation and requirements.

- Direct heat allows increased chamber size for a given footprint, while reducing the overall size and weight of the incubator.

- Air-jacketed incubators are lighter, and quicker to reach their temperature set-point than water-jacketed ones, but conversely lose heat faster in the event of a fan or electricity supply failure.

- Water-jacketed incubators are by nature much heavier, and require filling before use and draining before moving. While they take longer to reach their temperature set-point they can maintain it longer in the event of a power failure.

ii. Carbon dioxide concentration CO_2 can be monitored and controlled in various ways. Older-style incubators use a continuous flow of both CO_2 and air, which is mixed. This method can be wasteful of CO_2. Most modern incubators have an internal sensor that monitors the CO_2 in the chamber, and opens a valve to draw pure CO_2 in should the level drop below the set threshold. Infrared sensors are the most sensitive and accurate; they can respond extremely quickly and are unaffected by changes in humidity and temperature. Sensors may need to be removed from the incubator to allow thorough cleaning or fumigation; if in any doubt, check with the manufacturer.

iii. Humidity The atmosphere within the incubator is normally humidified by means of a tray of water placed in the bottom of the incubator, with a fan ensuring even distribution of humidity around the chamber. De-ionized or RO water should be used, with a regular and frequent preventative maintenance schedule in place to ensure the tray is emptied, thoroughly cleaned and refilled, as it can potentially be a source of bacterial and fungal contamination. It is therefore important that the tray can be easily accessed. Non-volatile cytostatic reagents such as thimerosal or a low concentration (1%) of a disinfectant detergent such as Roccal can be included.

- Alarms should be fitted to the incubator to alert the user if the chamber conditions fall outside pre-set limits with regard to any of the above. Some models are also able to keep error logs identifying such issues, which can aid the user should there be an intermittent problem.

- The incubator chamber should be designed in such a way as to minimize the risk of contamination. Ideally, there will be a smooth inner casing and shelves should be easily removable to allow cleaning and disinfection. Many manufacturers offer incubators with copper interiors and shelving, or automatic decontamination routines which can reduce the risk of contamination.

- An additional factor to consider is the culture capacity required by the laboratory – can the incubators be stacked on top of each other should the need arise at a later stage for more? The interior layout can also be varied with full or half shelves, multiple gas-tight inner doors, or roller rig adaptors. Most incubators also offer the option of left- or right-opening doors.

1.2.3.3 Centrifuges

A low speed benchtop centrifuge capable of generating at least 200 g is needed for pelleting cells during various culture operations. A variety of sizes and models are available, usually with assorted accessories such as interchangeable rotors (swing-out or fixed angle), different size buckets with compatible inserts to accommodate numerous tube types, and sealable lids for use in more hazardous applications where containment is essential. The requirement of different laboratories will vary, but most would need adapters for standard 50-, 30- and 15-ml tubes and perhaps holders for 96-well plates. Refrigerated versions of these centrifuges are also available.

Ideally, the centrifuge should be sited on its own table, bench or stand to prevent any vibration affecting nearby equipment. A small high-speed microfuge is also useful for centrifuging small tubes (<2 ml capacity). Laboratories handling large quantities of cells or cell-derived products may also require a large refrigerated floor-standing centrifuge, but the availability of an appropriately rated electricity supply should be checked before purchase (see Section 1.2.2).

Many modern centrifuges have the capacity for several programs (with varying speeds/forces and times) to be stored, allowing users to quickly select commonly used parameters. With all centrifuges it is worth checking that the whole of the bowl is easily accessible, and the rotor can be easily removed to allow thorough cleaning.

1.2.3.4 Microscopes

A good inverted microscope is essential in any tissue culture laboratory to allow visualization of cultures in flasks and plates, and also for cell counting. Several objective lenses are required such as $4\times$, $10\times$, $20\times$ and perhaps $40\times$, and phase contrast is a desirable feature to improve images. The siting of the microscope is important. Although it should be located in a convenient position not too far from the MSC and incubator(s), it must also be positioned away from sources of vibration. For operator comfort, it should be located at a convenient height with adequate space below the microscope to accommodate the operator's knees; a height-adjustable laboratory chair should also be supplied.

Microscopy is addressed in detail in Chapter 3.

1.2.3.5 Refrigerators and freezers

For storage of media and additives at least one refrigerator is needed. Ideally, unopened media should be kept separated from that in use – preferably in a cold room or refrigerator nearby. Once opened, each bottle of medium should be labelled with the user's name and date of opening, as well as the cell line it is to be used with, to ensure both that the medium is used within 1 month and also that it is not mixed between users and cell lines (in order to prevent cross-contamination of cultures). Refrigerators should not be glass fronted, as many media types are light-sensitive and may be degraded by constant exposure to UV.

Freezers at both $-20\,^{\circ}C$ and $-80\,^{\circ}C$ should be available for storage of media components that are unstable at higher temperatures. If possible, aliquots should be stored separately from unopened stocks, but must be clearly labelled with contents, date and lot numbers. Cells frozen under suitable conditions (see Chapter 6) may be stored short term (for a few days) at $-80\,^{\circ}C$ but will gradually lose viability thereafter. Liquid CO_2 back-up can also be added to low-temperature freezers, to maintain the temperature in the event of a power failure.

If possible, data logging and 24-hour monitoring of all freezers containing stock reagents and cultures is desirable to prevent expensive loss in the case of equipment failure. If this is not available then regular monitoring of equipment should be performed.

1.2.3.6 Liquid nitrogen refrigerators for cell storage

Cell stocks for long-term storage are usually kept in vials in either the liquid or vapour phase of LN (see Chapter 6). A wide range of Dewar vessel-type storage refrigerators are available from companies such as Taylor–Wharton. These range from small vessels with a capacity of around 90 1-ml vials, through the popular under-bench-sized vessels holding up to 4000 vials, to huge vessels that might be found in a cell repository and can hold over 33 000 vials.

There are two main methods of storage within these vessels. Vials can either be attached to aluminium canes, which are then placed within canisters in the vessel, or alternatively the vessel can contain tesselating towers of drawers, each of which contains many small compartments each designed to accept a single vial. Small vessels only tend to use the former system, and large vessels the latter, but intermediate-sized vessels can be purchased in either format. The drawer-type vessels tend to use more nitrogen, due to the larger opening required to access the drawers. Note also that LN vessels can hold cells either in the liquid or the vapour phase, the latter being preferable in order to minimize the risk of cross-contamination (see Chapter 6, Section 6.2.4).

As the cell stocks may be a laboratory's most valuable asset, it is important that the LN refrigerators never be allowed to run anywhere near dry. To ensure this, they must be filled on a regular and frequent rota (see Chapter 4, *Protocol 4.2*). For added protection, low level alarms can be fitted to most vessels, and larger vessels can be equipped with an auto-fill system that will add LN from an attached storage vessel if the level drops below a certain point. In addition, important cell banks should be split between vessels and locations whenever possible, in order to avoid total loss in the event of a vessel failure or a fire. For a detailed discussion of cell freezing and banking, see Chapter 6.

1.2.3.7 Miscellaneous equipment

Numerous other smaller pieces of equipment are desirable; some of these are listed below with brief descriptions of their use and key attributes.

- Water bath/dry incubator: for thawing and warming of reagents prior to use. Owing to the risk of contamination the water bath needs to be easy to empty and clean on a regular basis. Use of a cytostatic reagent (but not azide) in the water is advised. It should be noted that heat dispersal is less efficient in an incubator than in a water bath, but an incubator has the benefit of reducing the risk of contamination from water and associated aerosols. As it may be necessary to use these pieces of equipment at different temperatures (e.g. 37 °C to warm media, but 56 °C to inactivate complement) a working thermometer should always be present, and the current set temperature marked clearly and prominently on the outside of the equipment.

- Aspirator: to remove spent supernatants from culture vessels and plates. The aspirator jar should contain an anti-foam and small amount of detergent such as Decon (Decon Laboratories Ltd). A small volume of Chloros (sodium hypochlorite) solution should be drawn through the line between handling different cell lines and at the end of the working session.

- Automatic hand-held, battery- or mains-operated pipette aid: for use when pipetting with disposable volumetric pipettes (1–100 ml).

- Single- and multi-channel micropipettes covering volumes 1–20 μl, 20–200 μl, 100–1000 μl and 1–5 ml, preferably autoclavable. These are used for accurate pipetting of small volumes and reducing the manual handling involved in using 96-well plate formats. They are available in manual or rechargeable electronic form.

- Laboratory vortex mixer: for cell resuspension, particularly of cell pellets prone to clumping, for example following trypsinization. (Note that it may not be appropriate to use this method of resuspension with fragile cells.)

- Label printer: useful for clear and accurate identification of information on aliquots and tubes of frozen cells.

- Trolleys: may be useful for the transportation of large volumes of media and waste.

- Cell freezing device: providing controlled rate freezing for the long-term storage of cells (see Chapter 6).

- Weighing balance, magnetic stirrers: for preparation of media and reagents.

- pH meter and osmometer: to check pH and osmolarity of media.

- Automated cell culture instrumentation: for example cell counters, or spent medium analysers to assess culture growth.

1.2.4 Culture plasticware and associated small consumable items

Laboratories differ in their preference for the use of glass or plasticware for cell culture and related operations. However, whenever possible the use of single-use disposable plasticware is highly recommended over the alternative method of recycling and reusing glassware. Glassware has a propensity for adsorbing substances such as alkaline detergent onto its surface. If glassware is to be used it needs to be of a high-quality borosilicate and supported by the use of good washing facilities. There is no doubt that plastic containers are less prone to leaching trace elements that may be deleterious to cells.

1.2.4.1 Tissue culture plasticware

There are a number of companies that can offer the standard items of sterile cultureware required for tissue culture at comparable cost and ISO9001-certified quality. Each manufacturer treats their plastics in a slightly different manner, and some cell lines may adhere better to one than another. Comparative testing of samples from a number of suppliers can therefore be a worthwhile investment of time if dealing with demanding cultures. When assessing different products, some factors to be considered are handling benefits of different shapes/formats such as uniform footprint for ease of stacking – bad handling can itself be a source of contamination; alphanumeric identification – some products offer clearer labelling, for example individual alphanumerical codes for well identification; or other features such as (for multi-well plates) non-reversible lids with condensation rings to reduce the risk of contamination in multi-well plates, or culture flasks with built-in filters in the caps to aid gas exchange. In addition, certain products may be guaranteed to be non-pyrogenic and/or DNAse free, characteristics that may be essential for certain applications; these contaminants are known to have deleterious effects on certain cell cultures, assays and cell-derived protein products. It is much better to avoid possible issues from the beginning by scrupulous control of starting materials.

The basic procedures of cell culture that require plasticware fall into three categories.

i. Cell growth The most common vessels used for growing cells are 'tissue culture' treated for maximization of cell growth, be it in anchorage-dependent (adherent) or independent (suspension) mode. They are available in many formats depending on the scale of culture required including tubes, flat-bottomed flasks, spinner and Erlenmyer flasks, Petri and multiwell dishes, flasks, roller bottles (Chapter 10, Section 10.2.1.1) or stacked-plate vessels (Chapter 10, Section 10.2.1.2). Some manufacturers also offer non-tissue culture-treated versions, for use in applications where unwanted cell attachment is a problem.

Although uncommon, poor or uneven cell growth can occasionally be attributed to product quality issues – either poor design or moulding – and could be batch specific. Other possible causes of poor cell growth are discussed in Chapter 4, Section 4.3.

ii. Liquid handling and filtration This equipment, includes volumetric pipettes, Pasteur pipettes, centrifuge tubes, autoclavable micro-pipette tips, and filters for sterilizing tissue culture solutions.

The pore size of sterilizing filters should normally be 0.2 µm, or even 0.1 µm in order to exclude mycoplasma as well as other micro-organisms. Some filter membranes, for example those made of polysulphone, have low protein-binding properties and are essential where the protein concentration of the filtered solution is critical, particularly if the molecule is highly charged as are some of the polypeptide growth factors. Very small volumes of valuable reagents should be filtered through units where the dead space (hold-up volume) is minimal. It is cost-effective, therefore, to have some of these more expensive filters reserved for special purposes. Larger volumes can be filter sterilized using either pre-assembled, disposable units which attach to a vacuum line or, more economically, a washable, autoclavable, filter housing unit for repeated use (see Chapter 2, Section 2.2.5).

iii. Storage These include sample tubes, Eppendorf tubes, cryotubes and larger-volume screw-capped bottles or bags.

1.2.4.2 Miscellaneous small items

As in any other laboratory, items of general use, such as a calculator and test tube racks, are required and these items should be readily available, in sufficient quantity for the workload, and stored in the cell culture laboratory. This limits the transfer of materials in and out of the culture facility and thus helps to reduce contamination issues. If reusable pipettes are used, it is convenient to sterilize them in pipette cans, which should be marked as dedicated for use in one room, and once opened kept for exclusive use by an individual worker. Listed below are some of the small items most cell culture laboratories will require, but these lists are in no way intended to be comprehensive.

General items that may be required include:

* autoclave tape or bags
* pipette cans
* sharps- and waste-disposal units
* cryomarker pens
* tube racks
* cages or boxes for washing up
* personal protective equipment, for example gloves, safety glasses
* disinfectant spray bottles
* timers
* calculators.

Specialized items that may be required include:

* haemocytometer and/or other cell counting equipment (see Chapter 4, Section 4.2.6.3)
* cell dissociation grinders and sieves
* cell scrapers
* cloning rings (see Chapter 8)
* well membrane/cover slip inserts
* forceps
* scalpels and/or scissors
* microcarrier beads.

1.2.5 Washing reusable tissue culture equipment

All tissue culture equipment which can be washed, sterilized, and reused should go through the same general process, as described in *Protocol 1.1*. This can either be performed manually or by a tissue culture-dedicated automatic washing machine with both acid rinse and distilled water rinse facilities.

Over time a film of cell debris, protein and other materials can develop, and such glassware can become a source of endotoxin contamination. Therefore, periodic stripping by washing in sodium hydroxide (NaOH) is recommended. Recoating of certain vessels like spinner flasks with a silicone reagent to prevent cell attachment will be necessary. All reusable glassware will have to be sterilized after washing and drying. This topic is covered in Chapter 2.

PROTOCOL 1.1   Washing and sterilization of reusable labware

Equipment and reagents

- Ultra-pure water

- Chloros (hypochlorite) solution for soaking

- Phosphate-free detergent (e.g. Micro (International Products Corp.) for manual washing; low-foam for machine, as recommended by manufacturer)

- AnalaR (or similar high purity) HCl for manual washing; formic or acetic acid for machines

- Rigid plastic soak tanks, cylinders, and beakers (for small items and instruments)

Method

1 Soak items either in Chloros solution (except metals), or directly in detergent.

2 Wash with detergent (by soaking or machine).

3 Perform tap water rinses (continuous or sequential).

4 Rinse with acid (except for metals).

5 Repeat tap water rinses as in step 3 above.

6 Rinse with distilled/reverse osmosis/ultra-pure water (two or more times).

7 Dry with hot air.

8 Store temporarily, capped or covered, for example on preparation bench.

9 Prepare for sterilization (see Chapter 2).

10 Sterilize by autoclaving or in dry oven.

11 Store for use in dedicated area.

1.2.5.1 Pipette washing

Unless a pipette-washing facility is available in the automatic washing machine, in which case the supplier's instructions should be followed, this should be carried out using *Protocol 1.2*.

PROTOCOL 1.2 Washing glass volumetric pipettes

Equipment and Reagents

- Forceps
- Compressed air supply, or water jet
- Soaking cylinders and associated plastic pipette carriers
- Detergent suitable for use with cell culture equipment such as 7X or Decon 90
- Syphon-operated pipette washer
- Concentrated HCl
- Ultra-pure (tissue culture-grade) water (see Section 1.2.2.1)
- Drying oven

Method

1 Unplug the pipettes using forceps or a high pressure air or water jet applied to the opposite (tip) end.

2 Place them, tip upwards, in a plastic pipette carrier.

3 Insert the carrier into an outer cylinder containing detergent solution and leave overnight to soak.

4 Transfer the carrier to an automatic, water syphon-operated pipette washer[a].

5 After 3–4 h, add about 50 ml of concentrated acid (HCl or acetic) at the filling stage of the cycle and turn off the tap when full.

6 Leave for about 30 min to remove residual traces of detergent.

7 Run the tap water rinse cycle again for at least 1 h.

8 Immerse the pipettes in their carrier in two or three changes of ultra-pure water, in a clean plastic cylinder kept for the purpose.

9 Drain for a few minutes and transfer the carrier to a drying oven.

Note

[a] Pipette washers operate on tap water, the flow rate of which must be within certain limits, set empirically. The container fills and empties repeatedly, by a syphon mechanism.

Problems can frequently occur if equipment is inadequately cleaned. Such problems are largely avoided by attention to the following details.

- Handle all glassware with care, especially pipettes, and discard any broken or chipped items immediately using appropriate procedures.

- Rinse all items before soaking, remove labels and fully immerse.

- Soaking should be carried out as soon as possible after use, as many tissue culture solutions contain protein, which, if left for any prolonged period, can result in microbial growth and difficult-to-remove dried-on protein.

- Separate ready-filled soak tanks should be available near sinks for both easily broken delicate glassware and more robust vessels.

- The solutions in tanks need to be changed regularly. A convenient-to-use tablet form of sterilizing agent is always helpful as it speeds the preparation of soaking solutions.

- If washing is to be performed manually, further soak items overnight in detergent followed by an acid rinse to neutralize alkaline detergent residue.

- Ensure adequate rinsing to remove any residues of cleaning materials that can be toxic to cells.

1.2.6 General care and maintenance of the tissue culture laboratory

The proper care of equipment and attention to tidiness and cleanliness are especially important in a tissue culture laboratory. Consequently, it is useful to have a checklist of tasks that must be completed on a daily, weekly or monthly basis. The following lists can be used as a guide but should be modified and extended to suit the individual laboratory.
   Daily checklist:

- Where appropriate, check room air handling pressure differential(s) before entry.

- Record incubator, refrigerator and freezer temperatures.

- Check incubator temperature and CO_2 are at set points.

- Check CO_2 cylinder pressures.

- Check conductivity of purified water.

Protocols for laboratory start-up and shutdown can be found in Chapter 4, *Protocols 4.1 and 4.2.*
   Weekly checklist:

- Wash floor and bench work surfaces (if not done daily).

- Clean underneath MSC work surface.

- Change water in water baths.

- Change water in humidifier trays in incubators.

- Replenish stocks of routine reagents and plasticware. (Do not allow cardboard boxes to enter the sterile work area).

- Empty aspirator jars as necessary (if not done daily).

- Top up LN in cell freezers (or twice weekly).

- Change floor sticky mat (or more often if required).

- Change used laboratory coats for clean ones.

Monthly checklist:

- Cleanse and sterilize reverse osmosis unit.

- Check water ultra-purification cartridges.

- Check whether equipment services or safety tests are approaching.

- Defrost freezers (as necessary).

- Calibrate instruments and monitors – if not done more frequently or on a regular basis before use.

- Strip down and clean/sterilize the insides of incubators

- Perform environmental monitoring if required, and if not performed more frequently – see *Protocol 1.3*.

PROTOCOL 1.3   Environmental monitoring

Equipment and reagents

- 37 °C incubator – separate to culture facility

- Tryptone Soya Agar (TSA) settle plates (90 mm irradiated, Cherwell Labs)

- Sabouraud Agar (SDA) settle plates (90 mm irradiated, Cherwell Labs)

Method

1 Remove from outer packaging 2 × TSA and 2 × SDA plates and label each plate with details of the area to be monitored, for example MSC or bench position, and date.

2 Open one of each plate with agar exposed to the testing environment and with lid face down. The second plate is a negative control for the same area and remains closed.

(continued)

3 Leave for 4 h, then close.

4 Incubate (closed) with agar uppermost at 37 °C for 2 days.

5 Record number and type of colonies on the settle plates. Incubate further at room temperature for four more days and record.

6 Establish background colony count levels for specific areas and refer to troubleshooting section if above upper limit and decontamination is required. Typical acceptable counts for a general cell culture laboratory are MSC <5, bench <20, incubator <10.

7 More extensive testing, possibly including the use of swabs and contact plates, may be required in specialized areas, for example aseptic dispensing rooms.

If the laboratory is a multi-user facility, then one individual should be assigned the tasks of overseeing the running and maintenance of the laboratory and the training of staff, and ensuring that all users abide by an agreed code. Examples of good practice include:

- Clearing away promptly and effectively after finishing sterile work, i.e. vacating the MSC and cleaning up any spills, followed by swabbing down the work surface with alcohol and drawing disinfectant through the aspirator line.

- Update hood log records – details of use, for example time, date, operator, cell type handled, cleaning procedure adopted.

- Dealing promptly and appropriately with waste and any used glassware to be soaked.

- Labelling opened sterile equipment and cell line-specific solutions, for reuse only by the same named individual.

- Maintaining good communal records, for example of shared cell or reagent stocks.

- Checking incubated cells regularly (daily) for condition and early signs of contamination.

- Checking incubators regularly and discarding surplus or contaminated stock cultures according to approved safety procedures (e.g. by inactivation, autoclaving or soaking).

Further guidance on good practice in the cell culture laboratory can be found in reference 20.

1.3 Troubleshooting

In a well designed laboratory, where routine monitoring is kept up to date and best practice followed (as outlined above in Section 1.2.6 and in other chapters of this book) few problems should arise.

The major issues likely to be encountered are poor cell growth, or contamination; these are covered in greater detail in Chapters 4 and 9 respectively. There are three main routes whereby culture conditions and performance can be compromised:

- **Operator error** – poor aseptic technique
  These are normally sporadic events and easily identified and rectified by careful disposal and additional cleaning.

- **Quality issues** – introduction of inherent problems via raw materials (including cell lines) or consumables
  Examination of batch records can often identify the causative material – highlighting the necessity for good documentation. If the source of the problem is not easily identified, switch to alternative batches, and/or refer to the supplier if appropriate, to identify the cause via a process of elimination. Identity testing should be implemented, if relevant.

- **External source** – equipment failure
  This can be separated from the above two points by the global nature of the problems encountered, for example if the incubator is faulty all the cultures within it will be affected. Extra equipment checks can be implemented immediately to help pin-point the cause of a problem more accurately prior to calling a specialized engineer. For example, in the case of poor cell growth, check the incubator. Assuming the displays are showing the expected readings, check the actual parameter values by another means, e.g. place a thermometer inside so that you can take a reading without opening the door, and use a CO_2 monitoring device to check the accuracy of the displayed reading. Recalibrate the incubator's sensors as required.

1.3.1 Microbial contamination

This particular type of contamination can be subdivided into two basic categories: those easy to detect, usually by eye or by microscopic observation of the effects on the culture (such as a change in pH and/or turbidity) caused by bacteria, mould and yeast; and those more difficult to detect such as viruses, mycoplasma or cross-contamination by other cell cultures. For further details see Chapter 9 and reference 19.

Microbial contamination can result in persistent or recurrent problems for the whole laboratory, which are much easier to prevent than to cure. The length of time during which a culture contaminant escapes detection will be an important determinant of the extent of the damage it creates in a laboratory or on a research project. The maintenance of high standards is therefore fundamental, and best practice should be actively encouraged [20].

Each occurrence of microbial contamination should be recorded and investigated as far as possible: no incident should be simply ignored once the clean-up is complete, as this is only storing up problems for the future. With frequent checks and good record keeping contamination events can be dealt with far more effectively or avoided entirely, minimizing disruption to the laboratory.

In general, there are three factors to assess and act upon immediately: extent, containment and clean-up. These are crucial to the correct handling of the situation and are immediate considerations. At a later time it is important to try to identify the source of the contamination. This may not be so easy to pin down – but especially in the case of recurrent problems, this becomes increasingly important.

1.3.1.1 Extent

As soon as possible, all users should be alerted to the fact that there is microbial contamination in the laboratory. All cultures must be checked in a thorough manner, with any found to be contaminated being removed and disposed of appropriately as soon as possible. It is important to establish how widespread the problem is – this will help in the other steps of the investigation.

1.3.1.2 Containment

Often the contamination is restricted to one flask or plate – an isolated incident for one user. In this case it may be possible to trace the movements of that culture and operator so that any equipment involved can be selectively cleaned and decontaminated, causing minimal disruption but ensuring there is no further reservoir of contamination. This process can be aided when MSC logs are in use and work is segregated by type within incubators. If it is suspected that poor technique or practice is involved, it may be necessary to give the individual involved further training and/or supervision.

Sometimes the number of users and cultures involved indicate a larger problem, and a full shut down and decontamination of the whole laboratory is advisable.

1.3.1.3 Cleaning up

Once the spread of the contamination has been established, a course of action can be identified and undertaken.

i. Limited spread → selective cleaning

- Identify, remove, sterilize and dispose of any contaminated cultures.
- Trace the movement of users and cultures.
- Identify MSCs, incubators and other equipment used.
- Dispose of all opened consumables used within affected MSCs.
- Thoroughly clean and decontaminate the MSCs: including all internal and external surfaces (fumigate if possible – see Chapter 2, *Protocol 2.6*).
- Empty the affected incubator, remove water tray. Thoroughly clean and decontaminate the incubator and shelves; run a decontamination programme if the incubator has one.
- Clean and decontaminate any other equipment affected.

- Decontaminate all work surfaces.

- Discard opened media used by the operator(s) involved.

- Send operators' lab coats for laundering.

ii. Wide spread problem → full shut down

- Sterilize and dispose of all growing cultures.

- Dispose of all opened consumables within the laboratory.

- Discard all open media, and reagent aliquots.

- Clean and decontaminate all equipment.

- Thoroughly clean and decontaminate all MSCs: including all internal and external surfaces (fumigate if possible – see Chapter 2, *Protocol 2.6*).

- Thoroughly clean and decontaminate all incubators (including shelves); run a decontamination programme if the incubators have them.

- Decontaminate all work surfaces.

- Thoroughly clean and decontaminate the laboratory; both high and low level including the floor.

- Send all lab coats for laundering.

- If possible, fumigate the laboratory (see Chapter 2, *Protocol 2.5*).

1.3.1.4 Identifying the source

The source of the contamination should be identified – if possible. This will enable remedial action to be taken, and help to prevent a repeat.

There are four main potential sources of contamination to consider:

- consumables

- raw materials (cells, media or media component)

- operator error

- external sources.

i. Consumables   Has the sterility of a consumable been compromised?

This may be local (e.g. pipette wrappings damaged when opening a box), or more widespread, for example a batch of tips failing to be irradiated by the manufacturer. The identity of the contaminated consumables can often only be confirmed if batch numbers are recorded.

ii. Raw materials

- *Media*
  Has a medium or medium component been contaminated?

    Again this may be a local issue (such as contamination upon aliquotting, or while preparing a bottle of medium), or could be due to a manufacturing failure. Samples of a new batch of medium can be incubated prior to use to confirm sterility.

    With good record-keeping it should be possible to see who has used which aliquots, or batch numbers, to see if there are any patterns which reflect the spread of contamination.

- *Cells*
  Cells are a potential source of contamination, for example primary cells or incoming cell lines. Once again where possible cell lines should only be obtained from reputable suppliers (preferably culture collections such as the American Type Culture Collection (ATCC) or the European Collection of Cell Cultures (ECACC) – see Chapter 9 for contact details) and cells quarantined on receipt. Once cell growth is established, tests (e.g. for sterility and the absence of mycoplasma – see Chapter 9) can be performed to ensure acceptability for further use and storage.

1.3.1.5 Operator error

Is the contamination due to one operator?

    While occasional sporadic contamination can occur for any operator, it may become evident that some users are more prone to this than others. This reinforces the need for operator training, especially for those new to cell culture, and also when new techniques are being employed. Training can be 'in house', an experienced operator looking over the shoulder to spot any obvious errors and offer advice, or could be provided by an external agency (such as the courses offered by ECACC – see Chapter 9 for contact details).

1.3.1.6 External sources

Sometimes the source of the contamination comes from outside the laboratory, and may be beyond the control of the users. One issue for cell culture labs is nearby building work – it is well known for spreading fungal spores far and wide. There may be shared areas that can cause problems (nearby bacterial fermenters etc.), or common utilities that may fail – such as an autoclave or air handling unit.

1.3.2 Quality control testing

The best strategy for reducing contamination is to be proactive by routinely monitoring supplies, media and solutions, work areas and, most importantly, cell cultures for contaminants before they are used in critical applications and experiments. Unfortunately, there are no easy solutions: no single microbiological medium can detect all types of biological contaminants, and practical testing methods often miss low levels of contaminants.

The process of detection is made even more difficult by the use of antibiotics in culture media. However, it is recommended that laboratories, as a minimum, incorporate QC tests for sterility, mycoplasma, and cell culture identity into their monitoring process. Each of these is described in detail in Chapter 9.

Acknowledgements

Permission to reproduce extracts from BS 5726:2005 is granted by BSI. British Standards can be obtained from BSI Customer Services, 389 Chiswick High Road, London W4 4AL. Tel: +44 (0)20 8996 9001. E-mail: cservices@bsi-global.com.

References

1. Harrison, R.G. (1907) Observations on the living developing nerve fiber. *Proc. Soc. Exp. Biol. Med.*, **4**, 140–143.

2. Carrel, A. (1912) On the permanent life of tissues outside the organism. *J. Exp. Med.*, **15**, 516–528.

3. Sanford, K.K., Earle, W.R. and Likely, G.D. (1948) The growth in vitro of single isolated tissue cells. *J. Natl. Cancer Inst.*, **9**, 229–246.

★ 4. Gey, G.O., Coffman, W.D. and Kubicek, M.T. (1952) Tissue culture studies of the proliferative capacity of cervical carcinoma and normal epithelium. *Cancer Res.*, **12**, 364–365. *– First description of the successful culture of human cells* in vitro.

5. Abercrombie, M. and Heaysman, J.E.M. (1954) Observations on the social behaviour of cells in tissue culture, II. 'Monolayering' of fibroblasts. *Exp. Cell Res.*, **6**, 293–306.

6. Eagle, H. (1959) Amino acid metabolism in mammalian cell cultures. *Science*, **130**, 432–437.

7. Hayflick, L. and Moorhead, P.S. (1961) The serial cultivation of human diploid cell strains. *Exp. Cell Res.*, **25**, 585–621.

8. Buonassisi, V., Sato, G. and Cohen, A.I. (1962) Hormone-producing cultures of adrenal and pituitary tumor origin. *Proc. Natl Acad. Sci. USA*, **48**, 1184–1190.

9. Yaffe, D. (1968) Retention of differentiation potentialities during prolonged cultivation of myogenic cells. *Proc. Natl Acad. Sci. USA*, **61**, 477–483.

10. Sato, G., Fisher, H.W. and Puck, T.T. (1957) Molecular growth requirements of single mammalian cells. *Science*, **126**, 961–964.

★ 11. Köhler, G. and Milstein, C. (1975) Continuous cultures of fused cells secreting antibody of predefined specificity. *Nature*, **256**, 495–497. *– First description of the isolation of hybridomas producing monoclonal antibodies.*

12. Bernard, C.J., Connors, D., Barber, L. *et al.* (2004) Adjunct automation to the cellmate cell culture robot. *J. Assoc. Lab. Automation*, **9**, 209–217.

13. Zhou, H., Wu, S, Young Joo, J. *et al.* (2009) Generation of induced pluripotent stem cells using recombinant proteins. *Cell Stem Cell*, **4**, 381–384.

14. UK Government (2002) *Statutory Instrument No. 2677. The Control of Substances Hazardous to Health Regulations 2002*, HMSO, Norwich, UK.

15. Advisory Committee on Dangerous Pathogens (2004) *The Approved List of Biological Agents*, HMSO, Norwich, UK.

★★ 16. Davis, J. and Shade, K. (2010) Aseptic techniques in cell culture, in *The Encyclopedia of Industrial Biotechnology: Bioprocess, Bioseparation and Cell Technology* (ed. Michael Flickinger), John Wiley & Sons, Inc., Hoboken, NJ, pp. 396–415. – *Provides a good introduction to aseptic technique for cell culture, as well as HEPA filtration, MSCs, and cleanrooms.*

17. Whitehead, P. (2007) Water purity and regulations, in *Medicines from Animal Cell Culture* (eds G. Stacey and J.M. Davis), John Wiley & Sons, Ltd, Chichester, UK, pp. 17–27.

18. National Sanitation Foundation (1983). *Standard no. 49 for Class II (Laminar Flow) Biohazard Cabinetry*, National Sanitation Foundation, Ann Arbor, MI.

19. Fogh, J., Holmgren, N.B. and Ludovici, P.P. (1971) A review of cell culture contaminations. *In Vitro*, **7**, 26–41.

★★★ 20. Coecke, S., Balls, M., Bowe, G. *et al.* (2005) Guidance on Good Cell Culture Practice: A Report of the Second ECVAM Task Force on Good Cell Culture Practice. *Altern. Lab. Anim.*, **33**, 261–287. – *Essential reading for anyone performing cell culture.*

2
Sterilization

Peter L. Roberts

Research & Development Department, Bio-Products Laboratory, Dagger Lane, Elstree, Hertfordshire, UK

2.1 Introduction

Cell cultures are essential tools in research and diagnostics, and also for the production of various biological products. Unfortunately, they provide an ideal environment for the growth of microorganisms and it is only by excluding such agents that cells can be successfully cultured and meaningful experimentation can be carried out. The presence of contamination can have a wide range of deleterious effects. The most common effect is the complete destruction of the cell culture. However, more subtle effects such as inhibition of cell growth, reduction in the yield of a cell-derived product or loss of sensitivity in a biological assay may occur. Although it may be possible to remove contamination from a cell culture, the prevention of contamination by sterilizing equipment and components [1, 2] remains the first-line approach. Although sterility is an absolute term, in practice methods are used that give a high probability that any microorganisms are completely absent or inactivated.

When a cell culture is free of contaminating microorganisms, sterility can be maintained by the use of good aseptic technique as outlined in this and other chapters. This approach represents the best line of defence, but other 'insurance' approaches such as the use of antibiotics and the banking of important cell lines also have their place.

In the current chapter, the various physical and chemical methods used for sterilization are reviewed, and protocols for their application described. Particular attention is given to viruses and prions, which are of concern in biological products.

Animal Cell Culture: Essential Methods, First Edition. Edited by John M. Davis.
© 2011 John Wiley & Sons, Ltd. Published 2011 by John Wiley & Sons, Ltd.

2.2 Methods and approaches

2.2.1 Wet heat

Heat treatment under moist or wet conditions is a commonly used method of sterilization. Many microorganisms, including bacteria and viruses, are relatively easily inactivated at temperatures of 60–80 °C under these conditions, and this process is termed pasteurization. In fact, lower temperatures of 55–60 °C are often adequate. However, spore-forming bacteria and prions are more resistant to heat. Pasteurization is used in the food industry for treating certain food products. These procedures will kill vegetative microorganisms and any viruses that are likely to be present, but not bacterial endospores. Pasteurization at 60 °C for 10 h is also used in the pharmaceutical industry for certain products derived from human plasma as a method of inactivating any viral contaminants (see Section 2.2.6.1). Boiling, or steam treatment, at 100 °C for 5–10 min represents a simple means of sterilizing equipment and fluids. It is very effective and will even destroy some types of bacterial endospores. By the use of three cycles of heat treatment with intervening incubation periods at 37 °C during which the spores germinate, even resistant endospores can be inactivated. This type of procedure can be used in the cell culture laboratory for decontaminating incubators that have a heat sterilization option.

Although these approaches can be useful, it is preferable to use other more severe methods such as autoclaving (Section 2.2.1.1) or dry-heating at high temperature (Section 2.2.2) where possible.

2.2.1.1 Autoclaving

Steam under pressure at 121 °C or more is a highly effective sterilization method which will kill even bacterial endospores. To produce such conditions, an autoclave is required. Some of the standard cycles that are commonly used are given in Table 2.1. It should be noted that prions are particularly resistant to even this severe form of heat treatment and require more extreme conditions for inactivation (see Section 2.2.6.3).

The quality of the steam is critical for effective sterilization. The presence of too much air (i.e. non-condensable gas) in the steam supply will reduce the effective steam

Table 2.1 Standard autoclave cycles.

Temperature (°C)	Time (min)	Pressure (bar^a)	Endospore survival^b
115	30	0.7	1 in 10^4
121	15	1.0	1 in 10^8
126	10	1.4	1 in 10^{17}
134	3	2.0	1 in 10^{32}

^{a}1 bar = 10^5 Pa.
^{b}For a relatively heat-resistant bacterial endospore such as *Bacillus stearothermophilus*. Note that the cycle is more effective at higher temperatures. The 115 °C cycle is substantially less effective and is thus not generally used.

pressure in the chamber. The presence of air in the chamber, either from non-condensable gas in the steam supply or because of poor air removal, will lead to a lower temperature in the chamber and load and thus reduce the efficiency of sterilization. For pharmaceutical applications filtered 'clean steam' derived from water of conductivity less that 15 μS is used

The types of autoclave available vary widely in size, complexity and versatility. They range from small, simple pressure cookers, which generate their own steam, to large fully automated machines possessing multiple cycle options.

i. Portable benchtop autoclaves   Pressure cookers as used in the kitchen at home are the simplest form of autoclave. Steam is generated in the chamber base, either by an external or internal heat source, and an air–steam mixture is discharged from the chamber through a hole in the top until only steam remains. A range of additional features may also be present in more sophisticated models. There may be an automatic timer, safety features that seal the lid until the cycle has been completed and the contents have cooled to 80 °C, and cycle-stage indicators. A guide to the operation of a basic model is given in *Protocol 2.1*.

In addition a range of larger capacity (e.g. 100-l) floor-standing autoclaves, which are either top- or front-loading, are available. These are often essentially larger versions of the benchtop models, although more sophisticated types exist.

PROTOCOL 2.1   Operation of a small basic benchtop autoclave

Equipment and reagents

- Simple pressure cooker or small dedicated autoclave
- Autoclave paper bags or autoclave sheets, or foil for wrapping
- Temperature-sensitive autoclave indicator tape
- Items to be autoclaved

Caution: high temperatures and pressures are involved in this procedure

Method

1 Consult manufacturer's instructions and safety guidance before using instrument. The following protocol is for outline guidance only for a basic model and can vary considerably among models[a].

2 Remove lid and fill to required depth with water.

3 Prepare load[b] comprising equipment or fluid[c] items for sterilization.

4 Add a piece of autoclave tape to each item in the load.

5 Place the items to be sterilized into the machine on shelves or in a basket as required.

(continued overleaf)

6 Replace lid and seal it, heat water to boiling.

7 Allow steam to escape for several minutes before closing the pressure valve as appropriate.

8 Start timer when required pressure and temperature are reached[d].

9 At the end of the cycle, turn off the heat and allow the pressure and temperature to fall to below 80 °C, following the manufacturer's instructions, before removing the contents[e].

Notes

[a]More sophisticated models may include a choice of cycles such as 121 °C or 134 °C, a timer, a drying option, safety features such as a temperature lock, cycle-stage indicators and automatic timers.

[b]Wrap equipment items in aluminium foil or place in sterilization bags fastened with autoclave tape, but do not seal completely. Loosen the caps of bottles, and other containers, or wrap up the open tops. Some items, for example pipettes or micropipette tips, may conveniently be placed in containers such as tins or beakers, which are then covered with aluminium foil or a sterilization bag. However, remember to loosen the tops of tins. Micropipette tips are best sterilized in the boxed racks supplied by the manufacturer. The necks of autoclave bags containing mixed discard loads must be left open to ensure good steam penetration.

[c]For a fluid cycle, bottles may either be sealed or left unsealed. Check caps for their ability to withstand autoclaving. Note that with some types of cap the rubber liner may get sucked into the bottle.

[d]The temperature in the load is not usually monitored. There is no cycle record, i.e. a print-out or trace.

[e]The load only cools slowly after sterilization. The combination of wet steam and no, or limited, drying capability can lead to items being wet when removed from the autoclave. As microbes can grow through wet autoclave bags, such items should ideally be placed in a unidirectional airflow or microbiological safety cabinet to dry before being placed in storage.

ii. Gravity displacement autoclaves   More sophisticated autoclaves require an external steam supply. In gravity displacement autoclaves, air is displaced from the bottom of the chamber by the less dense steam entering at the top. These machines also possess a jacket that will contain steam or water at different stages of the cycle. Spray-cooling systems are used in some bottled-fluid autoclaves to cool the chamber contents more rapidly. A combination of heat in the jacket and a vacuum in the chamber can be used to aid in the drying of the load. This type of autoclave can be used with items that can be freely exposed to steam, for example glass or certain types of plasticware and also bottled fluids. However, it is not suitable for porous loads and wrapped items because air removal and steam penetration is not adequate.

iii. Multicycle porous-load autoclaves   Porous-load autoclaves (Figure 2.1) solve the problem of air removal from wrapped items and other loads by the use of a vacuum.

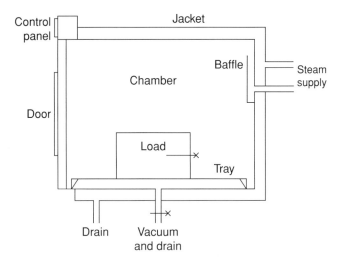

Figure 2.1 Representation of a multi-purpose laboratory autoclave. Steam or water supplied to the jacket is designed to warm or cool the chamber or load depending on the stage in the cycle. The control panel contains various indicators for temperature, pressure and cycle stage, as well as various recording instruments such as a chart recorder or printer. Thermocouple probes ($\rightarrow\times$) placed in the chamber drain or in the load are used to control the phases of the cycle and to monitor the conditions. A filter is usually placed on the air inlet (not shown) in order to prevent the load from becoming contaminated when air enters the chamber at the end of the cycle.

This, when used in pulses combined with steam, efficiently removes air from the load and chamber. The use of a vacuum also aids in the final drying stage by removing excess moisture. Such machines tend to be large with a minimum overall size of about $4\,\mathrm{m}^3$ and a chamber of at least $0.5\,\mathrm{m}^3$.

They can accommodate, for example 40×500-ml bottles, several large 5- or 10-l vessels, or several bags of discards. Some machines used in hospitals or the pharmaceutical industry are far larger and materials can be directly wheeled in, or loaded via dedicated trolley systems. In the laboratory a multicycle machine is commonly used, with a number of different cycles (Figure 2.2) available for use depending on the nature of the load. Pressure and temperature during the run are indicated by dials or digital meters. In addition a printer and/or chart recorder is fitted to allow the temperature throughout the cycle to be recorded. On more recent models an air-detector is usually fitted to confirm that all the air has been extracted from the chamber.

• *Preparation of the load and operation of the machine*

Before routine use, the sterilization of typical loads should be validated as described below under *iv. Autoclave testing*. The heat sensitivity of the items should be considered and tested before use. Many plastics can withstand 121 °C, but polystyrene cannot.

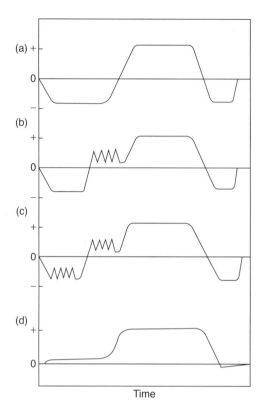

Figure 2.2 Examples of the pressure–time profiles of various autoclave cycles. A multipurpose laboratory autoclave possesses a number of cycles for different loads. (a) For porous loads, that is wrapped items or laboratory discards, the simplest cycle may use a single initial vacuum (−) pulse at the start of the cycle, before steam under positive pressure (+) is used. At the end of the sterilizing phase, a vacuum is used to dry the chamber contents while the load remains warm. (b, c) Cycles with a combination of negative and positive steam pressure pulsing stages are even more effective for air removal and steam penetration with porous loads. (d) A cycle in which the air is simply removed by gravity displacement as the steam enters the chamber. This is only suitable for bottled fluids and unwrapped items, and not porous loads. The exact details of pressures, times and resulting temperatures will depend on the specific nature of the cycle and its intended application.

Fluids such as water, salt solutions and buffers can also be autoclaved. The heat sensitivity of more complex formulations containing components such as vitamins, growth factors, antibiotics or heat-sensitive amino acids, for example glutamine, should be considered or tested. Cell culture media and other culture reagents are not usually autoclavable. However, autoclavable formulations of a few limited types of media exist, for example autoclavable minimal essential medium without glutamine. Temperature indicators such as autoclave tape, tubes (Browne's tubes) or indicator strips should be included in the load. *Protocol 2.2* outlines how the autoclave should be operated.

PROTOCOL 2.2   Operation of a multicycle porous load autoclave

Equipment and reagents

• Large multicycle autoclave

• Other equipment as in *Protocol 2.1*

Caution: high temperatures and pressures are involved in this procedure

Method

1 Follow the manufacturer's instructions, including safety aspects, and standard operating procedures[a].

2 Carry out routine or weekly tasks such as cleaning the door-seal, removing debris/broken glass from the chamber and cleaning the drain filter[b].

3 Perform the warm-up, leak-rate or performance tests as necessary.

4 Prepare load as in *Protocol 2.1*. Load the chamber but do not overfill.

5 Fill in details of run in the logbook.

6 Place the wander-probe into the load. In the case of a bottled-fluids cycle, place the probe into an identical control bottle with liquid which can be discarded later.

7 Seal the door, select the appropriate cycle and start the run.

8 On completion the cycle-end indicator will come on.

9 Check the chart-paper or printout for a satisfactory run[c].

10 Open the door and remove items using heat-resistant gloves to prevent injury.

11 Check autoclave tape or other in-load indicators. Label each item to indicate sterility if autoclave tape has not been applied to every item.

12 Complete logbook record[d].

Notes

[a] Before the machine is first used and at regular intervals thereafter, for example 6-monthly, arrange for the sterilization of critical worst-case loads to be validated by thermocouple studies carried out by sterilization/quality assurance personnel. Arrange for servicing to be carried out at regular intervals, for example every 3–6 months, and keep records of test results and machine performance.

[b] At regular intervals, for example daily or weekly, carry out a Bowie–Dick test (see *Protocol 2.3*), to assess steam penetration into a standardized difficult porous load. This may be less critical for modern machines fitted with an air detector.

[c] At the end of every run, check the chart and/or digital recorder for a satisfactory cycle. Relevant parameters include temperature, time and pulsing profiles.

[d] Record the details of all runs, including warm-up and leak-rate tests, in a log book with details of date, load, cycle, result (pass or fail), operator and other relevant information.

iv. Autoclave testing For an autoclave to perform satisfactorily and to ensure the sterility of the load, it is necessary for tests to be carried out by the user and by a sterilization engineer on a regular basis and for records to be kept [3]. One particularly important test for ensuring effective air removal from porous loads is the Bowie-Dick test (see *Protocol 2.3*). This test is designed to confirm that air removal and steam penetration into a standard test pack, during a porous load cycle, is effective, rapid and even.

PROTOCOL 2.3 Bowie–Dick and other tests for porous-load autoclaves

Equipment

- Cotton sheets 90 × 120 cm
- Autoclave sensor sheet or autoclave pack
- Tape
- Metal crate, *c.* 10 cm height

Caution: high temperatures and pressures are involved in this procedure

Method

This test is designed to confirm that effective air removal and steam penetration into a standard test pack, during a porous load cycle, is both rapid and even. It is performed within 1 h of a warm-up run. The standard method that is currently used, together with that originally described [4], are given below.

1 Take 25–36 dry cotton sheets (approx. 90 × 120 cm)[a]. Open them and allow to dry between tests. The sheets should be washed approximately every 2 weeks.

2 Fold each sheet in half three times and place one above another to form a stack approximately 25 cm in height. The test pack should weigh 7 ± 0.7 kg. The towels should be unfolded, dried and refolded opposite to the original crease after each test.

3 Fill in the relevant details on the autoclave sensor sheet[b] and place in the centre of the pack. The pack is then surrounded with a single sheet and secured with tape to keep the pack intact.

4 Place the pack on, for example, a metal crate approximately 10 cm high, in the centre of the autoclave. The wander-probe should be placed on the chamber hook if present, or on the base of the chamber. No other items should be present in the chamber during the test.

5 At the end of the run, check that the hold-time at sterilizing temperature was within the specified limits for the cycle used.

6 Remove the towels from the autoclave and interpret the results. An even distribution of darkening of the autoclave strips across the whole sheet indicates the run has passed.

(continued)

7 A variety of types of test sensor sheets or packs are available including pre-prepared disposable packs containing an indicator sheet surrounded by paper wadding to mimic the cotton sheets used in the original method.

8 In engineering validation tests using thermocouples placed in the pack, the temperature should be 115–117 °C for a 115 °C cycle, or 121–124 °C for a 121 °C cycle. The temperature indicated on the chart and digital read-out should all correspond. Other relevant loads should be tested including worst-case/difficult loads and items, for example discards, filters and wrapped items. In the case of liquid load tests, similar conditions must be met. All probes in the load should agree to within 1 °C.

Notes

[a] In the original method cotton huckaback towels (850 × 500 mm) were used.

[b] A sensor sheet is made using a sheet of A4 paper with a diagonal cross made of autoclave tape.

2.2.2 Dry heat

Heating in the dry state can also be used for sterilization. However higher temperatures and longer times are required for effective sterilization.

2.2.2.1 Incineration

Because of the risk that microbial contamination may be present in waste from the cell culture laboratory, it is recommended that such material should be autoclaved or treated by chemical methods in the worker's own laboratory or building prior to being sent for incineration [5]. For an incinerator to operate efficiently it should reach a temperature of 350°C and preferably have effective after-burners fitted to incinerate any unburned material that may be found in the smoke.

During routine aseptic technique, the openings of glass bottles and other containers may be briefly treated with the burning flame from a Bunsen burner. Metal items and scissors, used for example during the preparation of primary cell cultures, are re-sterilized between manipulations by being dipped in ethanol which is then ignited and allowed to burn off. **NOTE** – Make sure the flame is completely extinguished before doing **anything** further with the item.

2.2.2.2 Hot-air ovens

As an alternative to autoclaving, items can be heated in a hot-air oven. The main advantage of this method is that the equipment required is a lot less complex, and is simpler to maintain and cheaper to purchase. However, it can only be used with heat-resistant objects such as those made of glass or metal. Some plasticware can be treated in this way but requires a lower temperature and extended time, for example 120 °C for 18 h.

A hot-air oven comprises an insulated box with electrical heaters. Heat is distributed by simple convection or more effectively by a fan. For more critical applications, for example in the pharmaceutical industry, a filter may be fitted to the air-vent to prevent any possible recontamination of the load from occurring during cooling. For large-volume sterilization on an industrial scale, tunnel sterilizers are used in which items are continuously fed by conveyor through a hot-air tunnel.

The basic steps involved in preparing items for oven sterilization and for carrying out the process are given in *Protocol 2.4*.

PROTOCOL 2.4   Sterilization using a hot-air oven

Equipment and reagents

- Aluminium foil, or appropriate sterilization bags or sheets (these are available commercially)
- Heat sensitive indicator paper
- Sterilization oven able to reach circa 200 °C

Caution: high temperatures are involved in this procedure

Method

1 Wrap items in aluminium foil or sterilization sheets, or place in sterilization bags. Bottles or other containers need only have their openings wrapped or capped[a].

2 Add heat-sensitive tape, or other temperature indicator, preferably to every item in the load.

3 Fill the oven by placing items on shelves. To aid circulation of air and to promote warming up, do not over-fill.

4 Set timer[b], temperature and chart recorder controls. Various cycles can be used[c]. Add on additional time to allow for oven and/or load warm-up as necessary.

5 On completion, allow oven to cool. Some ovens have an automatic safety device which prevents the door from being opened until the temperature has dropped to a pre-set level.

6 Check chart and in-load sterilization indicators for a satisfactory cycle and then remove the load. Mark all items as sterile and complete the oven logbook.

Notes

[a]Check that the cap can withstand heat treatment.

[b]The required temperature and time is set and the actual temperature monitored by an in-built digital or dial thermometer and/or a thermometer placed through a port into the chamber. Some timers are only triggered when the required temperature is reached in the chamber, but note that the load itself may take longer to warm-up. To monitor the temperature continuously a chart recorder may be included.

[c]Various temperature–time combinations can be used for sterilization. The most commonly used conditions, are 160 °C for 2 h, 170 °C for 1 h and 180 °C for 0.5 h. These periods exclude the

(continued)

time required for the oven and load to warm up to the required temperature. Other conditions can be used where items cannot withstand these high temperatures, for example 150 °C for 2.5 h, 140°C for 3 h or 120 °C for 18 h, although these are generally less effective. The lower temperatures are useful for sterilizing certain types of plastic, for example polypropylene and polycarbonate as used in centrifuge tubes, micropipette tips and boxes. However, they are not suitable for polystyrene. The supplier's catalogue should be consulted for further details on the heat sensitivity of their plasticware.

i. Routine monitoring and testing Various indicators can be included as simple checks to confirm that the load has been heat treated. Most of these simply indicate that a specified temperature has been reached. However, as the time is not measured they do not indicate actual sterility. The most commonly used and cheapest method is the use of heat-sensitive tape on which dark stripes appear when a temperature of 160 °C is reached. Alternatively, indicator strips which change colour over specific ranges of, for example, 116–154 °C or 160–199 °C are available.

The performance and instrumentation of an oven should be tested after initial purchase and on a regular basis. Thermocouples placed in contact with the oven temperature indicator probe should confirm that the warm-up takes less than 135 min, overshoot is less than 2 °C, and drift over the complete cycle is less than 2 °C. The temperature in the load should be within about ±5 °C of the oven temperature indicator.

2.2.3 Irradiation

2.2.3.1 Ultraviolet light

Microorganisms and viruses can be inactivated by ultraviolet (UV) light at a wavelength of about 260 nm. However, because of its poor penetrating power, its usefulness is very limited. It can be used to sterilize surfaces and the air in cell culture cabinets and rooms, but only after they have been cleaned. As the power of UV lamps decreases with use, they should be checked on a regular basis. As found with heat, bacterial spores and prions are highly resistant. Also, many organisms possess DNA repair mechanisms that can overcome limited damage.

2.2.3.2 Gamma rays

Gamma irradiation is a commonly used method of sterilization. It is used with items that cannot be heat sterilized, and has the advantage that, because of its good penetrating power, items can be completely sealed and packaged. Various items of disposable plasticware can be bought pre-sterilized in this way, for example tissue culture flasks and dishes, filters, plastic pipettes, syringes, pipette tips and some chemicals such as antibiotics. A dose of 25 kGy from a cobalt-60 plant is the standard dose used by industry and is effective as long as only low levels of microbial contamination are present. Bacterial spores and viruses tend to be somewhat more resistant to irradiation and prions are extremely resistant.

2.2.4 Chemical sterilization

2.2.4.1 Fumigation

Gases such as formaldehyde or ethylene oxide can be used for fumigation. Both are effective against all types of microorganisms including viruses, although bacterial endospores are somewhat more resistant in the case of ethylene oxide. Their activity is greatest at higher temperatures, and humidity levels of 75–100%. However, conditions that reduce the accessibility of the microorganisms to the gas, for example being dried in organic or inorganic material, will decrease the effectiveness of fumigation.

Ethylene oxide, alone or in combination with steam, is commonly used for sterilizing items of clean equipment in so-called low-temperature autoclaves, particularly in hospitals. In the laboratory some items of plasticware may be purchased pre-sterilized with ethylene oxide, for example syringes and filters. However, this method of sterilization can leave behind residues that are toxic to cells, and other methods, for example gamma-irradiation, are preferable (see Section 2.2.3.2).

Hydrogen peroxide is being increasingly used as a new low-temperature sterilizing technology [6]. Treatment with this chemical is relatively rapid when compared with the use of ethylene oxide, that is 1–2 h versus up to 12 h. There is also the advantage that hydrogen peroxide is less toxic and only produces water and oxygen as by-products. This compound is commonly used for treating sensitive items of equipment, such as hospital endoscopes or sterile isolators, and can also be used for room decontamination. However, some items such as paper and cellulose are not compatible with this chemical.

Formaldehyde gas is used to decontaminate or sterilize unidirectional airflow cabinets and rooms used for handling cell cultures, as well as for small items of equipment. An outline of the procedure used when treating rooms is given in *Protocol 2.5*. Microbiological safety cabinets that are used for handling cell cultures may also be decontaminated with formaldehyde gas (*Protocol 2.6*), but this method can only be used if the cabinet can be sealed and the extract ducted to the exterior of the building.

PROTOCOL 2.5   Fumigation of a room using formaldehyde

Equipment and reagents

- Formaldehyde solution 40% (formalin)

- Sealing tape

- Hot-plate or formaldehyde kettle

- Safety notice

- Breathing apparatus

- Formaldehyde testing equipment

(continued)

Caution: formaldehyde is toxic by inhalation

Method

1 Remove all unnecessary items from the room, including those that must not be exposed to the gas, for example cell cultures, sensitive electrical equipment, and so on.

2 Clean the room to minimize the level of microbial contamination and to allow good gas penetration.

3 Turn off air-handling system and seal up room with sealing tape as far as possible.

4 Place a hot-plate with a large saucepan, connected to a timer, in the room (the time to boil off the liquid should be determined in advance). Alternatively use a dedicated formaldehyde-generating kettle[a].

5 Fill the container with formalin solution (20 ml per m^3 of room volume).

6 Turn on the hot-plate or kettle and leave the room immediately[b].

7 Lock the door, tape it up fully (including the keyhole) and attach safety warning notice. Leave until the next day.

8 Using a remote switch, turn on the room air extract and any cabinets ducted to the exterior of the building and vent to atmosphere. If a remote switch is not available, enter wearing breathing apparatus and turn on cabinet and/or air-handling system. Ensure there is an adequate supply of fresh air to the room to flush out the remaining formaldehyde.

9 Leave until the level of formaldehyde reaches an acceptably low (safe) level. Testing equipment should be used to confirm this.

10 It may take several days to fully remove the gas. The room may require cleaning to remove white residues of paraformaldehyde.

Notes

[a] Other methods for generating formaldehyde gas exist, for example heating paraformaldehyde (10 g per m^3 of room volume) or, for those situations where fumigation is carried out regularly such as in pharmaceutical manufacturing areas, equipment that generates a formaldehyde mist.

[b] It should be noted that formaldehyde gas is toxic and thus appropriate precautions, including the use of breathing apparatus, should be taken where necessary.

PROTOCOL 2.6 Fumigation of a microbiological safety cabinet

Equipment and reagents

- Formaldehyde solution 40% (formalin)

- Integral or small portable formaldehyde generator

- Sealing tape

- Safety notice

(*continued overleaf*)

Caution: This procedure must only be used with a cabinet vented to the exterior of the building. Also note that formaldehyde is toxic by inhalation

Method

1 This procedure should be carried out on a regular basis, for example once a week or once a month (depending on frequency of use of the cabinet and the nature of the organisms being used), or after it has been found that contaminated cultures have been handled.

2 Clean the cabinet to ensure good gas penetration.

3 Fill the integral formaldehyde generator attached to the exterior, or a small portable generator placed inside, with *c.* 25 ml of formalin solution.

4 Also, place in the cabinet any items of equipment used in cell culture procedures, for example pipette-aids and so on. Check for chemical resistance to formaldehyde.

5 Replace the door and seal with sealing tape.

6 Switch on formaldehyde generator and place warning notice on the cabinet[a].

7 Leave overnight before turning on air-exhaust while opening cabinet door.

8 Leave to vent before cleaning cabinet to remove residues of white paraformaldehyde.

Notes

[a] It should be noted that formaldehyde gas is toxic and thus any necessary precautions must be taken. However, the use of breathing apparatus should not be necessary for this application.

2.2.4.2 Liquid disinfectants

Diluted disinfectants can be used in the cell culture laboratory for routine hygiene and disinfection of used items of equipment and surfaces in rooms and cabinets. The main properties of some of the principal types of liquid disinfectant are summarized in Table 2.2. This approach is also useful for the treatment of used or contaminated cell culture media before disposal.

There are a number of different types of disinfectant available and these are considered below. There are also a number of methods available for applying the disinfectant. These include using a cloth, a hand-held sprayer, or paper towels applied to liquid spills and then soaked in disinfectants. For a full 'wet' disinfection of a room, a 'knapsack' sprayer, or large-volume garden sprayer, with a long lance should be used. In the case of items of glassware or plasticware, they must be fully immersed, that is with no air pockets, in the disinfectant in order for the procedure to be fully effective.

i. Aldehydes Formaldehyde and glutaraldehyde disinfectants are not generally influenced by the presence of high levels of organic matter and they will inactivate all types of conventional microorganisms including bacterial endospores. Formaldehyde is used at about 4%, and is conveniently prepared by diluting 40% formaldehyde solution (formalin). Glutaraldehyde is used at 2%. A number of commercial aldehyde formulations are

Table 2.2 Use of common disinfectants.

Type	Standard conditions[a]		Effect[b]			
	Concentration	Time (min)	Bacteria	Endospores	Fungi	Viruses
Alcohol	70%	30[c]	+	−	−	±
Hypochlorite	10 000 ppm	30	+	+	+	+
Aldehydes	2–4%	30	+	+	+	+
Phenolics	2–5%	30	+	−	+	±

[a]Many disinfectants are available as proprietary products. In these cases the specific manufacturer's recommendations should be followed. Note that organic matter neutralizes the antimicrobial activity of some disinfectants and thus items should be cleaned prior to disinfection. In the case of contaminated items, this may simply involve an initial cleaning step with the disinfectant itself.
[b]Effective inactivation (+) depends on the microorganism involved. In the case of viruses, the agent may be effective against enveloped viruses but only partially (±) effective against some non-enveloped viruses.
[c]Because alcohol tends to evaporate rapidly, it may require repeated applications to achieve a sufficient exposure time.

available. Treatment times are typically of the order of at least 30 min, with longer times required to obtain full sporicidal activity. Appropriate safety precautions must be taken with these toxic chemicals.

ii. Hypochlorite   Hypochlorite is a very widely used disinfectant but is not effective in the presence of high levels of organic matter. Also, it corrodes metals and only has a shelf life of 24 h after dilution. The concentration recommended for use when high levels of organic matter are present is 10 000 ppm of available chlorine, although lower concentrations have been recommended for routine hygiene, for example 2500 ppm. An exposure time of at least 30 min, and preferably overnight, is recommended. For treating spills, sodium dichloroisocyanurate powder can be used as an alternative to absorb the spill.

iii. Phenolics   Phenolic disinfectants such as Hycolin are commonly used at a concentration of 2–5%, or as recommended by the manufacturer. Although not inactivated by organic matter, they have little activity against bacterial endospores. They may also leave sticky residues when used for cleaning surfaces.

iv. Alcohol   Ethanol, or propanol, is commonly used for disinfecting surfaces and gloved hands. Its effect is optimal at a concentration of 70–80%. After spraying, the ethanol/water mixture is then left to evaporate off naturally. Although a convenient and easy disinfectant to use, and of low toxicity, it is not very effective against fungi, bacterial endospores or non-enveloped viruses. Caution must be exercised in its use because it is flammable. It is best used as a disinfectant for less critical applications or in combination with other disinfectants.

v. Others   There are a wide range of other disinfectants that can be used in the cell culture laboratory including hydrogen peroxide (5–10%), acids, alkalis, ethanol (70%) mixed with 4% formaldehyde or 2000 ppm hypochlorite, or ethanol/propanol/aldehyde mixtures. One proprietary disinfectant, Virkon (Antec International), uses a combination of peroxide, low pH and detergent that has been shown to inactivate a wide range of viruses and other microorganisms.

2.2.5 Filtration

Membrane filtration using filters with a porosity of 0.2 μm will remove bacteria and fungi, and is the method commonly used for sterilizing solutions that cannot be sterilized by methods such as autoclaving. Uses in the cell culture laboratory may include sterilizing culture media, sera and other cell culture supplements, and any biological products made by the cells.

2.2.5.1 Types of filter

A range of filter materials is available that can be used for removing unicellular microorganisms from liquid by filtration [7]. These include materials made of cellulose acetate, cellulose nitrate, or a mixture of both, nylon, polysulphone or polycarbonate. Other materials with particularly low protein-binding properties, for example polyvinylidene difluoride, can be used for critical applications, although these are not generally needed. Principal filter manufacturers include Sartorius, Pall and Millipore, as well as some of the manufacturers of cell culture consumables.

Membrane filters come in a range of pore sizes, but 0.2 μm is considered the standard for removing bacteria and fungi. Larger pore sizes and depth pre-filters, for example of glass fibre, are useful in serial filtration systems to increase the filtration capacity where significant levels of particulate material may exist, such as in serum, complete medium containing serum, or cell culture harvests. For effectively removing mycoplasma, filters with a pore size of 0.1 μm are required and these are now used by most commercial processors of serum products. In addition, filtration to this level may also be useful for removing some of the larger viruses that may be present in such products, for example infectious bovine rhinotracheitis virus and parainfluenza-3.

Apart from filtering liquid, membrane filters are also used for filtering gases, for example CO_2 or O_2 supplied to cell cultures growing in mass-culture systems. A membrane filter is an integral part of the cap of some types of cell culture flasks. Various types of filter, ranging from membrane filters to simple cotton wool plugs, are also used to protect liquid-handling equipment, such as pipettes, automatic pipetting devices and micropipette tips, from air- or liquid-borne contamination.

2.2.5.2 Types of filtration unit

Units range from the simple syringe-filter of 11 or 25 mm diameter, which will filter small volumes, to cartridges of various sizes and shapes, often pleated to maximize filter area, for filtering large volumes. Disposable units designed for filtering liquids directly into a bottle (i.e. bottle-top filters), some with an integral receiving vessel, are also available.

Filter units comprising the filter housing and the filter itself can be purchased either as pre-assembled, sterilized and disposable units or as a reusable housing which, after fitting the required filter(s), can be sterilized by autoclaving. Disposable units, despite their higher costs, are increasingly being used because of their convenience and assured integrity/quality.

2.2.5.3 Filtration set-up

Filtration is generally carried out under positive pressure, for example by depressing a syringe plunger for small volumes up to c. 50 ml, or by air pressure from a pump or pressure-line, or by or a peristaltic pump. However, negative pressure can also be used, for example with bottle-top filters and filter units with integral receiving vessels. However, the latter method has the disadvantage that with tissue culture medium containing bicarbonate the pH rise during filtration is more pronounced. In addition, when protein is present, frothing and protein denaturation can occur. Filter, cell culture and plasticware manufacturers can provide various complete filtration systems designed for use in particular situations.

In *Protocol 2.7* the typical steps involved in liquid filtration are described, and a typical setup is shown in Figure 2.3.

PROTOCOL 2.7   Sterilization of liquids by positive-pressure membrane filtration[a]

Equipment and reagents

- Pre-filter depth or membrane filter discs of 0.45, 0.8 or 1.2 μm pore-size (if required)

- Sterilizing membrane filter discs of 0.2 μm pore size

- Stainless steel filtration apparatus

- Pressure source

Method

1 A diagram of a typical filtration system is shown in Figure 2.3. Place any pre-filters required and a 0.2-μm sterilizing grade filter in the disassembled filter housing.

2 Assemble the unit taking care to install all supports and O-rings in their correct positions. Take care not to damage the filter(s). Attach tubing, if necessary, to inlet/outlet ports.

3 Wrap the items and sterilize by autoclaving at 121 °C for 15 min using a porous-load cycle (see Section 2.2.1.1 *iii*, and *Protocol 2.2*). Higher temperatures, or dry-heat sterilization, must not be used unless recommended by the filter manufacturer.

4 Place any accessories that are required in a microbiological safety cabinet or vertical unidirectional flow cabinet.

(continued overleaf)

5 Remove the sterilized filter from its packaging using aseptic techniques, and assemble the complete filtration system.

6 Tighten all connections and attach to a suitable pressure source containing the liquid to be filtered[b,c].

7 Ensure pressure is adjusted to fall within the range recommended by the filter manufacturer. Start filtering into sterile container(s), after having bled off any air in the system via the air-vent (if present).

8 Test the integrity of the filter unit (see Section 2.2.5.4) and/or carry out sterility tests on the filtered liquid as necessary.

Notes

[a]The method described assumes the use of a non-sterile, reusable filter housing employed to contain a non-sterile, single use filter disc(s). If a filter disc or cartridge unit, pre-sterilized by gamma irradiation and supplied in a sterile package, is to be used, then the protocol should be started at step 4.

[b]If necessary, pre-filter the solution, using depth or membrane filters of large pore size. This will increase the capacity of the sterile filtration stage.

[c]Alternatively, positive pressure can be provided by pumping the liquid using a peristaltic pump.

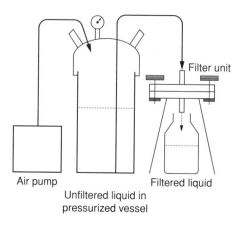

Figure 2.3 Diagram of a positive-pressure liquid filtration setup. A number of different systems may be used to sterilize liquids by filtration. In this example, designed for filtering volumes of about 1–5 l, a pump or pressure line is used to pressurize a vessel holding the liquid to be sterilized. This is then filtered through a filter disc contained in a stainless steel filter housing. Alternatives include the use of filters in cartridges which can be purchased presterilized.

2.2.5.4 Testing of bacterial filters

To ensure the correct functioning of the filtration system, the assembled unit, that is filter and housing, needs to be tested for integrity. The manufacturers carry out on their filters various physical tests that are related to the pore size, for example bubble-point or air-flow/diffusion tests. These are then correlated with the actual performance of the filter or filter unit in tests designed to challenge the filters with high levels of an extremely small test bacterium, for example *Pseudomonas diminuta* in the case of a 0.2-μm filter. A number of procedures can be carried out by the user to confirm the integrity of a wetted, hydrophilic filter unit. The tests should be carried out before and/or after using the filter. The methods are the following.

(a) The bubble-point method, which is based on determining the pressure required to force air through the wet filter. In its simplest form this involves confirming that a significant resistance is felt when trying to force air through a wet filter using a syringe.

(b) The diffusion airflow or forward-flow method, which is based on determining the air-flow across the wet filter at a specified pressure (*c.* 80% of the bubble-point pressure).

(c) The pressure-hold test, which is based on measuring pressure decay after pressurizing the upstream of the filter.

In all cases, the filter manufacturers provide information on the pass limits for specific types of filter and housing when using a specified liquid. In addition to these physical tests, sterility testing of the final product, by inoculating a range of microbiological growth media, can be carried out. Both integrity testing and sterility testing are routinely carried out in the pharmaceutical industry.

2.2.5.5 HEPA filters

Large volumes of air in sterile or clean rooms, and in laminar flow or microbiological safety cabinets [8,9] can be filtered by HEPA (high efficiency particulate air) filters. Such filters act as depth filters and are capable of removing >99.97% of particles of 0.3 μm or larger. They are thus effective against not only bacteria but also viruses which are present as aerosols in the atmosphere attached to dust particles and liquid droplets. In the pharmaceutical industry, standards exist for the quality of air, that is the maximum level of particles and viable microorganisms permitted [8]. Cell cultures are best considered as a potential source of viruses and are thus best handled in a Class II MSC in a verti-cal laminar flow of HEPA filtered air (see Chapter 1, Figure 1.2). The air in the cabinet is exhausted through HEPA filters to remove microorganisms. These conditions provide protection to both the operator and the cell cultures, as well as the environment. Class I MSCs (see Chapter 1, Figure 1.1) offer no protection to the work but are recommended because they give a high and consistent level of operator/environment protection when handling dangerous pathogens (Hazard Group 3) [5]. This is because the air is totally ex-hausted through a HEPA filter. Alternatively, a fully enclosed Class III cabinet (see Chap-ter 1, Figure 1.3) would provide complete operator/environment protection and some level of protection to the work because all incoming and outgoing air, although not sub-jected to laminar flow, is HEPA filtered. Simple horizontal unidirectional flow cabinets, in which air is directed directly at the operator, should only be used for assembly of sterile equipment, as they provide no protection to the operator or environment. Similarly, simple vertical unidirectional flow cabinets, in which air is directed downwards towards the operator's hands, should only be used for the filtration or dispensing of non-hazardous liquids, as again such cabinets provide no protection to the operator or environment. Further details on the use of a microbiological safety cabinet are given in *Protocol 2.8*.

PROTOCOL 2.8   Use and maintenance of a microbiological safety cabinet

Equipment and reagents

• Microbiological safety cabinet

• 70% ethanol, or similar liquid disinfectant

Method

1 Turn on airflow while removing door panel. Leave the door in a suitable clean location (do not place on the floor). Turn on the cabinet's internal light.

2 Spray base and sides of cabinet with 70% ethanol[a] and leave to evaporate.

3 Leave cabinet to run for at least 5 min.

4 Check airflow dial reading is correct and in 'safe' position.

5 Work in cabinet using standard microbiological technique and limit rapid movements, especially of the arms.

6 After use, remove all items from the cabinet and clean the interior with disinfectant. Finally, spray with 70% alcohol.

7 Turn off airflow and light, and replace the door panel.

8 At regular intervals or after handling contaminated cultures, fully clean and disinfect the interior, including underneath the work surface[b]. Fumigate the cabinet, if possible (see *Protocol 2.6*).

9 Have the performance of the cabinet tested at regular intervals, at least once a year.

Notes

[a] Other alcohols such as isopropanol may also be used. Crude industrial-grade chemicals, which are less costly, are suitable.

[b] Perform this step with the fan switched off.

HEPA filters and cabinets incorporating them should be tested at least once each year, or more frequently for critical applications. Testing should include challenging the HEPA filter with dioctyl-phthalate smoke (this contains particles of about 0.3 μm diameter) and confirming the airflow velocities meet appropriate standards, for example BS5726 in the UK. For MSCs, an operator protection test, using potassium iodide generated within the cabinet (KI-Discus test) should also be carried out when the cabinet is first installed and at regular intervals. The efficiency of the HEPA filters in a sterile room should also be tested. The level of particles within a sterile room can be determined using a particle counter. Additional biological tests for product protection in rooms and cabinets used for sterile filling can be carried out by exposing bacteriological agar plates for the detection of viable organisms.

2.2.6 Viruses and prions

2.2.6.1 Virus elimination

Viruses pose a particularly difficult problem in cell culture due to the ability of many to be latent or persistent without causing any obvious effects. The first approach for controlling such agents is the screening of the cell line. The limitations of the screening approach include the fact that detection methods may be of limited sensitivity and be too specific with regard to the range of viruses they can detect. A complementary approach is to inactivate or remove any virus that may be present [10–13] in the biological additives added to cell cultures or in the products produced by the cultures. (However, this approach cannot be applied to the cell cultures themselves.) The methods that are commonly used for this purpose are not as universally effective as standard sterilization methods such as autoclaving. For this reason specific validation using relevant test viruses may be necessary for any specific product and process and the incorporation of multiple virus-reduction steps should be considered.

A wide range of methods has been specifically developed for inactivating and/or removing viruses from biological products. Terminal or in-process heat treatment in the liquid state, that is pasteurization at 60 °C for 10 h, is one commonly used method. However, the addition of specific stabilizers (e.g. sodium octanoate in the case of human albumin) or general stabilizers (e.g. sugars, amino acids) is required in all cases to prevent protein denaturation. The use of less severe treatment conditions, that is 30 min at 56 °C, has been used for treating bovine serum [14]. An alternative approach is to heat the protein in the dry state after freeze-drying. Treatment conditions of 80 °C for 72 h or even 100 °C for 24 h have been used with coagulation factors. Chemical methods can also be used to inactivate viruses, the most widely used being solvent/detergent treatment [15] for the inactivation of enveloped viruses. The product is treated with the solvent tri-*n*-butyl phosphate, combined with a suitable non-ionic detergent, for example sodium cholate, Tween 80 or Triton X-100. Low pH, for example pH 4, is effective for the inactivation of enveloped viruses and some non-enveloped viruses.

In the case of serum used for cell culture, the most commonly used method is gamma irradiation at a dose of 30 kGy. This method is recommended by the regulatory agencies [16]. Treatment can be carried out on the bulk serum or on the final bottled material after sterile filtration. This procedure has been shown to be effective for inactivating a wide range of viruses without affecting the growth promoting properties of the serum [17].

2.2.6.2 Filters for viruses

Filters have been used for removing viruses from biological products [18–21]. Such virus filters, or nanofilters, have a very small pore size and are available in a range of formats, as shown in Table 2.3. Such filters have been shown to remove viruses in a size-dependent fashion. Some filter types come in a range of pore sizes and the most appropriate type that will maximize virus removal without removing significant levels of any important medium component or product should be selected and further tested. The efficiency of virus removal can be increased by using multiple units in series.

Virus-removing filters have found application, in the cell culture field, in eliminating viruses, particularly those of a large size such as bovine viral diarrhoea virus (BVDV), infectious bovine rhinotracheitis virus (IBRV) and parainfluenza-3 from bovine serum. Virus filters are currently being used in the pharmaceutical industry in the production of a wide range of cell culture-derived or plasma-derived biological products.

As with sterilizing filters that are designed to remove bacteria, it is essential that integrity testing is carried out on viral filters. Different test methods are available to the user depending on the type of filter. The integrity tests used are essentially similar to those used with standard bacterial sterilizing filters, although viral filters need to be pre-treated in order to lower the bubble point. In all cases the performance of the filter in the integrity test has been correlated with the removal of a representative virus, usually a bacteriophage which has been selected to have a size near to that of the filter pore size. A pass value is recommended by the filter manufacturer. In the case of Planova filters, a gold particle removal test is used. This method involves evaluating the removal by the filter of gold particles slightly larger than the pore size. The absorbance of pre- and post-filtration samples are measured on a spectrophotometer and used to calculate a gold particle reduction factor which is correlated with virus removal.

2.2.6.3 Prions

Prions are the agents that cause transmissible spongiform encephalopathies (TSEs) [27, 28] such as Creutzfeldt–Jakob disease (CJD), bovine spongiform encephalopathy (BSE) and variant CJD (vCJD). CJD can be transmitted by human-derived tissue, for example cornea, or by products like human growth hormone.

Prions differ considerably from conventional viruses, with infectivity being associated with a specific protein, that is the abnormal form of prion protein (PrP^{Sc}), rather than a nucleic acid. In addition, prions are extremely resistant to inactivation by conventional sterilization methods [29]. Furthermore, suitable screening methods have yet to be developed that are sensitive and specific enough for routine use. Currently the most sensitive detection methods rely on infectivity titration using mice or hamsters and take about a year to complete. Health screening of donor animals (or humans) and careful sourcing of biological materials is the approach recommended for controlling the transmission of such agents [30]. Thus, in the case of cell culture, it is recommended that alternatives to bovine-derived serum are used. Where this is not possible, material from countries that are free of BSE should be used (see www.oie.int).

Table 2.3 Filters for removing viruses.

Manufacturer[a]	Filter type	Material[b]	Grade	Pore size[c]	Protein cut-off (kDa)[d]	Virus removal (nm)[e]	Operation[f]	Removal Mechanism	Integrity Testing
Millipore	Viresolve	PVDF	70	—	70	≥30	T	Membrane	Liquid/Liquid Intrusion
		PVDF	180	—	180	≥30	T		
		PVDF	NFP[g]	—	160	≥20	D		
		PES	NFR[h]	—	700	≥80	D		
Asahi	Planova	Cellulose Fibres	15N	15 nm	160	≥20	D or T	Depth	Gold Particle or Leakage
			20N	20 nm	350	≥20			
			35N	35 nm	800	≥50			
Pall	Ultipor VF	PVDF	DV-20	—	70–160	≥20	D	Membrane	Forward-Flow
			DV-50		160–200	≥50			

[a] Additional information can be found in refs 20, 22 (Viresolve); 23, 24 (Planova) and 25, 26 (Ultipore VF).

[b] Membranes are constructed of polyvinyl deoxyfluoride (PVDF) or polyethersulphone (PES) or cuprammonium regenerated cellulose fibres.

[c] The nominal or approximate pore-size and molecular weight cut-off, as provided by the filter manufacturer, are given.

[d] Note for reference; the molecular weight of immunoglobulin is about 150 kDa and interferon and other interleukins <35 kDa.

[e] Viruses of ≥20 nm include minute virus of mice (MVM) and B19, of ≥30 nm encephalomyocarditis virus (EMC) and hepatitis A virus (HAV), of ≥50 nm BVDV and hepatitis C virus (HCV), and of ≥80 nm IBRV, murine leukaemia virus (MLV) and human immunodeficiency virus (HIV).

[f] Different types of filter are designed to operate in different modes, that is in recirculating/tangential (T) or dead-end (D) mode.

[g] Designed for the removal of small viruses such as parvoviruses.

[h] Designed for the removal of large viruses such as retroviruses.

Even those methods that have been recommended for destroying prions [31] may only be partially effective and do not produce a level of sterilization similar to that expected with conventional microorganisms. Treatment with hypochlorite (20 000 ppm for 1h), or 1 M sodium hydroxide for 1 h are the standard methods used. Autoclaving at high temperatures, i.e. 132 °C for 1h in a gravity displacement autoclave, or 134–138 °C for 18min given in one or more steps using an effective porous load cycle, will reduce but not totally eliminate infectivity and are routinely used. However, incinerating is effective. In the case of biological products, filtration using a virus filter with a pore size of *c.* 15 nm is also effective [23, 32, 33].

2.3 Troubleshooting

- When contamination does occur, the destruction of the cell culture is the likely outcome. However, it may be possible to rescue critical cultures by treating them with suitable antibiotics or antimycotics. In the case of mycoplasma, specific elimination agents are available. These issues are covered in Chapter 9.

- Phosphate-buffered saline formulations are likely to go cloudy on autoclaving due to the inclusion of calcium and/or magnesium in the solution. Autoclaving at 115 °C will, to some extent, prevent the formation of any precipitate. However buffer formulations lacking these components completely can be autoclaved at 121 °C without any problem.

- Autoclaved items can be wet at the time they are removed from the autoclave. If this occurs repeatedly then (where possible) the drying stage conditions, that is temperature, vacuum and time, should be adjusted.

- Low levels of various extractable materials have been detected in the past in membrane filters, for example surfactant used during manufacture or residues of the ethylene oxide gas that has been used by some manufacturers as a sterilizing agent. This has been reported to affect cell growth [34]. For critical cell culture applications, where the complete absence of such residues is essential, filters with very low extractable levels, and sterilized by gamma irradiation, can be used. Alternatively, where practical, simply discard the first sample of the filtrate.

- Although contamination from bacteria and fungi is readily prevented, the removal of mycoplasma by conventional membrane filtration can be more of a problem. If this is a concern then biological reagents may be filtered to 0.1 μm as a precaution.

- The removal of formaldehyde after fumigation can be difficult. This process can be speeded up by placing dishes of ammonia in the area after treatment to help neutralize the gas.

- When decontaminating equipment with disinfectants, it is essential to submerge the object fully. This can be a problem with items that tend to float. In this case the use of weights or rods to hold the objects under the surface will be required, but care will still be necessary to eliminate all air bubbles.

- Although there are a number of techniques available to eliminate viruses from cell culture additives and products, these are not generally as effective as those used for the elimination of cellular microorganisms. For this reason, and to meet the requirement of regulators, specific evaluation using test agents and the specific product of concern may be required.

• Further guidance and requirements for sterilization, disinfection and cleaning in critical hospital or pharmaceutical settings is available [35] and should be consulted where appropriate.

References

★★★ 1. Block, S.S. (ed) (2001) *Disinfection, Sterilization, and Preservation*, 5th edn, Lippincott, Williams and Wilkins, Philadelphia. – *Individual chapters covering specific relevant topics on sterilization.*

★★★ 2. Fraise, A.O., Lambert, P. and Maillard, J-Y. (eds) (2004) *Russell, Hugo and Ayliffe's principles and practice of disinfection, preservation and sterilization*, 4th edn, Wiley-Blackwell Publishing, Oxford. – *Comprehensive coverage of the full topic of sterilization.*

3. National Health Service Estates. Health Technical Memorandum 2010 (1994–1997) *Good Practice Guide: Sterilization Parts 1–6*. NHS Estates, London, UK.

★ 4. Bowie, J.H., Kelsey, J.C. and Thompson, G.R. (1963) The Bowie and Dick autoclave tape test. *Lancet*, **1**, 586–587.– *The original description of the Bowie-Dick test for autoclaving porous loads.*

5. Advisory Committee on Dangerous Pathogens. Health and Safety Executive (2004) *The Approved List of Biological Agents*, HMSO, London.

6. Rutala, W.A. and Weber, D.J. (2001) New disinfection and sterilization methods. *Emerg. Infect. Dis.*, **7**, 348–353.

★★ 7. Brock, T.D. (1983) *Membrane Filtration*, Springer-Verlag, Berlin. – *Review on the use of filters for removing cellular microorganisms.*

8. Davis, J. and Shade, K. (2010) Aseptic techniques in cell culture, in *Encyclopedia of Industrial Biotechnology: Bioprocess*, vol. **1** (ed. M. C. Flickinger), *Bioseparation and Cell Technology*, John Wiley & Sons, Inc., New York, pp. 396–415.

9. Collins, C.D. and Kennedy, D.A. (eds) (1999) *Laboratory-Acquired Infections; History, Incidence, Causes and Prevention*, 4th edn, Butterworths Heinemann, Oxford.

★★ 10. Roberts, P. (1996) Virus safety of plasma products. *Rev. Med. Virol.*, **6**, 25–38. – *General review of the various methods for removing and inactivating viruses from biological material.*

★★ 11. Lynch, T.J. and Fratantoni, J.C. (2000) Viral clearance methods applied to blood products, in *Scientific Basis of Transfusion Medicine: Implications for Clinical Practice*, 2nd edn (eds K.C. Anderson and P.M. Ness), Saunders, Philadelphia, pp. 599–617. – *Good general review on virus elimination from biological material.*

12. Burnouf, T. and Radosevich, M. (2000) Reducing the risk of infection from plasma products: specific preventative strategies. *Blood Rev.*, **14**, 94–110.

★★ 13. Foster, P.R. and Cuthbertson, B. (1994) Procedures for the prevention of virus transmission by blood products, in *Blood, Blood Products and HIV*, 2nd edn (eds R. Madhok, C.D. Forbes and B.L. Evatt), Chapman and Hall, London, pp 207–248. – *Good general review on virus elimination from biological material.*

14. Danner, D.J., Smith, J. and Plavsic, M. (1999) Inactivation of viruses and mycoplasmas in fetal bovine serum using 56°C heat. *Bio. Pharm. Int.*, **June**, 50–52.

★★ 15. Horowitz, B., Prince, A.M., Horowitz, M.S. and Watklevicz, C. (1993) Viral safety of solvent-detergent treated blood products. *Dev. Biol. Stand.*, **81**, 147–161. – *General review on virus inactivation in biological material by solvent/detergent.*

16. European Medicines Agency (2005) Committee for Medicinal Products for Veterinary use. Requirements and controls applied to bovine serum used in the production of immunological veterinary medicinal products. EMEA/CVMP/743/00 Rev2, EMA, London, UK. This document can be downloaded at http://www.ema.europa.eu/docs/en_GB/document_library/Scientific_guideline/2009/10/WC500004575.pdf.

17. Purtle, D.G., Festen, R.M., Etchberger, K.J. *et al.* (2006) SAFC Biosciences Research Report, Validated Gamma Radiation Serum Products, SAFC Biosciences, Kansas, USA.

★★ 18. Roberts, P. (2000) Application of virus filters to biological products: practical and regulatory considerations. *Eur. J. Parenter. Sci.*, **5**, 3–9. – *The general strategy for the application of virus filters.*

★★ 19. Burnouf, T. and Radosevich M. (2003) Nanofiltration of plasma-derived biopharmaceutical products. *Haemophilia*, **9**, 24–37. – *Review of the nanofiltration systems available and their application.*

20. Levy, R.V., Phillips, M.W. and Lutz, H. (1998) Filtration and removal of viruses from biopharmaceuticals, in *Filtration in the Biopharmaceutical Industry* (eds T.H. Meltzer and M.W. Jornitz), Marcel Dekker, New York, pp 619–646.

21. Carter, J. and Lutz, H. (2002) An overview of viral filtration in biopharmaceutical manufacturing. *Eur. J. Parenter. Sci.*, **7**, 72–78.

22. Hughes, B., Bradburne, A., Sheppard, A. and Young, D. (1996) Evaluation of anti-viral filters. *Dev. Biol. Stand.*, **88**, 91–98.

23. Burnouf-Radosevich, M., Appourchaux, P., Huart, J.J. and Burnouf, T. (1994) Nanofiltration, a new specific virus elimination method applied to high-purity factor IX and factor XI concentrates. *Vox Sang.*, **67**, 132–138.

★ 24. Manabe, S.-I. (1996) Removal of virus through novel membrane filtration method. *Dev. Biol. Stand.*, **88**, 81–90. – *Early description of the use of the Planova hollow fiber cellulose filters.*

★★★ 25. Roberts P. (1997) Efficient removal of viruses by a novel polyvinylidene fluoride membrane filter. *J. Virol. Methods*, **65**, 27–31. – *Evaluation of a PVDF filter for virus removal.*

★★ 26. Oshima K., Evans-Strickfaden, T.T., Highsmith, A.K. and Ades, E.W. (1996) The use of a microporous polyvinylidene fluoride (PVDF) membrane filter to separate contaminating viral particles from biologically important proteins. *Biologicals*, **24**, 137–145. – *Virus removal and protein recovery by a PVDF membrane filter.*

27. Prusiner, S.B. (1991) Molecular biology of prions causing infectious and genetic encephalopathies of humans as well as scrapie of sheep and BSE of cattle. *Dev. Biol. Stand.*, **75**, 55–74.

★★★ 28. Weissmann, C. (2004) The state of the prion. *Nat. Rev. Microbiol.*, **2**, 861–871. – *Properties of TSEs and their role in disease.*

★★ 29. Taylor, D.M. (2004) Resistance of transmissible spongiform encephalopathy agents to decontamination. *Contrib. Microb. Basel*, **11**, 136–145. – *Inactivation of TSEs by physical and chemical methods.*

30. European Medicines Agency (2004) CHMP position statement on Creutzfeldt-Jakob disease and plasma-derived and urine-derived medicinal products. EMEA/CPMP /BWP/2879/02/rev 1, EMA, London, UK.

31. Advisory Committee on Dangerous Pathogens TSE Working Group (2003 and later supplements and amendments) *Transmissible Spongiform Encephalopathy Agents: Safe Working and the Prevention of Infection.* Department of Health, London, UK. www.dh.gov.uk/ab/ACDP/TSEguidance/index.htm (Accessed November 2010).

32. Tateishi, J., Kitamoto, T., Mohri, S. *et al.* (2001) Scrapie removal using Planova virus removal filters. *Biologicals*, **29**, 17–25.

33. Van Holton, R.W., Autenrieth, S., Boose, J.A. *et al.* (2002) Removal of prion challenge from an immune globulin preparation by use of a size-exclusion filter. *Transfusion*, **42**, 999–1004.

34. Knight, DE. (1990) Disposable filters may damage your cells. *Nature*, **343**, 218.

35. Microbiology Advisory Committee (2006) *Sterilization, Disinfection and Cleaning of Medical Equipment: Guidance on Decontamination from the Microbiology Advisory Committee to Department of Health*, Medical Devices Agency, London.

3
Microscopy of Living Cells

Colin Gray[1] and Daniel Zicha[2]

[1] School of Medicine and Biomedical Sciences, University of Sheffield, Beech Hill Road, Sheffield, UK
[2] Cancer Research UK, London, UK

3.1 Introduction

3.1.1 Chapter structure

This chapter presents an overview of live cell microscopy, from simple tissue culture applications to specialized techniques using high resolution, sensitivity and quantification. Section 3.1.2 contains historical background information. Section 3.2.1 contains general information that is relevant to all remaining sections.

3.1.2 Historical context

The principle of how to make powerful compound microscopes has been known since the sixteenth century. However, the quality of devices from this period was very poor due to optical aberrations. In fact, the first observation of individual bacteria and protozoa – by Anton von Leeuwenhoek – was not reported until 1674 and this was only achieved using a single-lens microscope.

A method for combining lenses was introduced by Joseph Jackson Lister to improve resolution by reducing aberrations. This opened up the field of microscopy and enabled Matthias Schleider and Theodor Schwann, in 1838, to propound the cell theory: the idea that the nucleated cell is the unit of structure and function in plants and animals. Later progress in basic optical principles, such as Ernst Abbé's theory of light diffraction in 1873, allowed Carl Zeiss to achieve resolution at the theoretical limits of visible light in 1886.

Animal Cell Culture: Essential Methods, First Edition. Edited by John M. Davis.
© 2011 John Wiley & Sons, Ltd. Published 2011 by John Wiley & Sons, Ltd.

Following the development of tissue culture by Ross Harrison in 1907, microscopy was soon applied to the observation of living cells in culture, and by the 1920s Ronald Canti had established the use of time-lapse cinemicroscopy for recording the behaviour of cultured cells [1]. However, the application of the compound microscope provided very limited contrast with living cells. Consequently, much effort was expended to develop suitable contrast enhancement techniques, leading to the development of interference microscopy by Aleksandr A. Lebedev in 1930, phase contrast by Fritz Zernike in 1935 and differential interference contrast by Francis H. Smith in 1955 (with further improvements by Georges Nomarski in 1969).

The availability of these contrast enhancement techniques finally allowed live cells to be studied in detail using both qualitative and quantitative techniques. In the 1950s, Michael Abercrombie pioneered the application of rigorous quantitative methods to the analysis of cell behaviour [2], at a time when microscopy images were still on celluloid film. The introduction of video microscopy greatly improved the acquisition of dynamic data [3] and the advent of computers and digital imaging technology has contributed to progress.

The increasing popularity of light microscopy has been largely due to developments in fluorescence microscopy, which has complemented dynamic observation with specific information on intracellular structures in live cells. This has been allied to powerful techniques in molecular biology to provide important information about the relationship between specific intracellular structures and the molecular mechanisms of cellular functions. Such studies are becoming even more important and new techniques are continually emerging. Two important directions in the development and wider application of specialized light microscopy techniques for live cells illustrate this trend:

- the enhancement of 3D observation by confocal microscopy and deconvolution techniques;

- techniques for the visualization of specific molecular interactions, such as fluorescence resonance energy transfer (FRET) assay and fluorescence correlation spectroscopy.

3.2 Methods and approaches

3.2.1 General imaging methodology

Light microscopes are produced in upright or inverted configuration according to the position of the objective lens in relation to the specimen. Inverted microscopes are most frequently used for live cell imaging; they are more convenient for use with a culture dish because the objective lens is underneath in close proximity to the cells. Upright microscopes can also be used for live cell studies but require water-dipping objective lenses or special culture chambers. In this chapter we have focused on the more common inverted microscope and this configuration is shown diagrammatically in Figure 3.1. A photograph of an inverted microscope in an environmental enclosure, with its important components labelled, is shown in Figure 3.2.

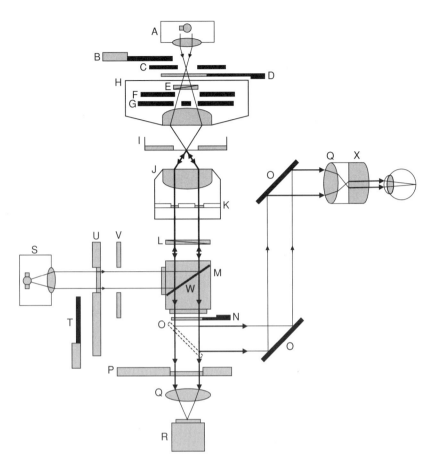

Figure 3.1 Diagram of an inverted compound microscope with common additional component. The components of an inverted compound microscope in the diagram are not shown to scale, in order to allow the path of the light rays through the microscope to be shown. The diagram shows the light path from the transmission lamp house (A), past a transmission shutter (B), through the field iris (C), polarizer (D), condenser (H) containing Wollaston prism (E), condenser diaphragm (F) and phase annulus (G), and onto the specimen (I) in a glass bottomed dish. Light transmitted through the specimen is collected by the objective lens (J) and passes through the phase plate (K), Wollaston prism (L), filter block (M) and DIC analyzer (N). It may be reflected by the mirrors (O) onto the tube lens (Q) and eyepiece (X). Alternatively, it may pass straight though a fluorescence emission filter wheel (P), a tube lens (Q) and be focused onto a camera (R). Components D, E, L and N are only present with DIC. Components G and K are only present with phase contrast. With fluorescence microscopy, light from the epifluorescence lamp house (S) passes the excitation shutter (T), excitation filter wheel (U), field diaphragm (V), excitation filter in the filter block (M) also containing a dichroic mirror (W) where it is reflected towards the objective lens (J) focusing it onto the specimen (I). The light emitted by the specimen (I) is collected by the objective lens (J), and passes through the dichroic mirror (W) and emission filter in the filter block (M). The emitted light can then be reflected to the eyepiece as above or pass through the emission filter wheel (P) to the tube lens and onto the camera. Normally only one excitation and one emission filter are used either in the filter block or in the wheels.

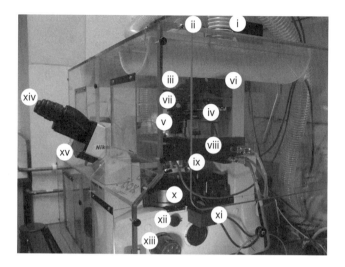

Figure 3.2 TE2000 Nikon microscope in an environmental enclosure. An environmental enclosure (Life Imaging Services) fitted to a TE2000 Nikon microscope with transmission illumination lamp house (i), field iris (ii), polarizer (iii), Wollaston prism/ phase annulus selector (iv), condenser (v), vertical condenser adjuster (vi), horizontal condenser adjusters (vii), motorized stage (viii), objective lens (ix), filter blocks (x), analyzer (xi), intermediate magnification lens (xii), focus control (xiii), eyepieces (xiv) and Bertrand lens (xv).

The modern inverted light microscope is a compound microscope [4] and consists of four main parts:

- a light source with a collector lens and a diffuser (A in Figure 3.1, i in Figure 3.2),

- a condenser (H in Figure 3.1, v in Figure 3.2) to focus the light on a specimen (I in Figure 3.1),

- an objective lens (J in Figure 3.1, ix in Figure 3.2, also called an objective) to collect the light from the specimen and project it through a tube lens (Q in Figure 3.1),

- an eyepiece (X in Figure 3.1, xiv in Figure 3.2) to produce a focused image in the observer's eye.

This section is devoted to use with any culture vessel and low magnification dry objective lenses but the principles described here are also important for more advanced and specialized techniques.

3.2.1.1 Magnification and resolution

The magnification of an image formed by a simple lens (Z in Figure 3.3) can most easily be understood by a geometrical construction tracing the rays from the object (Y in Figure 3.3) to its image (Y′ in Figure 3.3). The magnification is then given by the ratio of the size of the image (Y′) and the size of the object (Y), which is the same as the ratio of the distances from the lens to the image (d′) and from the lens to the object (d). The combinations of lenses used in microscope objectives work in a similar manner but

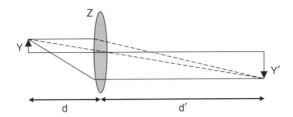

Figure 3.3 Magnification. The magnification of a simple lens (Z) is given by the ratio of the sizes of the image (Y′) and the object (Y) as well as by the ratio of the distances from the lens to the image (d′) and from the lens to the object (d).

produce a higher magnification and eliminate some of the aberrations (see Section 3.2.1.4) introduced by a single lens system.

The magnification in a compound microscope is achieved in two steps. The objective lens with the tube lens produces a magnified intermediary image, which is further increased by the eyepieces. The final magnification of the image is the product of these two steps and can usefully be as high as 2000×.

Additional magnification, usually 1.5× or 2.5×, can be achieved by the insertion of an intermediate magnification lens (xii in Figure 3.2; called Optovar by Carl Zeiss Inc.) located between the objective and the eyepiece. The intermediate magnification reduces the field of view but does not reveal more detail.

The level of detail that can be seen in an image is not solely governed by the magnification but mainly by the resolution of the objective. The image is formed when the lenses in the microscope recombine light diffracted by the specimen. In order to resolve small objects the objective must collect the diffracted light from a larger range of angles.

The numerical aperture (NA) is a convenient measure of the cone of light that an objective (i in Figure 3.4a) or a condenser is able to collect or emit. It is defined as the sine of the half angle (α) of the cross-section of the cone of light multiplied by the refractive index (n; for air $n \approx 1.0$) which quantifies the relative optical density of the medium between the specimen (ii in Figure 3.4a) and the objective. An objective with a larger angle α can resolve smaller objects. In order to achieve optimal resolution in applications with transmitted light, it is necessary to use a condenser with a NA matched to the objective.

At high angles of incidence, light will not travel from the glass coverslip (*iii* in Figure 3.4b) into air but will be reflected (iv in Figure 3.4b). This means that the NA achievable in air is severely limited (left half of Figure 3.4b). In order to achieve a high NA, a substance (v in Figure 3.4b) with a higher refractive index must be used between the coverslip and the objective lens. Ideally, if the refractive index of the immersion liquid matches that of the glass, then no reflection will occur at the interface between the two (right half of Figure 3.4b). Specialized immersion oils (e.g. Cargille) are designed to have a refractive index of 1.515 (at 37 °C or other temperature as required), matched to that of glass. Immersion oils are also selected for low auto-fluorescence to reduce background in fluorescence microscopy. Such immersion oils are suitable for high-resolution observation of cells on coverslips (see Section 3.2.3). When live cells are observed at some distance into culture medium, which is water-based and has a lower refractive index than immersion oil, a water immersion objective should be used to avoid image distortion.

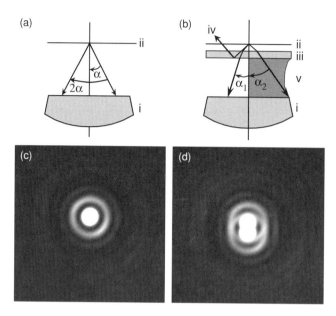

Figure 3.4 Numerical aperture and resolution. (a) The objective lens (i) collects light from the specimen (ii) in a cone with cross section angle of 2α. The left half of (b) shows an air objective lens (i) collecting light from the specimen (ii) in a cone of light with half cross section angle α_1. Light falling outside this cone is being reflected off the coverslip (iii) and lost (iv). The right half of (b) shows the effect of an immersion oil (v) of similar refractive index to that of glass, where light is collected from a larger cone with cross-section angle α_2. (c) The microscope does not produce sharp edges to an image of a small object (tiny hole in a metal foil) but a series of concentric rings, decreasing in brightness as the diameter increases, called an Airy pattern with an Airy disc at its centre. (d) Resolution is defined by the Rayleigh criterion, which states that two small objects can be distinguished as long as their Airy discs are arranged such that the centre of the second Airy disc falls on or beyond the intensity minimum between the first Airy disc and its first surrounding ring.

Note that objective lenses designed for oil immersion are designated 'Oil', for water immersion 'W' and for glycerol immersion 'Glyc' (glycerol having an intermediate refractive index that can be useful for special applications). Multi-immersion objective lenses are available, allowing adjustment for different immersion liquids via a collar on the objective.

What does 'resolution' actually mean? It refers to the minimum distance at which two small objects can just be distinguished from each other rather than merging into a single object. First, it is necessary to understand that the microscope image of a point object is not a point with sharp edges but a blurred spot with diffraction rings surrounding it (Figure 3.4c). The spot is called an 'Airy disc'. Whether two such objects can be seen separately in the resultant image depends on how close the centres of their Airy discs are. The resolution is usually defined by the Rayleigh criterion, which states that two small objects can be distinguished as long as the centre of the second Airy disc falls on or beyond the intensity minimum between the first Airy disc and its first surrounding ring (Figure 3.4d). The distance to the first minimum (d_0) gives the resolution and can be calculated according to the following equation

$$d_0 = \frac{1.22\lambda}{NA_{obj} + NA_{cond}}$$

where λ is the wavelength, which determines the colour of the light, and NA_{obj} and NA_{cond} are the numerical apertures of the objective and condenser respectively. Practically, the maximum resolution achievable by light microscopy is about 200 nm.

3.2.1.2 Illumination

To achieve the best possible resolving power of a microscope, correct illumination is essential and should provide sufficient intensity with an even distribution across the field of view without any stray light. There are several main types of light sources used for microscope illumination. Halogen (also called tungsten halogen) lamps are used for most transmission applications. In order to achieve good microscopy images it may be necessary to illuminate with high-intensity light, which is often poorly tolerated by cells. To protect the cells, the transmission illumination can be restricted to a single wavelength using a green interference filter. Adjustable lamp houses require occasional alignment as described in *Protocol 3.1*, and will always need realignment when the bulb is replaced.

PROTOCOL 3.1   Halogen lamp alignment[a]

Equipment

- Microscope with a lamp house (A in Figure 3.1, i in Figure 3.2)
- Centring telescope or a Bertrand lens (xv in Figure 3.2).

Method

1 Remove the lamp house and the diffuser (sometimes in a holder on the front of the lamp house) from the microscope and project an image of the filament onto a wall about 3 m away.

2 Adjust the focus and position screws so that the filament and its mirror image overlap and are in focus. The two filament images should be adjusted so that they each fill in the gaps in the other, to cover a central rectangular region.

3 Mount the lamp housing and perform final adjustment by observing the images of the filament using a centring telescope or Bertrand lens. The filament image should be central, in focus and interlaced with its mirror image.

4 Remove the lamp housing, replace the diffuser and remount the housing on the microscope.

Note

[a] Follow the manufacturer's instructions to replace the bulb. Not all microscopes require tungsten lamp adjustment. Zeiss Axiovert is given as an example.

Virtually all modern microscopes are designed to use Köhler illumination when operated in transmission mode. This can be set up as described in *Protocol 3.2.*

PROTOCOL 3.2   Setting Köhler illumination[a]

Equipment

• Microscope

• Specimen

Method

1 Switch on the light source, open the field iris (C in Figure 3.1, ii in Figure 3.2) as far as possible and fully open the condenser diaphragm[b] (F in Figure 3.1) to maximize the light available.

2 Put the condenser exit lens at the approximate working distance from the specimen.

3 Look down the eyepieces (xiv in Figure 3.2) and bring the specimen into the best focus (Figure 3.5a) using the focusing knob (xiii in Figure 3.2). When setting Köhler illumination with higher magnification objective lenses, it is often easier to start with a low-magnification objective.

4 Close the field iris as much as possible while the specimen is still visible (Figure 3.5b). Bring the specimen into focus by adjusting the condenser vertically (vi in Figure 3.2) and move the condenser horizontally with its adjusters (vii in Figure 3.2) until the field iris appears centred (Figure 3.5c). As the iris image nears the centre, it is helpful to open the field iris so that it almost fills the field of view, making it easier to judge how central it is.

5 Open the field iris so that it is just larger than the field of view (Figure 3.5f).

6 Adjust the condenser diaphragm[b] so that the image is bright but has good contrast. It may be helpful to remove the eyepiece from one tube and look down the tube from about 20 cm. You should see a bright circle, which is an image of the condenser diaphragm. This circle should be set to about 80% of its maximum size[c].

Notes

[a] This protocol can only be carried out if the lamp has first been aligned as in *Protocol 3.1.*

[b] There may not be a condenser diaphragm present, for example on a microscope using phase contrast.

[c] In order to generate the best image, it is important to readjust the illumination for every change of objective. Usually this will just require the repetition of steps 4–6.

Mercury and xenon arc lamps are used for most fluorescence applications, and lasers are mainly used for scanning confocal and multiphoton fluorescence microscopy. More recently, metal halide lamps and LEDs that provide stable illumination with a longer life span have become available. In contrast, the mercury or xenon arc lamps have to be replaced and realigned relatively often. Arc lamp alignment is described in *Protocol 3.3.*

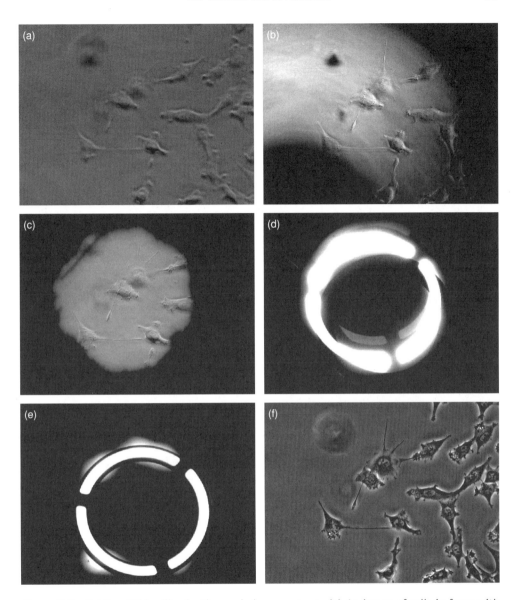

Figure 3.5 Setting Köhler illumination and phase contrast. (a) An image of cells in focus with the condenser misaligned; (b) the field iris is first closed, and then (c) focused and centred. For phase contrast, a Bertrand lens is used to view the phase annulus and the phase plate which are shown (d) misaligned and (e) following adjustment. Finally (f) the field iris is opened to just over-fill the field of view. Images were taken using phase contrast on a Zeiss Axiovert 135TV microscope with a 20× NA 0.5 Ph2 objective and a Hamamatsu Orca ER cooled CCD camera.

PROTOCOL 3.3   Arc lamp alignment[a]

Equipment

- Fluorescence microscope

- White paper

- UV protective glasses

- Fluorescent plastic slide (Chroma Technology Corp.)

Warning: Arc lamps emit a large amount of UV radiation. This can seriously damage your eyes or cause severe skin burns so wear UV protection. When carrying out the following protocol, arc lamps should be handled cold since there is an explosion hazard when hot. Mercury arc lamps contain liquid mercury and its vapour, which are considered to be hazardous; in the event of lamp breakage, evacuate the room and consult your safety officer.

Method

1 Follow manufacturer's instructions to replace the arc lamp within the lamp house (S in Figure 3.1). The lamp should not be touched with bare hands, but if necessary, it may be cleaned with ethanol.

2 Follow the Köhler illumination alignment procedure (*Protocol 3.2*) as far as centring the closed-down field iris and then increase the transmitted lamp brightness to maximum. This will produce a bright spot used as a target for centring the fluorescence illumination.

3 Remove one of the objectives (J in Figure 3.1, ix in Figure 3.2). Place the sheet of white paper over the microscope stage and illuminate it with the epifluorescence light. Fully open the fluorescence field diaphragm (V in Figure 3.1).

4 Focus the direct image of the arc on the paper using the lamp adjusters. The lamp housing contains a reflector, which produces a mirror image of the arc. Unless the manufacturer instructs otherwise, move the mirror image of the arc next to the direct image using the mirror adjusters. Centre the arc images evenly around the spot of light from the condenser.

5 Replace the objective lens and check for even illumination using a fluorescent plastic slide. Alternatively, use a fluorescent specimen which is moved around the field, or a solution of a fluorophore in a dish.

Notes

[a]Some light sources do not require alignment, such as Lambda LS (Sutter Instruments), which uses pre–aligned arc lamp cartridges.

3.2.1.3 Contrast

The straightforward application of Köhler illumination in transmission mode, described above, is known as bright field microscopy. It is of limited use for living cells since they are transparent and have little effect on the intensity of light passing through them. Therefore, various contrast enhancement techniques have been developed [5], some of which are described below.

Phase contrast is the most popular contrast enhancement technique for live cells because it is relatively inexpensive, robust, easy to configure and for most purposes produces high-contrast images. However, it is not suitable for the accurate determination of cell edges, especially by automatic image processing, since thick edges produce a bright 'halo', obscuring them. With phase contrast, the specimen illumination is a hollow cone of light provided by a narrow annular aperture in the condenser called the phase annulus or ring. The contrast is produced when a matching phase plate is located within the objective lens [4]. The procedure for setting up phase contrast is described in *Protocol 3.4*, and an example of cells photographed using phase contrast is shown in Figure 3.5f.

PROTOCOL 3.4 Setting phase contrast

Equipment

- Microscope
- Phase contrast annulus for the condenser (G in Figure 3.1, iv in Figure 3.2)
- Phase contrast objective lens (J in Figure 3.1, ix in Figure 3.2)
- Centring telescope or Bertrand lens (xv in Figure 3.2)
- Thin specimen

Method

1 Adjust the microscope for Köhler illumination as described in *Protocol 3.2* leaving the microscope focused on the specimen.

2 Ensure that you are using a phase objective with a matching phase annulus in the condenser, each designated as Ph1, Ph2 or Ph3.

3 Insert the centring telescope or Bertrand lens, which enables observation of the phase annulus (G in Figure 3.1) and phase plate (K in Figure 3.1). Focus the centring telescope or Bertrand lens so that the bright phase annulus and the dark phase plate are in focus (Figure 3.5d).

4 Move the phase annulus using its adjusters so that it is concentric with the phase plate (Figure 3.5e).

The simplest form of contrast enhancement is to use oblique (anaxial) illumination. This can be achieved by partially closing and off-centring the field iris using the condenser adjusters (vi and vii in Figure 3.2). Although oblique illumination suffers from reduced resolution and undesirable optical effects resulting from regions of the specimen that are not exactly in focus, it easily provides a shadow-cast effect.

In dark field microscopy, a specimen is illuminated by a hollow cone of light produced by a special condenser, although a close approximation to dark field illumination may be achieved by selecting a Ph3 condenser 'phase ring' with a low NA objective lens. The incident angle of the illumination is so great that none of the light directly enters the objective. Light scattered from the specimen is collected and used to form the image. The background appears dark, and diffracting features of the specimen appear bright with high contrast even if their size is smaller than the microscope's limit of resolution. Dark field produces high-contrast images with clean specimens, but any contamination, such as particles in the background, appears extremely bright and spoils the image. Therefore, this technique is less commonly used with live cells.

In differential interference contrast (DIC) microscopy (see *Protocol 3.5*), the illumination light is polarized by a polarizer (D in Figure 3.1, iii in Figure 3.2) and split by a condenser Wollaston prism (E in Figure 3.1) into two spatially shifted components. The two components pass through the specimen and objective lens, and contrast is generated when the two components are recombined by the objective Wollaston prism (L in Figure 3.1) and pass through another polarizing filter called the analyzer (N in Figure 3.1, xi in Figure 3.2). DIC produces a shadow-cast image that is especially good for picking out small variations in refractive index such as the edges of cells, vesicles or narrow filaments (in Figure 3.6g–j).

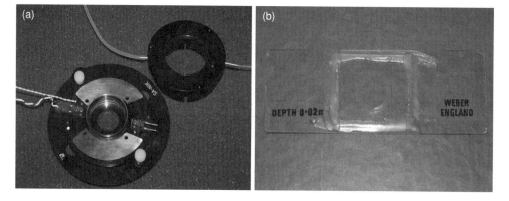

Figure 3.6 Environmental control for microscopy and example micrographs. (a) A glass bottomed culture dish (MatTek) in a Micro-incubation System DH-35i (Warner Instruments; the lid of the DH-35i is in the top right corner) which allows medium replacement, and temperature and gas control. (b) A Z80000 Bacteria Evaluation Chamber (Hawksley, formerly made by Weber Scientific) containing culture medium and sealed with a coverslip using a wax mixture.

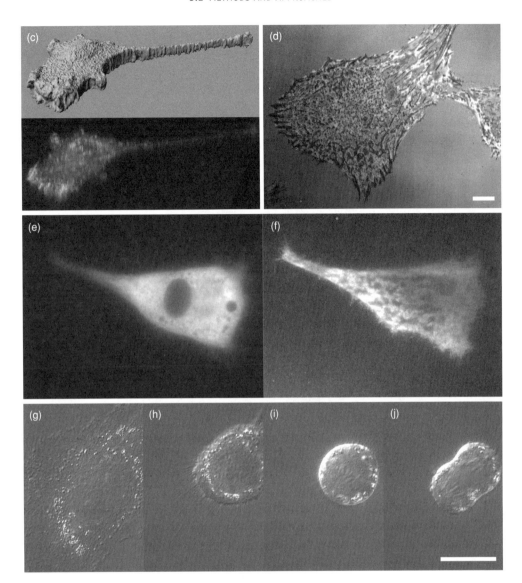

Figure 3.6 (*continued*) (c) A surface rendering (top image) and a maximum projection (bottom image), produced with Imaris (Bitplane), of the 3D distribution of actinin–GFP fusion protein in a T15 sarcoma cell imaged with a laser scanning confocal microscope LSM 510 (Zeiss) equipped with a 63× NA 1.4 objective lens. Imaging volume was 92.5 × 92.5 × 12.5 µm. (d) An interference reflection microscopy (IRM) image of rat sarcoma T15 cells acquired by the LSM 510 using laser light of 488 nm wavelength with no emission filter. The scale bar represents 10 µm. (e) Epifluorescence and (f) total internal reflection fluorescence (TIRF) images of a transfected T15 cell expressing 4.1B–GFP fusion protein acquired by a TE2000 microscope (Nikon) equipped with a 100× NA 1.49 objective using Xenon illumination and 488 nm laser light respectively. Four frames from a time-lapse recording of a dividing T15 cell acquired at (g) 0 min, (h) 20 min, (i) 44 min and (j) 55 min using DIC microscopy on a TE2000 microscope (Nikon) equipped with 100× NA 1.4 objective. The scale bar represents 10 µm.

PROTOCOL 3.5   Setting up DIC

Equipment

- Microscope
- Polarizer (D in Figure 3.1, iii in Figure 3.2)
- Condenser Wollaston prism (E in Figure 3.1)
- Objective Wollaston prism (L in Figure 3.1)
- Analyzer (N in Figure 3.1, xi in Figure 3.2)
- Centring telescope or Bertrand lens (xv in Figure 3.2)
- Thin specimen on glass coverslip[a]

Method

1 Set up Köhler illumination as described in *Protocol 3.2* using bright field illumination, having fitted a DIC objective lens[b].

2 Remove the specimen. Insert the polarizer and the analyzer into the light path. Place an objective Wollaston prism that matches the objective lens in designation (e.g. magnification, NA or Oil/Glyc/W) into the light path.

3 While utilizing the centring telescope or Bertrand lens, rotate the polarizer or the analyzer[c] until as dark a diagonal stripe as possible runs across the field of view.

4 Remove the centring telescope or Bertrand lens and replace the specimen. If using a glass-bottomed dish remove any plastic lid[a]; a special stage incubator with a glass lid can be used instead (for example Figure 3.6a).

5 Engage the correct condenser Wollaston prism, usually by rotating a wheel in the condenser (iv in Figure 3.2). Some microscopes, for example, allow 'M' for medium NA objective lenses between 0.4 and 0.8, or 'H' for high NA objectives between 0.8 and 1.4.

6 Make small adjustments by rotating the polarizer or the analyzer to achieve the desired level of contrast. Some manufacturers allow small lateral adjustments to the position of the objective Wollaston prism in order to achieve additional fine control of the image contrast.

Notes

[a]DIC cannot be used with plastic cultureware since it interferes with polarization of the light essential for this technique.

[b]A DIC objective lens is made from special strain-free glass in order to perform well with polarized light. Cheaper ordinary bright field objective lenses may still produce acceptable DIC images.

[c]Most microscopes allow rotation of the polarizer or the analyzer but not both.

A more economical solution than DIC is the Hoffman modulation contrast (HMC) system which can produce images of acceptable quality. HMC uses an illumination slit in the condenser, and the objective aperture holds a second complementary masque (modulator) which consists of three parallel segments: a dark segment, an intermediate segment in the centre and a bright segment at the other side. Light deviated by the specimen to one side is thus blocked whereas light deviated to the other side is unchanged and the background is partially suppressed by the middle segment. A shadow-cast image, similar to DIC, is thus produced. HMC illumination is available from several microscope manufacturers (e.g. Nikon Instruments Inc. or Olympus America Inc.).

Interference reflection microscopy (IRM) uses reflected light backscattered from the cells to form an image. Illumination as well as observation is achieved by means of the objective. In order to eliminate unwanted interference, the objective needs to have a highly efficient anti-reflection coating, or confocal microscopy (only systems compatible with reflection imaging) can be used instead. Zeiss designates these objectives 'antiflex' and produces a polarization filter set for separation of the illumination and observation beams. IRM with live cells is especially useful for the visualization of focal adhesion contacts where the light reflected from the ventral cell surface recombines with light reflected from the coverslip substrate. Regions of cells in direct contact with the coverslip appear dark and brighter regions indicate a small separation from the coverslip (Figure 3.6d).

Varel contrast is another technique for imaging live cells and is more economical than DIC. It combines phase contrast and inclined unilateral illumination. The technique is especially useful where phase contrast fails, for example because of the curved bottom of culture vessels. Varel contrast is available from Zeiss. A similar system called relief phase contrast is available from Olympus.

The direct imaging of optical density in live cells can be achieved by dual beam interference microscopy [6] or quadrature interferometry [7]. These techniques are not currently commercially available but have a considerable potential for computer-assisted imaging and quantitative analysis of live cells. Similar information can be derived by digital holographic microscopy available from Lyncée Tec (www.lynceetec.com).

3.2.1.4 Aberrations and their corrections

Magnification, as described earlier (see Section 3.2.1.1), applies to very thin lenses and only to objects near the lens axis. Real lenses are thicker and introduce variations in the focus with both position in the field of view and wavelength. These are called aberrations. Practical modern objectives contain combinations of lenses designed to minimize aberrations. A list of common aberrations, their causes and specially corrected objectives follows.

Spherical aberration occurs when light rays, originating from one point and passing through different regions of a lens, do not all focus at the same distance from the lens (Figure 3.7a). This aberration affects images of objects placed anywhere in the observation field including the lens axis. Some objective lenses are equipped with a coverslip thickness correction collar, which can be used to minimize this defect.

Field curvature is an aberration where the periphery of the observation field focuses in a different plane from the centre, resulting in images where the centre and the periphery are never in focus at the same time. The defect becomes more severe for points of an

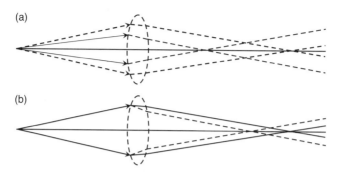

Figure 3.7 Spherical and chromatic aberration. Rays originating at a single point, on the lens axis, going through different regions of the lens do not focus at the same distance from the lens and so produce spherical aberration (a). Rays of different colour originating from a single point are brought to focus at different distances from the lens, producing chromatic aberration (b).

object located away from the centre of the image. Objectives designated 'Plan' are corrected for this aberration.

Chromatic aberration is due to the refractive index of a lens being wavelength dependent. Therefore, light rays of different wavelengths are focused at different distances from a lens (Figure 3.7b). In objectives, this effect is corrected using a combination of lenses made from different types of glass. Chromatic aberration is corrected at two wavelengths in an objective designated 'Achromat' and at four wavelengths in one designated 'Apochromat'. Some manufacturers now produce VC (violet corrected) Apochromatic objective lenses that are corrected for most of the visible range. Objectives designated 'NeoFluar' (Zeiss) or 'Fluor' (Nikon) are corrected at three wavelengths but have increased transmission especially for UV light, and these are essential for some fluorescence applications. These objective lenses may not be fully corrected for field curvature aberration at the same time.

Other aberrations, such as *coma, astigmatism* and *geometrical distortions* should not be detectable in modern high-quality objective lenses.

In general, objectives with greater correction for aberrations have more lens elements and can have reduced transmission. They are also usually more expensive. Corrections for several aberrations can be combined in one objective, for example 'Plan-NeoFluar'. Some objectives are selected as strain free and designated 'Pol' or 'DIC', and are essential for microscopy using polarized light such as DIC.

Choosing the right type of objective for a particular experiment is important since this can significantly influence image quality. Furthermore, even objectives with identical specifications can vary in their performance. For critical high-resolution studies, additional benefit can be gained by making a selection of an especially good objective from a set. In particular, chromatic aberration may lead to errors in high-resolution, colocalization studies especially those using confocal microscopy, where greater reliance is placed upon the results. Chromatic aberration can be checked by evaluating the co-localization of different fluorescence emission wavelengths from small (e.g. 0.2 μm diameter) TetraSpeck beads (Invitrogen), which fluoresce at multiple wavelengths.

3.2.1.5 Environmental control for microscopy specimens

Living cells in culture can respond rapidly to changes in their environment. Therefore, critical microscopy, especially long-term observation, requires a tightly controlled environment particularly with regard to temperature, osmolarity and pH.

Temperature-controlled stage heaters can be purchased from microscope manufacturers. Heated stages may have a hole under the specimen position to allow access for the objective. A consequence of this is unwanted cooling of the specimen at the very position being observed. This is more dramatic when using immersion liquid between the objective and the specimen, since it also acts as a heat conductor. Such cooling can be prevented by using an objective heater (e.g. from Bioptechs Inc.). Alternatively, a temperature-controlled plastic box with access ports, for the controls, can be constructed to house the entire microscope. Such environmental enclosures are available commercially from several manufacturers, for example Life Imaging Services (Figure 3.2), or can be built in-house if workshop facilities are available. This solution has the additional benefit of focus stability, since a variation of temperature ≈ 1 °C can lead to a significant shift in focus due to expansion and contraction of the microscope body.

If a heated box is used with open culture dishes, then the atmosphere should be humidified to minimize evaporation of the culture medium and prevent changes in osmolarity. Alternatively, an open-top culture dish can have a layer of mineral oil (designed for tissue culture, e.g. from Sigma) poured on top of the culture medium to prevent evaporation.

In order to maintain pH, it is often easier to work with a sealed chamber (for example b in Figure 3.6) and a CO_2-independent medium such as a Hanks-based culture medium. However, if a CO_2-rich atmosphere is required, a stage incubator can be used with a mixed air/CO_2 supply. Commercial solutions are available, for example from Life Imaging Services.

3.2.2 Imaging of intact cultured cells in flasks, dishes and plates

The visualization of live cells in plastic flasks, dishes and plates requires objectives and condensers with long working distances. Standard dry $4\times$ and $10\times$ objectives will generally be suitable. However, at higher magnification long working distance objectives (which tend to have lower NA and are able to resolve less detail than normal working distance objectives) will usually be required. These objectives often have an adjustable collar to ameliorate optical effects due to the thick bottoms of culture vessels. Setting these collars is often overlooked, leading to severely degraded images. Plastic vessels are generally unsuitable for any techniques involving polarized light such as DIC; HMC is a possible alternative. Another solution, called PlasDIC, is commercially available from Zeiss.

When using multiwell plates with small wells, the meniscus interferes with Köhler illumination and phase contrast alignment, resulting in images with low contrast especially at the well edges. For example, with 96-well plates, only the very centre allows reasonable image quality as a result.

When CO_2 needs to be provided with multiwell plates, without a special stage incubator, the lid should be raised using spacers to allow gas mixing.

3.2.3 High-resolution imaging of cells on coverslips

As discussed in Section 3.2.1.1, high-resolution imaging requires an objective lens with a high NA. This is achieved using oil, glycerol or water immersion objective lenses usually designed to work with 0.17-mm-thick glass coverslips (thickness No. 1.5).

It is preferable to culture the cells directly on the coverslips and then mount these coverslips on a chamber for imaging. Cells can easily be cultured on a coverslip in a 35-mm culture dish (e.g. Corning) by adding cell suspension and allowing the cells to settle and adhere. For this to be successful, the coverslips must be properly cleaned before use (*Protocol 3.6*).

In addition, it may be beneficial to pre-coat the coverslips with, for example, fibronectin or gelatin to aid adhesion with some cell types. A convenient, though more expensive, alternative is to use a Glass Bottom Culture Dish (MatTek Corp.), which is a plastic Petri dish that has a glass coverslip glued over a hole in its base. These glass bottomed dishes are also available pre-coated with poly-d-lysine or collagen. An example of the Glass Bottom Culture Dish is shown (a in Figure 3.6) in a Micro-incubation System DH-35i (Warner Instruments LLC), which also allows medium replacement.

PROTOCOL 3.6   Cleaning coverslips

Warning: Take care when handling the acids as they are highly corrosive and dangerous. Use appropriate protection.

Equipment and Reagents

- Coverslips (e.g. from RA Lamb)

- Lens tissue (e.g. from Whatman)

- 1% solution of 7X Cleaning Solution For Laboratory Use (MP Biomedicals, Inc.) in clean water

- Mixture of concentrated HCl–ethanol (60 : 40)

- 70% (v/v) ethanol in ultra-pure water produced, for example, by a Milli-Q Plus system (Millipore Corp.) from de-ionized water

- Teflon coverslip racks (Invitrogen)

- Acid-resistant tweezers

- Sterile tweezers

Method

1 Rub the coverslips with lens tissue and a small amount of 7X solution.

2 Rinse thoroughly using clean water.

3 Place the coverslips in a Teflon rack. From this point onwards use acid-resistant plastic tweezers to handle the rack. Place the rack in a bath containing 60:40 HCl–ethanol and leave for 20–60 min (this should be done in a fume hood).

(continued)

4 Dip the rack in a bath of ultra-pure water, then remove. Rinse out the bath and replace with fresh ultra-pure water. Place the rack in the bath and leave for 10–60 min. Repeat two more times (three times in total).

5 Place the rack in a bath of fresh 70% ethanol–ultra-pure water for at least 20 min. It is possible to leave the coverslips stored in this mixture.

6 Thoroughly dry the coverslips in the rack in a unidirectional flow cabinet or microbiological safety cabinet, ideally using filtered compressed air to prevent drying marks. Place in a sterilized glass beaker and cover with foil and leave for several hours or overnight to ensure that the coverslips are completely dry before moving on to step 7.

7 Using sterilized tweezers, place the clean, aseptic and dry coverslips in either a multiwell plate or individual 35-mm plastic dishes. If the dishes or plates are wrapped with Parafilm they can be stored for months before use.

3.2.3.1 Cell culture chambers

Once cells are cultured on coverslips it is necessary to use a chamber to observe them on the microscope.

For basic observation it is possible to use a very simple Z80000 Bacteria Evaluation Chamber (Figure 3.6b; Hawksley, formerly made by Weber Scientific), which is flooded with culture medium and a coverslip seeded with cells inverted and sealed using a wax mixture made, for example, from beeswax (Fisher), soft yellow paraffin (Fisher) and paraffin wax (melting point 46 °C; Fisher) in the ratio 1 : 1 : 1. Such a chamber contains ample medium to support low-density culture for at least 24-h observation if properly sealed. This arrangement produces particularly good images since microscopes are usually optimized for slides that have the same overall thickness. A modification of this chamber is the DCC Dunn chamber (Hawksley), which has been specifically designed for the assessment of chemotaxis [8,9].

A home-made chamber slide can be constructed using two coverslips acting as 'spacers' on a cleaned slide, or for longer term experiments a 15-mm hole can be drilled through a slide and a coverslip permanently glued to one face of the slide with Sylgard 184 (Dow Corning Corp.) or Araldite epoxy resin (RS Components Ltd.) to create a well which is then filled with culture medium. A second coverslip with adherent cells is inverted over the well and sealed down using the wax mixture described above.

If the medium needs to be changed, a number of commercial products are available. For example, a Confocal Imaging Chamber RC-30HV (Warner Instruments LLC) allows perfusion of cells on coverslips. Simple multichannel slides (Ibidi GmbH) can also be used for high-resolution imaging. However, the substrate is made of thin high-quality plastic and is unsuitable for DIC.

3.2.4 Imaging of fluorescently labelled cells

Fluorescence is a phenomenon where a fluorophore (also called fluorochrome) is illuminated by high-intensity monochromatic light (e.g. blue with a wavelength of 488 nm), which stimulates the fluorophore molecules to briefly absorb the light energy before emitting it at a longer wavelength (e.g. 510 nm, producing green colour).

3.2.4.1 Fluorophores and cell labelling

To make the best use of fluorescence with live cells, they need to be labelled using suitable fluorophores.

Most live cell fluorescence studies now use green fluorescent protein (GFP) or one of the newer fluorescent biomolecules [10]. These are vital research tools, since when used alone they can be markers of particular cells, but when used as fusion proteins they allow the direct visualization of the fusion protein and its interactions within living cells. However, to achieve this, cDNA constructs must be introduced into the target cell. This may be done using transfection techniques with agents such as Lipofectamine (GIBCO-BRL) or Effectene (QIAGEN), or by the direct microinjection of the cDNA construct into cells [11]. Microinjection has advantages for microscopy since viable expressing cells, in sufficient quantity for most applications, are available within hours rather than next day. As an example, images of a microinjected cell expressing a GFP fusion protein are shown (Figure 3.6e,f).

Other approaches for labelling specific cellular structures with fluorescence include the microinjection of fluorescently labelled molecules, which are available for example from Cytoskeleton, and special fluorophores, such as Mitotracker (Invitrogen) for mitochondria or CellTracker Orange CMTMR (Invitrogen) for the plasma membrane.

3.2.4.2 Microscope configuration for fluorescence

Fluorescence microscopy is generally used in the so-called epifluorescence mode, where illumination and observation both occur through the objective. This has the great advantage that only the reflected excitation light (a relatively small amount) needs to be filtered out in order to observe the emitted fluorescence [12]. Usually excitation light from a high-intensity arc lamp (S in Figure 3.1) passes through an excitation filter, which only transmits the required wavelengths, and is then reflected by a dichroic mirror (W in Figure 3.1) through the objective onto the specimen. Emitted light from the specimen is collected by the objective and passes through the dichroic mirror (W in Figure 3.1), rather than being reflected as the excitation light before, and on through an emission filter to further reduce unwanted light before observation at the eyepieces or camera. The excitation filter, the dichroic mirror and the emission filter are normally housed in a filter block (M in Figure 3.1, x in Figure 3.2) usually found beneath the objective turret in the microscope. Suitable filters and dichroic mirrors allow viewing of single or multiple fluorophores simultaneously, using multiband filters, or consecutively by replacing the filter block. Filters are usually designated by the wavelengths they allow through and are known as 'band pass' where wavelengths in a narrow range are transmitted and 'long pass' where all wavelengths longer than that chosen are transmitted. A band pass filter will generally be more specific for a particular fluorophore whereas a long pass is less specific but allows more light to pass through (useful with weakly fluorescent specimens).

A different approach, now common, is to fit the excitation filter into a filter wheel (U in Figure 3.1), controlled by the microscope or a computer, which when used with a multiband dichroic mirror (W in Figure 3.1) and emission filters allows fast automated switching between colours. Because the dichroic mirror is not moved, this avoids any shift in the image that would result from replacing single filter blocks. This means that the resulting images can be accurately overlaid or valid colocalization analysis performed.

Excitation using a broader range of wavelengths may also excite other fluorophores present (to a lower extent) during multilabelled experiments. In this situation, using a multiband (or long pass) emission filter with the chosen excitation filter will allow unwanted signal, usually called crosstalk, to be collected. For example, a dual-labelled experiment using GFP and DsRed2 (a red fluorescent protein) when illuminated using a 488-nm band pass filter will maximally excite the GFP but also results in a 40% stimulation of the DsRed2. Therefore, the resulting image cannot be relied upon as an accurate localization of the GFP signal since it contains some DsRed2 as well. This problem can be overcome by the use of a computer-controlled emission filter wheel (P in Figure 3.1) with single band emission filters, although the image acquisition can be slower. In the above example, although the excitation filter has achieved 40% stimulation of the DsRed2, a 510- to 520-nm-band GFP emission filter does not pass the DsRed2 emitted light which has much longer wavelengths greater than 550 nm (and only 2–3% emission at 510–520 nm). When fluorescence is combined with DIC imaging, the analyzer (N in Figure 3.1, xi in Figure 3.2) can be installed in the emission filter wheel to prevent loss of fluorescence intensity. An important disadvantage with this two-filter wheel approach is that the fluorescence cannot be viewed via the eyepieces since the emission filter (P in Figure 3.1) is only in the camera light path. Major manufacturers of fluorescence filters and complete filter sets are Chroma Technology Corp., Omega Optical Inc. and Semrock Inc., who all have useful websites to aid filter selection.

It is possible to buy fibre-optic scramblers (e.g. Technical Video Ltd.) that allow the mounting of the arc lamp away from the microscope. These systems remove the large heat source of the arc lamp from direct contact with the microscope, allow filter wheels and shutters to be mounted remotely, reducing vibration at the microscope, and are able to provide a more even field of illumination. However, the intensity of the illumination is diminished.

Fluorescent molecules can usually be stimulated several times before irreversible change occurs and fluorescence ceases; this is called photobleaching. The chemical products of the photobleaching process are often highly reactive chemicals that can interfere with cellular processes. It is therefore important to minimize light exposure to preserve both fluorescence signal and normal cellular activity, so modern fluorescence microscopes have either manual or computer-controlled shutters (T in Figure 3.1). Another way of reducing the intensity of excitation is the use of neutral density filters between the lamp house and the microscope.

Fluorescence microscopy often needs extra precautions since the light emitted by the fluorophore is often weak and live cells are sensitive to the excitation light. It is frequently advantageous to culture cells directly onto a coverslip and observe them using an immersion objective since its higher NA gathers significantly more light. It should be noted that phenol red pH indicator, commonly used in culture media, is fluorescent and should be avoided in order to reduce background. When selecting fluorescent markers for experiments it is worth remembering that, in general, exposure to longer wavelengths causes less damage than shorter ones.

Recent technological development has resulted in objective lenses with extremely high NA that are still designed to work with standard immersion oil and coverslips. These objective lenses achieve NA of 1.45 or 1.49 and allow easier implementation of a specialized microscopy technique based on total internal reflection fluorescence (TIRF). An objective with a NA of at least 1.45 allows through-the-lens laser excitation at an incidence angle smaller than the critical angle (the laser light reflects off the coverslip surface rather than passing through it) thus achieving total internal reflection and generating a narrow excitation field (called an evanescent wave) only in the immediate vicinity ($\approx$150 nm) of the coverslip. This type of illumination permits visualization of the distribution of the fluorophore next to the substrate with enhanced sensitivity (Figure 3.6f) in comparison to the bulk distribution visualized by standard epifluorescence (Figure 3.6e). The sensitivity of TIRF microscopy can be so high that it allows the imaging of individual molecules in live cells [13]. TIRF microscopes are available from all major microscope manufacturers.

Although the direct usage of objective lenses in a microscope limits the resolution as described in Section 3.2.1.1, recent development has allowed super-resolution with live cells, using for example the stimulated emission depletion (STED) principle [14] (Leica).

3.2.4.3 Three-dimensional imaging

Confocal microscopy [15] provides three-dimensional (3D) information about cells by optical sectioning. Illumination is restricted by a pinhole projected onto a limited region in the specimen. Observation is restricted by a second pinhole, placed in the position where the image of the illuminated region is formed. This arrangement of the illumination and observation pinholes is said to be confocal and eliminates light originating from features above and below the focal plane in the specimen, producing optical sectioning and at the same time increasing the maximum achievable resolution. An image is produced by scanning the specimen with the illumination and observation pinholes. The original confocal microscopes used a spinning disc with small holes in it for both illumination and observation. A version of the original design, which enhances the excitation intensity, has been implemented by Yokogawa Electric Corp. (available from Zeiss, Andor Technology, PerkinElmer Inc., Intelligent Imaging Innovations Inc.) but is not compatible with reflection imaging. This approach allows the entire field of view to be imaged at once, increasing speed. Most confocal microscopes work by slower raster scanning of a single laser spot across the specimen using galvanometer mirrors. When coupled to computer-controlled focusing (normally supplied with confocal microscopes) a stack of images of thin optical sections can be acquired and 3D structure reconstructed using software (Figure 3.6c). The main manufacturers of laser scanning confocal microscopes are Leica, Nikon, Olympus and Zeiss.

In multiphoton fluorescence confocal microscopy, fluorescent molecules are excited by an extremely high-intensity light of longer wavelength (e.g. infrared light). This high-intensity excitation light can only be tolerated by the specimen when applied as very short pulses (lasting pico- or femto-seconds) from special titanium–sapphire lasers (e.g. from Coherent Inc., Newport Corp.). The extremely high-intensity excitation is only achieved at the point of focus and consequently no out-of-focus material is excited, lowering the background and reducing overall photobleaching. It is also possible to excite UV fluorophores using visible wavelengths, eliminating the necessity for UV lasers and UV optics. In addition, multiphoton microscopy allows deeper penetration into thick specimens. Multiphoton confocal microscopes are available, for example, from LaVision Biotech GmbH, Leica, Prairie Technologies Inc., Olympus and Zeiss.

3.2.4.4 Direct imaging of protein dynamics and specific molecular interactions

Fluorescence recovery after photobleaching (FRAP) is a specialized technique that utilizes photobleaching for evaluating the dynamics of molecules. The molecule to be studied is fluorescently labelled and a region of interest within a cell is photobleached using high-energy excitation light. Photobleaching is rapidly followed by normal fluorescence imaging to record the recovery of fluorescence as unbleached molecules move into the photobleached region. The speed of recovery reflects the mobility of the labelled molecule. Rapid recovery indicates fast diffusion or active transport of free molecules, whereas slow recovery indicates that the molecule participates in larger complexes. Fluorescence localization after photobleaching (FLAP) is a technique related to FRAP but in addition can follow the translocation process of the photobleached molecules, with absolute as well as relative quantification [16, 17]. Another technique providing the absolute quantitation of the steady-state translocation process of specific molecules is photoactivation [18]. These techniques can be conveniently implemented using biofluorescent fusion proteins.

Fluorescence techniques also permit the sensitive measurement of molecular interactions. In the fluorescence resonance energy transfer (FRET) assay the presence of close molecular interaction can be inferred by the transfer of energy from one fluorophore to another [19, 20]. This process only occurs where the emission spectrum from one fluorophore significantly overlaps the excitation spectrum of the other and if they are in close proximity (up to ≈ 10 nm). Two interacting molecular species are separately labelled with different fluorophores. One fluorophore is then excited in the normal way and the emission of the other is measured, showing how much energy has been transferred between them. The separation and relative orientation of the two fluorophores strongly affects the amount of energy transferred.

In fluorescence correlation spectroscopy [21, 22] a single fluorescently labelled species is monitored, via a confocal pinhole, in a focus-limited spot. Time correlation between changes in emitted intensity is used to calculate the rate at which single fluorescent molecules enter and leave the volume of interest. These correlations can then be used to estimate the size of the molecule based on its diffusion rate. Two fluorophores can be monitored simultaneously to produce cross-correlations and hence information about the interactions between the two labelled molecules. Commercial systems are available, for example from Zeiss or Becker & Hickl GmbH.

3.2.5 Recording of microscopy images

Image recording using celluloid film has become obsolete because of the awkward development process and difficult handling of images for reproduction, presentation and publication, whereas digital image acquisition techniques have, in fact, superseded even 35-mm film in quality [23].

The choice of camera is determined by the microscopy technique. Standard charge-coupled device (CCD) video cameras are adequate for contrast enhancement techniques such as phase contrast and can be purchased from Cohu Inc., Hitachi Ltd, JAI Inc. (formerly Pulnix) or Sony Corp. Weak fluorescence signals require a sensitive scientific low-light-level CCD camera such as those available from Andor Technology plc, Hamamatsu Photonics K.K. and Photometrics. A standard CCD video camera can be connected to a video printer, video recorder or to a computer equipped with a digitization board. Scientific CCD cameras are designed to be directly connected to computers by their digital interface. Commercially available computer-based recording systems are listed in Section 3.2.4.6.

In computer recording of microscopy images, it is important to use correct spatial sampling determined by the size of a pixel (PICture ELement). In order to record an image at the resolution level provided by the microscope, the pixel size should be smaller than half of the resolution according to the Nyquist–Shannon sampling theorem [24]. The pixel size can be increased by binning, where adjacent pixels on the camera are merged. Another way of changing the pixel size is by the intermediate magnification lens. An important aspect of digital imaging is bit depth, which determines the numbers of intensity levels recorded in each pixel. For quantification or presentation, where high-quality data are required, images should be acquired and stored at the highest possible bit depth, often 12 (defining 4096 intensity levels). It is also important to choose a suitable image format, particularly its compression. In order to keep all the detail in an image, a non-lossy (where all original image information is preserved) compression has to be used. Tagged image file format (TIFF), for example, allows saving without any compression or with non-lossy Lempel–Ziv–Welch (LZW) compression, which usually reduces the file size but makes saving slower. Joint Photographic Experts Group (JPEG), on the other hand, often introduces considerable loss of detail and should be used with care.

An important consideration when choosing a microscope is the camera port on the microscope. Maximum light efficiency is achieved by an arrangement that does not require mirrors and minimizes the number of optical components in the light path to the camera. The bottom port of an inverted microscope can be used in this direct way. A disadvantage of this arrangement is that the camera detector tends to collect more dust.

3.2.5.1 Time-lapse imaging

The equipment required for recording individual images of live cells is also essential for time-lapse imaging, namely a suitable microscopy technique, specimen environmental control and computer image acquisition [25]. Advances in digital imaging and computing have dramatically simplified time-lapse recording (*Protocol 3.7*). The environmental control of tissue culture conditions needs to be provided as described in Section 3.2.1.5. For long-term recording this becomes much more critical, as does the elimination of unnecessary illumination. An automatic illumination shutter is recommended for the observation of live cells, whereas for time-lapse recording using high-intensity illumination, such as fluorescence, it is a necessity. Suitable software is then used to acquire images in time-lapse mode and to operate light shutters and wheels with fluorescence filters for multichannel recording. Temperature control of the microscope should be provided since variations even of less than 1 °C can lead to significant shifts in the focus or image due to thermal expansion and contraction of the microscope body or stage. Alternatively, automatic focus can be applied using software solutions (e.g. Metamorph from Molecular Devices Inc.) or specialized hardware (e.g. Perfect Focus, Nikon). Vibration may also be a destabilizing factor and can be minimized by an anti-vibration table (e.g. CVI Melles Griot). Complete imaging systems are available from Andor Technology plc, Applied Precision Inc., Molecular Devices Inc. and Photometrics Inc. ImageJ with the MicroManager plugin or LabVIEW (National Instruments Corp.) can be used to control bespoke or user-assembled systems.

PROTOCOL 3.7   Time-lapse imaging

Equipment

- Microscope

- Anti-vibration table (if vibration is a problem)

- Heater

- Stage incubator allowing delivery of humidified CO_2/air for open tissue culture dishes (for example a in Figure 3.6)

- Automatic light shutters (B and T in Figure 3.1)

- Automatic filter wheels for fluorescence multichannel recording (P and U in Figure 3.1)

- Camera (R in Figure 3.1)

- Computer with software for time-lapse imaging

Method

1 Set up the microscope as required (*Protocols 3.1* to *3.5*) with environmental control (see Section 3.2.1.5) and choose the observation field with cells in a suitable vessel (see Section 3.2.2 or 3.2.3).

2 Focus the image displayed on the monitor, using the focusing knob (*xiii* in Figure 3.2).

3 Choose the exposure, time-lapse interval and number of frames for recording.

4 Start the time-lapse program.

3.2.5.2 Multidimensional imaging

The development of computer technology combined with digital imaging has vastly increased the amount of data and the speed at which it can be captured and stored. When combined with computer-controlled motorized stages (viii in Figure 3.2), focusing, filter wheels and shutters, it became possible to collect five-dimensional data sets (x, y, z, time and colour). Reconstruction of these data sets allows the movement of multiple structures or cellular components to be followed through time and space at different locations. Recent developments using robotics allow the imaging of cells in multiple plates (e.g. IncuCyte from Essen BioScience Inc., ImageExpress Micro from Molecular Devices Inc. or BioStation from Nikon).

3.2.6 Analysis of microscopy images

Studies of the molecular mechanisms of cellular functions require reliable detection of subtle changes in cell behaviour against a background of high intrinsic heterogeneity and variability. Quantification is an obvious solution, especially combined with automatic image processing providing the necessary large amount of data required for eliminating the effects of heterogeneity.

Pre-processing often enhances images, although it can be detrimental to quantitative analysis.

3.2.6.1 Image enhancement

Acquired 3D fluorescence information can be processed by deconvolution in order to eliminate blurring resulting from structures that are out of focus. The procedure can be used to improve confocal images but it gives more dramatic improvement with images acquired from standard microscopy. Globular and filamentous structures are especially suitable for this technique. There are numerous commercially available programs for this purpose, all of which usually require considerable computing power. Huygens (Scientific Volume Imaging B.V.) is a flexible deconvolution program allowing many parameters to be defined for optimum performance. Autodeblur (AutoQuant Imaging Inc.) is a simpler program for personal computers.

3.2.6.2 Image processing – extraction of morphometric data

Vertebrate cells in tissue culture visualized using standard contrast enhancement techniques pose serious problems for automatic image processing since it is difficult to distinguish a cell from the background or other cells in contact because of inconsistent variations of light intensity along the cell outline or inside the cell. Interactive semi-automatic techniques can be used for the acquisition of limited amounts of data. Many image-processing programs, such as ImageJ [26], allow measurements of, for example, position or spread area of cells [27] or fluorophore co-localization. Automatic algorithms can be used in special cases such as for highly refractive cells, for example neutrophil leucocytes or *Dictyostelium discoideum*. Fluorescence images with good contrast can also be used for automatic image processing. Special interferometric techniques are ideally suited for automatic analysis of cell images [28] but these are not currently commercially available. Measurements based on cell outline can be used to measure various parameters such as the total amount of fluorescence intensity, displacement in a time-lapse recording and morphometric parameters such as elongation or polarity.

3.2.6.3 Data analysis – statistical evaluation

Data derived from observations of live cells usually form a hierarchical structure. Typical levels in the hierarchy might include treatment, cell culture, observation field or cell. Statistical tests on such data require analysis of variance (ANOVA) with a nested model [29, 30]. In a general case, the nested model also has to be unbalanced because the numbers of data points at different levels are not equal since they depend on chance, for example number of cells in the observation field. The ANOVA test calculates variability in the data introduced at individual levels and assigns statistical significance to the variability (at individual levels) taking into account interactions between the levels. The level of control/treatment is then usually expected to show a significant additional difference on top of the variability at the lower levels. When experimental conditions are well controlled then the level immediately below, the culture level, does not introduce significant variability. This means that individual cell cultures are reproduced with the same internal variability. The cell level usually gives a high significance since individual cells are highly variable even within individual cultures.

Observation of live cells reveals a great deal of information about dynamic intracellular structures and their interactions with cell behaviour. Because of the complexity of these interactions it is difficult to determine their consistency without computer simulation. Successful models of well-understood biological systems have been developed using established simulation techniques (see for example reference 29). More complicated systems with partial understanding require the development of a powerful simulation environment (see for example reference 31) to achieve theoretical predictions that can be tested by further experiments.

3.2.7 Maintenance

Microscopes that are not used frequently should be put through a maintenance programme prior to use. Frequently used microscopes should receive routine maintenance about once a week.

Maintenance consists of checking the alignment (*Protocols 3.1, 3.2, 3.3, 3.4*, and *3.5* if required for immediate use). Any exposed optical surfaces, such as objectives, eyepieces and condensers, should be examined for cleanliness during maintenance. An eyepiece, removed from the microscope and used backwards as a magnifier, allows examination of the small front lenses of objectives. Dust can be removed by clean compressed air. Contamination that cannot be removed in this way may be cleaned with a lens tissue. More serious contamination, such as grease or immersion oil, requires lens tissues and a solvent recommended by the microscope manufacturer, for example diethyl ether, ethanol or xylene (note that some manufacturers strictly prohibit the use of some solvents as they may dissolve the cement between the lens elements). The lens tissue can be wrapped around a cotton bud and dipped in solvent. Alternatively, for cleaning the front lens of an objective, a drop of solvent is placed on a lens tissue held horizontally above the objective. The tissue is then lowered so that the solvent contacts the lens and the tissue is dragged sideways, drying the lens. This process can be repeated several times using a clean piece of tissue each time.

3.3 Troubleshooting

- *No light is present in the eyepieces or on the camera.* Check that the light is on, follow the light path from the light source and look for obstructions.

- *The microscopy image is of poor quality.* Check the front element of the objective lens for cleanliness or damage, suitability of the objective specification, setting its correction collar if present, and illumination alignment.

- *The fluorescence image is of low intensity.* Check that the correct fluorescence filters have been selected, that the DIC analyzer is not in the light path, and that no neutral density filters are in use.

- *The computer-recorded image is of degraded quality.* Check the pixel size and compression in use.

- *The focus drifts during time-lapse recording.* Check whether the specimen is firmly attached to the stage and whether the microscope body may be exposed to vibration or variations in temperature. If temperature seems to be the problem, it may be possible to improve the room air conditioning and/or use a stable environmental incubator for the whole microscope. Alternatively, automatic focusing may solve the drift problem. In the case of unacceptable vibration, use an anti-vibration table.

- *The background is not clean.* Consider filtering the culture medium.

- *Cell viability is a problem with long-term observations.* Check temperature control, pH of the culture medium, humidification, and potential excess exposure to high-intensity light.

Acknowledgements

We thank Deborah Aubyn, Frédéric Bollet-Quivogne, Lucy Collinson, Ian Dobbie and Alastair Nicol (Cancer Research UK London Research Institute, Lincoln's Inn Fields Laboratories) and Sarah Whitehouse (University of Sheffield, UK) for their comments and contributions.

References

★ 1. Canti, R.G. (1928) Cinematographic demonstration of living tissue cells growing in vitro. *Arch. Exp. Zellforsch.*, **6**, 86–97. – *Original description of time-lapse microscopy with cells in vitro.*

2. Abercrombie, M. and Heaysman, J.E. (1953) Observations on the social behaviour of cells in tissue culture. I. Speed of movement of chick heart fibroblasts in relation to their mutual contacts. *Exp. Cell. Res.*, **5**, 111–131.

★★ 3. Inoue, S. and Spring, K.R. (1997) *Video Microscopy: the Fundamentals*, Plenum Press, New York & London. – *Comprehensive review.*

4. Jenkins, F.A. and White, H.E. (1976) *Fundamentals of Optics*, McGraw-Hill, New York, London.

5. Slayter, E.M. and Slayter, H.S. (1992) *Light and Electron Microscopy*, Cambridge University Press, Cambridge.

6. Dunn, G.A. and Zicha, D. (1993) Phase-shifting interference microscopy applied to the analysis of cell behaviour. *Symp. Soc. Exp. Biol.*, **47**, 91–106.

7. Hogenboom, D.O., Dimarzio, C.A., Gaudette, T.J. *et al.* (1998) Three-dimensional images generated by quadrature interferometry. *Opt. Lett.*, **23**, 783–785.

8. Zicha, D., Dunn, G. and Jones, G. (1997) Analyzing chemotaxis using the Dunn direct-viewing chamber. *Methods Mol. Biol.*, **75**, 449–457.

9. Zicha, D., Dunn, G.A. and Brown, A.F. (1991) A new direct-viewing chemotaxis chamber. *J. Cell Sci.*, **99** (Pt 4), 769–775.

10. Shaner, N.C., Steinbach, P.A. and Tsien, R.Y. (2005) A guide to choosing fluorescent proteins. *Nat. Methods*, **2**, 905–909.

11. Pepperkok, R., Saffrich, R. and Ansorge, W. (1994) Computer-automated capillary microinjection of macromolecules into living cells, in *Cell Biology: a Laboratory Handbook* (ed. J.E. Celis), Academic Press, San Diego, pp. 22–29.

12. Ploem, J.S. (1971) A study of filters and light sources in immunofluorescence microscopy. *Ann. N. Y. Acad. Sci.*, **177**, 414–429.

13. Mashanov, G.I. and Molloy, J.E. (2007) Automatic detection of single fluorophores in live cells. *Biophys. J.*, **92**, 2199–2211.

14. Willig, K.I., Rizzoli, S.O., Westphal, V. *et al.* (2006) STED microscopy reveals that synaptotagmin remains clustered after synaptic vesicle exocytosis. *Nature*, **440**, 935–939.

15. Pawley, J.B. (2006) *Handbook of Biological Confocal Microscopy*, Springer, New York, NY.

16. Dunn, G.A., Dobbie, I.M., Monypenny, J. *et al.* (2002) Fluorescence localization after photobleaching (FLAP): a new method for studying protein dynamics in living cells. *J. Microsc.*, **205**, 109–112.

17. Zicha, D., Dobbie, I.M., Holt, M.R. *et al.* (2003) Rapid actin transport during cell protrusion. *Science*, **300**, 142–145.

18. Patterson, G.H. and Lippincott-Schwartz, J. (2002) A photoactivatable GFP for selective photolabeling of proteins and cells. *Science*, **297**, 1873–1877.

19. Gu, Y., Di, W.L., Kelsell, D.P. and Zicha, D. (2004) Quantitative fluorescence resonance energy transfer (FRET) measurement with acceptor photobleaching and spectral unmixing. *J. Microsc.* **215**, 162–173.

20. Wouters, F.S., Bastiaens, P.I., Wirtz, K.W. and Jovin, T.M. (1998) FRET microscopy demonstrates molecular association of non-specific lipid transfer protein (nsL-TP) with fatty acid oxidation enzymes in peroxisomes. *EMBO J.*, **17**, 7179–7189.

21. Mustapa, M.F., Bell, P.C., Hurley, C.A. *et al.* (2007) Biophysical characterization of an integrin-targeted lipopolyplex gene delivery vector. *Biochemistry*, **46**, 12930–12944.

22. Schwille, P. (2001) Fluorescence correlation spectroscopy and its potential for intracellular applications. *Cell. Biochem. Biophys.*, **34**, 383–408.

23. Entwistle, A. (1998) A comparison between the use of a high-resolution CCD camera and 35 mm film for obtaining coloured micrographs. *J. Microsc.*, **192**, 81–89.

24. Shannon, C.E. (1949) Communication in the presence of noise. *Proc. Inst. Radio Eng.*, **37**, 10–21.

25. Zicha, D. (2000) in *Culture of Animal Cells: a Manual of Basic Technique* (ed. R.I. Freshney), Wiley-Liss, New York, pp. 429–431.

26. Abramoff, M.D., Magelhaes, P.J. and Ram, S.J. (2004) Image processing with Image *J. Biophotonics Int.*, **11**, 36–42.

27. Cammer, M., Wyckoff, J. and Segall, J. (1997) Computer-assisted analysis of single-cell behavior, in *Basic Cell Culture Protocols* (ed. J.W. Pollard and J.M. Walker), Humana Press, Totowa, NJ, pp. 459–470.

28. Zicha, D. and Dunn, G.A. (1995) An image processing system for cell behaviour studies in subconfluent cultures. *J. Microsc.*, **179**, 11–21.

29. Milliken, G.A. and Johnson, D.E. (1992) *Analysis of Messy Data*, Chapman & Hall, London.

30. Snedecor, G.W. and Cochran, W.G. (1989) *Statistical methods*, Iowa State University Press, Ames, IA.

31. Zicha D and Vesely P (1989) The use of a production system for simulation analysis of tumour cell migration in vitro - Development of a specialized control strategy, in *Artificial Intelligence in Medicine* (eds J. Hunter, J. Cookson and J. Wyatt), Springer-Verlag, Berlin, pp. 269–275.

4

Basic Techniques and Media, the Maintenance of Cell Lines, and Safety

John M. Davis

School of Life Sciences, University of Hertfordshire, Hatfield, Hertfordshire, UK

4.1 Introduction

This chapter provides an introduction to some of the more general and routine aspects of cell culture methodology. For more detailed procedures and critical analysis of other general methods, as well as details of many specialized cell culture techniques, there are numerous other excellent publications that can be consulted [1–21]. In addition, the EC Guidance on Good Cell Culture Practice [22], drawn up by a team of respected cell culture scientists from both Europe and America, should be read by any worker intent on performing useful work using the techniques of mammalian cell culture.

4.1.1 The terminology of cell and tissue culture

The following definitions help explain how cultured cell populations are derived, and how they come to express some of their distinguishing characteristics. For additional information, see reference 23.

Animal Cell Culture: Essential Methods, First Edition. Edited by John M. Davis.
© 2011 John Wiley & Sons, Ltd. Published 2011 by John Wiley & Sons, Ltd.

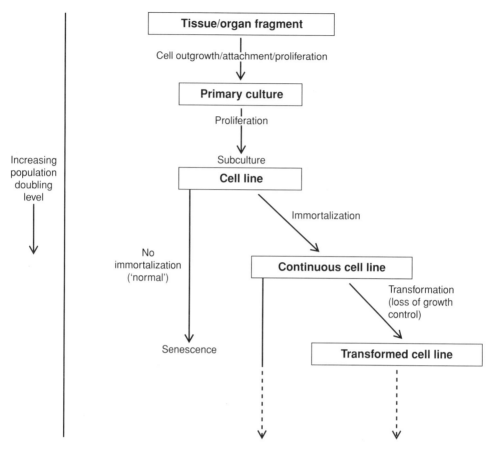

Figure 4.1   Scheme depicting the origin and progression of cell lines with population doubling level. (Note, however, that in some cases transformation can occur in the absence of immortalization - see Section 4.1.1.2 *vii.*)

4.1.1.1 The origin of cell lines

i. Primary culture   Primary cultures are derived from intact or dissociated tissues, or from organ fragments. A culture is generally considered to be a primary culture until it is subcultured (or passaged), after which it is termed a cell line (Figure 4.1). The product of the first subculture can also be called a secondary culture. The term tertiary culture is hardly ever used.

ii. Subculture (also called passage)   To subculture is to transfer or transplant cells of a culture into a new culture vessel in order to propagate the cell population or set up replicate cultures.

iii. Cell line   A cell line is a cell population derived from a primary culture at the first subculture. Adjectives such as finite (indicating a finite *in vitro* lifespan) or continuous (indicating an assumed unlimited generational potential) should be used only when data arc available to support them. The term cell line does *not* imply homogeneity or reflect the degree to which a culture has been characterized.

iv. Cell strain   This term is used to describe a subcultured cell population selected for the expression of specific properties, functional characteristics or markers. When describing a cell strain, its specific features should be described.

v. Clonal culture (clone)   Clonal culture, clonal selection or cloning is the establishment of a cultured cell population derived from a single parental cell. Although one might expect that a culture derived by the successive mitoses of a single cell would consist of identical cells, this is not necessarily the case. Clonally derived cell populations frequently (and often rapidly) come to express some degree of heterogeneity, and thus cloning does *not* imply absolute homogeneity or long-term stability of characteristics within the cell population (see Chapter 8).

4.1.1.2 The characteristics of normal and transformed cells

i. Normal cell   Although the properties of 'normal' cells can be described, the term evades precise definition. For primary cultures, 'normal' implies that the cells were derived from normal, healthy tissue, not from a pathological lesion. Following subculture and the generation of cell lines, normal cells commonly (although not invariably) exhibit a set of characteristic properties including (but not limited to) anchorage dependence, density-dependent inhibition of growth and a finite lifespan (in terms of the number of cell doublings attainable). Traditionally, the converse of normal is 'transformed' (but see section *vii* below). These terms are usually defined by considering properties of cell behaviour commonly expressed by cells of one category and not the other. Such features (Table 4.1) are helpful in characterizing cultures, but it must be stressed that no single property is necessarily diagnostic of the normal versus transformed condition, and exceptions do exist.

ii. Anchorage dependence   Anchorage-dependent (also termed substrate-dependent) cells need to attach to a surface (substrate) in order to survive, multiply or express their characteristic or differentiated function(s). In common usage, the term implies the requirement for attachment in order to support proliferation, a characteristic of most normal, non-haemopoietic cells. The alternative to anchorage dependence is the ability to proliferate in fluid suspension or within a three-dimensional, semi-solid medium such as one incorporating agar or methylcellulose ([24, 25] and Chapter 8). Many transformed cell lines form colonies in soft agar, and this ability plus the lack of density-dependent inhibition of proliferation is considered to be an expression of 'altered growth control'. Growth on a rigid substrate by itself, however, does not suggest that a cell line is normal, as most transformed cells also grow well on such surfaces.

iii. Monolayer culture   The term monolayer culture describes the distribution of cells as a single layer on a substrate such as glass or plastic. The ability of cells to grow as a monolayer does not necessarily imply that they are normal or that they are anchorage dependent. The phrase 'substrate-dependent monolayer culture' should only be used

Table 4.1 Comparison of the characteristics of normal and transformed cells.

Normal cells

- Are anchorage dependent (except haemopoietic cells)
- Show population density-dependent inhibition, and often contact inhibition, of proliferation
- Have a finite lifespan (*c.* 10–100 population doublings, dependent on species, cell type, and age of donor organism)
- Display altered characteristics with increasing age *in vitro.*

Transformed cells

- Are genetically unstable (heteroploid/aneuploid)
- Show altered growth characteristics:
    - Reduced requirements for exogenous growth factors (serum)
    - Generally, a reduced population doubling time
    - Reduced dependence on substrate adhesion/increased ability to form colonies in agar
    - Reduced population density inhibition of proliferation
    - Reduced contact inhibition of proliferation.

---

when it has been demonstrated that the cells are indeed attachment dependent and that the culture is composed of a single layer of cells.

iv. Density-dependent inhibition (or limitation) of proliferation Most normal cell lines exhibit a marked reduction in proliferative activity upon the attainment of confluency, that is occupancy of all available attachment surface. In most cases this is a population density effect, not a cell–cell contact effect (i.e. contact-inhibition of growth). Density-dependent inhibition of proliferation can occur before confluence is attained, due to a diminished nutrient supply and/or the release of cell-derived factors (including waste products) into the medium.

v. Saturation density The saturation density of an attached culture is its population density (in cells/cm^2) at the point when it reaches density-dependent inhibition of growth. This is normally reported in terms of cells/cm^2, but is sometimes given in cells/ml; in either case, the conditions (medium volume, culture surface area, etc.) under which the saturation density was measured must be defined. Transformed cell lines commonly have a higher saturation density than normal cells.

vi. In vitro cell senescence (cell ageing in culture) Normal cell lines usually have a finite lifespan, that is they do not grow beyond a finite number of cell generations (population doublings). For example, the lifespan of normal diploid human fibroblasts is in the range of 50–70 population doublings [26]. The process of cell ageing in culture involves progressive alterations in a number of cell characteristics. Table 4.2 lists some of those exhibited by normal diploid fibroblasts.

Table 4.2 Changes in the characteristics of normal fibroblasts with increasing generation number *in vitro*.

- Increased cell doubling time
- Decreased proportion of cells in the cell cycle/reduced DNA synthesis
- Decreased adhesion to substrate/changes in cell surface molecules
- Altered cytoskeletal organization
- Increased level of aneuploidy
- Decreased rate of protein synthesis and degradation (turnover)
- Decreased amino acid transport
- Decreased sensitivity to growth factors
- Shortened telomeres

vii. Immortalization and transformation   A transformed cell (or cell line) is one that has a reduced dependence on extracellular factors for continued growth, usually along with other properties such as elevated saturation density and anchorage-independence that distinguish it from a normal cell line (Table 4.1). Transformation was once thought to occur only after immortalization, but more recently it has become clear that the two processes are actually independent, and transformation can be induced without immortalization, for example by the transfection of cells (particularly human cells) with *RAS* or *RAF* genes. Similarly, immortalization can occur without transformation – 3T3 cells, for example, appear to be immortalized but not transformed.

The process of immortalization is composed of a number of steps [29, 30], the first of which is escape from normal (p53- and p16-mediated) senescence. This can occur apparently spontaneously in mouse cells, but human cells must be treated with exogenous agents (e.g. ionizing radiation, chemical mutagens/carcinogens, viruses) to induce this change [4, 27, 28]. However, in the absence of telomerase expression the cells' telomeres become extremely short, and the cells enter a period of proliferative arrest ('crisis') often accompanied by abnormal morphology and degenerative loss in cell number [5]. Should telomerase expression then occur spontaneously or be induced, this can lead to immortalization, which is characterized by the emergence of a rapidly dividing cell population that overgrows the culture. Note that some cells, notably mouse fibroblasts, appear to have natural telomerase expression and never undergo crisis prior to immortalization.

It should be noted that *in vitro* transformation is not equivalent to neoplastic transformation, which is the ability of cells to form tumours when injected into an immuno-incompetent host. However, the belief appears to be emerging that cells that are both transformed *and* immortal are probably potentially tumorigenic.

4.1.2 Basic components of the cell culture environment

To promote cell survival and proliferation, it is essential that the *in vitro* culture environment meets the fundamental physiological requirements of the cell. Components that one can control include factors associated with the medium such as its composition, pH, osmolality and the volume and frequency of replenishment. In addition, incubation conditions such as temperature, relative humidity and gas composition can be regulated, as can the form and composition of the physical substrate for cell attachment.

4.1.2.1 Culture medium

The culture medium must provide all the essential nutrients, vitamins, cofactors, metabolic substrates, amino acids, inorganic ions and trace elements needed to support cellular functions and the synthesis of new cells. Additionally, buffering is required in order to maintain the cells within their normal pH range and limit the effects of acidic waste products generated by the cell, most notably CO_2 and lactic acid. Most media also contain non-essential ingredients that facilitate the work of the cell culture scientist, for example Phenol Red, which gives an instant visual readout of the approximate pH of the medium. All of the above components are generally supplied by the basal medium (Section 4.1.2.1.*i*). However, this is not in itself sufficient to support cell viability for more than a few hours, and a source of growth factors and certain other substances is required to induce cell growth and multiplication. For many decades this was supplied almost without exception by animal sera (Section 4.1.2.1.*ii*). Fetal bovine serum (FBS – also known as fetal calf serum), a by-product of the beef industry, was the most popular of these, and is still used extensively today. However, a variety of factors (including its batch-to-batch variability, lack of definition, expense, the potential for carrying adventitious agents and the presence of unwanted substances) has driven a move to replace serum with more defined materials. This is discussed in depth in Chapter 5.

i. Basal medium   There are four main categories of basal medium for mammalian cells:

(a) Eagle's minimal essential medium (EMEM) and its derivatives – these include alpha minimal essential medium (αMEM), Dulbecco's modified Eagle's medium (DMEM), Glasgow minimal essential medium (GMEM) and Joklik's minimal essential medium (JMEM).

(b) Media designed at the Roswell Park Memorial Institute (RPMI). The most popular of these is probably RPMI 1640, but others include RPMI 1629 and RPMI 1630.

(c) Basal media designed for use with sera, for example Fischer's, Leibovitz's L-15, Trowell's, and Williams' Medium E.

(d) Basal media designed for use in serum-free formulation, for example Iscove's modified Dulbecco's medium (IMDM), CMRL 1060, Ham's F10 and derivatives such as F12, TC 199 and derivatives, the MCDB series, NCTC series, and Waymouth's.

These categories are in fact inter-related, as the history of medium development has been one of step-wise modification and development. This is clearly illustrated by IMDM, which was developed from DMEM, itself derived from EMEM.

Basal media are composed of the following sets of components

- *A balanced salt solution*

Balanced salt solutions comprise a mixture of inorganic salts designed to:

- maintain osmotic pressure

- buffer the medium at physiological pH

- maintain membrane potential

- act as cofactors in enzyme reactions and in cell attachment.

The main inorganic ions used are Na^+, K^+, Mg^{2+}, Ca^{2+}, Cl^-, SO_4^{2-}, PO_4^{3-} and HCO_3^-. (Trace elements, such as selenium, are also required by cells [31], but these are often omitted from media intended for use with serum, as the serum will generally supply sufficient of these.) In addition, glucose is often added as an energy source for the cells.
   The four main categories of balanced salt solution are:

- Dulbecco's phosphate-buffered saline (DPBS)

- Hanks' balanced salt solution (HBSS)

- Eagle's spinner salt solution (ESSS)

- Earle's balanced salt solution (EBSS).

The last two require an atmosphere of CO_2 in air to maintain their intended pH.

- *A buffering system*

Media need to be buffered for the reasons mentioned at the start of Section 4.1.2.1. Bicarbonate is frequently used for this purpose, but requires an atmosphere of 5–10% CO_2 in order to maintain the pH that the medium was designed to have (at least until the cells themselves produce significant quantities of CO_2 and/or lactic acid). Each of the basal media has a recommended bicarbonate concentration and CO_2 tension to achieve the correct pH and osmolality (Table 4.3). Some media include other buffers such as HEPES. This has a pK_a of 7.3 at 37 °C, and thus buffers very effectively without CO_2 in the pH 7.2–7.6 range, unlike bicarbonate with its pK_a of 6.1. However, bicarbonate is also essential to cells as a nutrient, so there must always be sufficient in the medium for this purpose.

- *An energy source*

Carbohydrates are generally the major energy source used by cultured cells and glucose is the one most frequently supplied in basal media, but sometimes other sugars such as

Table 4.3 Recommended gas-phase CO_2 concentrations for use with some common media.

Basal Medium	NaHCO$_3$ concentration (mM)	Percentage CO_2 in gas phase
Eagle's MEM (Hanks' salts)	4	0
Ham's F12	14	5
RPMI-1640	24	5
TC199	26	5
DMEM/Ham's F12	29	5
IMDM	36	5
DMEM	44	10[a]

[a]Although DMEM was designed for use with 10% CO_2, many methods are published that, for whatever reason, employ DMEM with 5% CO_2. It should be recognized that this will lead to the initial pH being a little more alkaline.

maltose, sucrose, fructose, galactose and mannose may also be included. Glucose is catabolized by a balance of both aerobic and anaerobic pathways, with the balance depending on environmental factors such as incubation temperature and pH (J. Davis, unpublished results). This leads to the production of CO_2 and lactic acid, respectively, both of which will tend to acidify the medium. Aside from acidification effects, it is often claimed that lactic acid is toxic to cells, but in fact the level of toxicity varies greatly between cell lines. Some studies indicate that use of other sugars, including galactose and fructose, can substantially reduce the generation of lactate [32]. The amino acid glutamine is also an important energy source, and in some cell types may supply over 50% of the ATP generated [33, 34].

- *Amino acids*

Although most animal cells have a high requirement for glutamine, both as an energy source (see above) and as an amine donor in biosynthesis [35], they also require a supply of the essential amino acids (those not synthesized in the body). In humans these are arginine, cystine, histidine, isoleucine, leucine, lysine, methionine, phenylalanine, threonine, tryptophan and valine. Cysteine and tyrosine are also added to media to compensate for inadequate synthesis. Other amino acids may be added to compensate for a particular cell type's inability to make them or to counteract loss into the medium. Balancing the concentrations of the various amino acids in a medium is a complex process (see Chapter 5).

It is important to note that glutamine is heat-labile, and degrades both during storage at 4 °C and during use at 37 °C. The rate of degradation will depend not only on the temperature but also on the pH of the medium. Thus glutamine is usually stored frozen, then thawed and added to bottles of medium just before first use. Derivatives of glutamine that are more stable can be used instead in many cultures, and are commercially available (e.g. GlutaMAX™ [L-alanyl-L-glutamine] supplement from Invitrogen).

- *Vitamins*

Several vitamins of the B group are essential for cell growth and multiplication. Those most frequently found in basal media are *para*-amino benzoic acid, biotin, choline, folic acid, nicotinic acid, pantothenic acid, pyridoxal, riboflavin, thiamine and inositol.

- *Other components*

Depending on the demands to be placed on the medium, a variety of other components may be added. Trace elements such as selenium, iron, zinc and copper may be important, particularly where the medium is intended for use without serum supplementation. Some media may contain tricarboxylic acid cycle intermediates such as pyruvate, or polyamines such as putrescine.

- *Antibiotics*

Antibiotics are added to media in many laboratories, but this is poor practice unless it can be justified on a scientific basis (for example, where materials that cannot be sterilized, such as virus preparations, have to be added to the culture, or in primary cultures where the risk of microbial carry-over from the animal of origin may be very great). Antibiotic use tends to encourage sloppy aseptic technique, cryptic microbial contamination, and in the longer term the emergence of antibiotic-resistant organisms (see also Chapter 9, Section 9.2.1.2).

A full list of the components of the various commercially available media are available in the suppliers' catalogues, or from their websites (e.g. www.lifetechnologies.com, www.sigmaaldrich.com, etc.).

ii. Serum This is the liquid excluded from the clot when blood is allowed to coagulate. Because it is rich in growth factors, and has a low antibody concentration, the most widely used serum in mammalian cell culture is FBS, followed by newborn calf serum, calf serum and adult bovine serum. Horse, human and other sera are also used for some purposes.

- *Constituents*

The typical concentrations of some of the constituents in bovine sera (and some related parameters) are shown in Table 4.4. Along with these components, serum also contains

- a cocktail of growth factors – among these are platelet-derived growth factor (PDGF), fibroblast growth factors (FGFs), epidermal growth factor (EGF), vascular endothelial growth factor (VEGF), insulin-like growth factors-I and -II (IGF-I and IGF-II);
- proteins – these include albumin (which protects cells against shear forces, binds heavy metals and pyrogens [36, 37], and carries lipids, fatty acids and hormones (see below)); fibronectin (which helps cells bind to culture surfaces); transferrin (which transports iron); and protease inhibitors (especially α_1-antitrypsin and α_2-macroglobulin). The

Table 4.4 Typical composition/parameters of bovine sera. (adapted from reference 38).

Parameter	Fetal	Newborn	Calf	Adult
Source Animal				
Age	Gestational	<10 days	<12 months	1–6 years
Weight (kg)	9–32	23–41	160–340	550–1600
Diet	Varying	Nursing from mother only	Varying: primarily grass, some grain	Varying: hay, grain and grazing
Serum				
Total Protein (g/dl)	3.0–4.5	4.5–6.5	5.8–7.5	5.8–8.2
pH	7.1–7.5	7.2–7.8	7.5–8.0	7.6–8.0
Osmolality (mOsm/kg)	298–325	285–302	268–313	268–287
Endotoxin (EU/ml)	<0.5–1.1	3.8–176.3	<0.4–240	<0.1–0.5
Sodium (meq/l)	133–142	134–143	131–144	135–142
Potassium (meq/l)	9.7–12.7	5.4–8.0	5.4–7.7	4.5–5.9
Calcium (mg/dl)	13.5–15.0	8.5–11.8	9.6–11.8	7.8–9.4
Phosphorus (mg/dl)	9.1–11.6	8.1–10.3	7.7–11.0	6.6–7.4
Chloride (meq/l)	93–101	93–102	91–105	94–102
Uric acid (mg/dl)	1.5–3.2	0.1–2.1	0.6–1.6	0.2–0.6
Alk. Phosphatase (U/l)	165–247	122–220	137–329	45–63
LDH (U/l)	409–605	667–1102	950–2388	747–980
SGOT/AST (U/l)	22–69	27–66	47–98	16–63
SGPT/ALT (U/l)	0–8	12–20	8–23	13–20
GGT (U/l)	4–7	76–425	22–47	14–26
Cholesterol (mg/dl)	27–58	53–112	111–174	69–205
Bilirubin, total (mg/dl)	0.1–0.2	0.3–0.9	0.1–0.4	0.1–0.2
Glucose (mg/dl)	83–132	63–193	86–946	17–100
Iron, serum (µg/dl)	170–194	69–100	35–144	80–169
BUN (mg/dl)	13–17	13–15	6–14	4–15
Creatine (mg/dl)	2.6–3.5	0.9–1.4	0.9–1.8	1.1–1.6
Triglyceride (mg/dl)	17–91	34–48	32–53	31–38
Haemoglobin (mg/dl)	6.0–17.9	12.0–20.2	6.6–23.0	3.1–22.9
Albumin/globulin ratio	0.6–4.1	0.4–0.9	0.4–1.1	0.3–0.8
Albumin (g/dl)	1.2–2.9	1.5–2.8	1.9–3.7	· 1.9–2.6
Alpha-1-globulin (g/dl)	<0.1–0.8	<0.1–0.8	<0.1–0.1	<0.1
Alpha-2-globulin (g/dl)	0.1–1.3	0.5–1.3	0.7–1.6	0.6–1.5
Beta-globulin (g/dl)	0.2–1.5	0.8–1.2	0.9–2.0	0.8–1.4
Gamma-globulin (g/dl)	<0.1	0.5–2.0	0.6–2.7	1.6–4.1

BUN, blood urea nitrogen; GGT, gamma-glutamyl transferase; LDH, lactate dehydrogenase; SGOT/AST, aspartate aminotransferase; SGPT/ALT, alanine aminotransferase.

protease inhibitors have an important role to play in neutralizing trypsin following the treatment of cells with trypsin-EDTA to remove them from a culture surface;

- hormones – these include thyroxin (which can be carried by albumin), triiodothyronine, insulin, growth hormone and hydrocortisone.

- *Storage*

All commercially obtained sera should be received from the supplier frozen, and should be kept frozen at $-20\ ^\circ$C or below until it is necessary to open a bottle. It should then be thawed in a 37 $^\circ$C water bath for the minimum time required to thaw it completely. The bottle should be swirled every 10–15 minutes until completely thawed to avoid the build-up of pH and solute gradients that can lead to damage to serum components. If the whole amount is not to be used immediately, it can be stored at 4 $^\circ$C for 2–3 days. For longer storage, it should be aliquotted into suitable volumes for use (e.g. 25 ml or 50 ml) and again stored frozen at $-20\ ^\circ$C or below. Repeated freezing and thawing must be avoided in order to maintain the levels of growth factors. See reference 38 for further details on serum stability and storage.

- *Risks associated with sera*

It is important to note that animal sera are a potential source of adventitious agents including, but by no means limited to, mycoplasma (sera are probably the greatest source of mycoplasma contamination in the laboratory – see Chapter 9), foot and mouth (hoof and mouth) disease virus, bluetongue virus and (at least theoretically) BSE. Thus it is essential to source sera carefully from reputable suppliers who can fully document the serum all the way back to the point of collection. This becomes critical where FBS must be sourced from BSE-free countries, for example for use with cell lines involved in the preparation of therapeutic products [39]. Such sera are particularly expensive, and this has attracted counterfeiters who pass off material sourced from non-BSE-free countries and as coming from BSE-free locations [40].

Human sera are, of course, a potential source of pathogens and should be tested for those pathogens of greatest concern (HIV, HTLV, hepatitis B and C, etc.) before use.

For a very useful review of the composition, production, processing, testing, storage and stability of animal media, see reference 38. For more details of its applications in cell culture, see references 4 and 41.

iii. Complete medium Unless it is serum-free, protein-free or chemically defined (see Chapter 5) complete medium will be a mixture of a basal medium and serum, for example DMEM + 10% FBS. The actual basal medium, the type of serum and the concentration of the serum will depend predominantly on the cells to be cultured, along with the nature of the work to be undertaken. Advice should be sought from the supplier of the cell line, or failing that from the published literature.

It is important to note that different cells may have different nutrient requirements, and that most media were designed for very specific applications, such as the optimal growth of one particular cell line. Thus a given medium cannot be expected to provide optimal support for any cell line other than the one for which it was developed. Additionally, certain media were designed for the growth of cells at low population densities, for example some of the MCDB series of media, and may not have the levels of certain components necessary for cells at 'normal' population densities. Similarly, media for use at very high population densities, for example in bioreactors, will require further modification, not least in their buffering capacity.

Clearly, enormous effort is required to establish the medium composition and culture conditions for optimal cell proliferation, or for maximal expression of a differentiated function, in any particular cell line. Most workers do not actually attempt to do this, but rather find conditions that support 'adequate' function and growth. Nevertheless, there are very significant advantages to be gained by using well-characterized, fully defined media [10], and approaches and methods to achieve this are covered in depth in Chapter 5.

4.1.2.2 Physicochemical Factors

Several features of the culture environment combine with the medium in use to define the conditions experienced by the cells. As these can fluctuate over time, they should be monitored. Generally, optimal culture conditions for most mammalian cells fall within the following parameters (although exceptions do exist):

- **pH**: 7.0–7.4

- **osmolality**: 280–320 mOsmol/kg

- **CO$_2$**: 5–10% in air

- **temperature**: 35–37 °C.

However, just as different cell lines have different nutrient requirements, certain cells may have unique physicochemical demands for optimal growth or function. A good example is temperature: whereas most mammalian cells show optimal growth in the range 35–37 °C (although this may depend on the tissue of origin), cells from cold-blooded vertebrates often grow best at lower temperatures. The A6 toad kidney epithelial cell line, for example, grows best at 26–28 °C [42].

4.1.2.3 Stationary versus dynamic media supply

Media delivery is another important factor in a culture system. Cells must be supplied with sufficient medium to meet physiological demands and the medium must be replenished at an appropriate frequency and volume. If the volume of medium in which the cells are growing is too low, or the medium is not replenished frequently enough, nutrients may be depleted too rapidly and metabolic products become too concentrated, resulting in deleterious effects on cell function or proliferation. Conversely, if the initial volume of medium is too great or the frequency of replenishment is too rapid, cells may be unable to condition their environment adequately (see Section 4.2.5.2 and Chapter 8,

Section 8.1.3.4). In general, the recommended volume of medium for routine substrate-dependent culture is 0.2–0.3 ml/cm^2. Since the medium overlaying a layer of cells acts as a barrier to gas diffusion, an excessive volume of medium can restrict cell access to CO_2 and O_2.

Automatic delivery and replenishment of medium is an integral feature of many suspension and microcarrier culture systems (see Chapter 10, Section 10.2.1.3), and particularly those used at the industrial scale [43–45]. Methods have also been devised to improve the delivery and mixing of medium in routine substrate-dependent cultures. One of the simplest techniques is to place the culture vessel on a rocking platform within the incubator. As the platform tilts from one side to the other (3–6 cycles/min) the cells are gently washed by the medium. This mixes the medium, thereby minimizing the local build-up of potentially harmful waste products at the cell surface. Once in each tilt cycle the cells are covered by only a thin film of medium, which helps improve O_2 and CO_2 delivery and may be beneficial to some cell populations. Roller bottles used for large-scale culture accomplish the same end (see Chapter 10, Section 10.2.1.1).

4.2 Methods and approaches

4.2.1 Sterile technique and contamination control

Fastidious aseptic technique is an essential prerequisite for successful cell culture. In order to keep cultures free from both microbial contamination and cellular cross-contamination, one must understand the factors that can lead to contamination, and must adopt well thought-out procedures for handling cells and reagents, paying constant attention to even the most routine manipulations. While laboratory errors can be costly in any discipline, cell culture has the added difficulty that not all problems are immediately obvious. Some forms of contamination, particularly low-level mycoplasma infection, the occurrence of cellular cross-contamination, or the presence of persistent virus infection, can go undetected for years.

Sloppy cell culture technique can be very disruptive to the laboratory. It is true that occasional isolated microbial contaminations can happen to anyone, and are somewhat in the nature of an occupational hazard for the cell culture scientist. When such problems only occur once in a blue moon they are normally no more than a nuisance, but recurrent microbial contamination of working cultures is more troublesome and can become very time-consuming and expensive. Even more serious is the contamination of cell line stocks. This can wipe out an irreplaceable model system or, particularly if cellular cross-contamination is involved, can completely invalidate the results of all studies performed with the suspect lines. Consequently, it is absolutely essential that the cell culture specialist adopts the best possible working habits and is constantly on the lookout for problems and their causes.

4.2.1.1 Working within the microbiological safety cabinet

Many cell culture procedures can, in principle, be carried out at the open bench (see Section 4.2.1.3). However, working in the open increases the chance of introducing external contaminants to cultures and, in cases where the work involves potential pathogens, places the worker at increased risk of infection. Thus it is recommended that, as far as possible, the handling of cultured cells is always carried out in a MSC, preferably a Class II cabinet unless microbiological safety considerations demand the use of a Class I or Class III cabinet – see Section 4.2.8.2.*ii*. As described in Chapter 1, only MSCs offer protection to the operator from aerosols generated during culture work, and as almost all cell lines have the potential to carry viruses (see Section 4.2.8.2.*ii*), and as it is essentially impossible to prove that any cell line is entirely virus-free, it is strongly recommended that cells are only manipulated in an MSC, and never in a horizontal or vertical UDAF (laminar flow cabinet) – see Chapter 1, Section 1.2.3.1.

There are a number of points to keep in mind regarding the use of MSCs for cell culture.

(a) The MSC creates a *clean* work environment, *not* a sterile one.

(b) The MSC is the most critical work area in the cell culture laboratory and thus, despite the filtered air, is the most likely site where cultures can become contaminated.

(c) Perform all activities in an MSC with the understanding that air flow can carry contaminants to the culture system.

(d) Hands and arms (or gloves and sleeves) can be a serious source of contaminant-laden particulates, such as dry skin, talc and lint. Thus when working in a Class II MSC never reach over an open dish or bottle.

i. Organizing the hood for routine work   Ideally one should initially stock the MSC with every item needed (but bear in mind point (a) below), so that once cultures are transferred to the hood and work has begun, the chance of introducing external particulates to the working area is minimized. This includes having a supply of pipettes, and possibly a pan or tray for waste pipettes, within the hood. If sterile pipettes are kept in pipette canisters, they should be kept closed except when in use. In practice, many workers find it more convenient to have separate waste buckets for reusable pipettes, disposable plastics and sharps (e.g. glass Pasteur pipettes) on the floor outside the MSC. Similarly, many who use individually wrapped sterile pipettes keep the supply outside the hood, but open them within the hood. The risk of contamination can be minimized by following a good work routine:

(a) Minimize clutter. Include in the hood only those items (equipment and supplies) essential for the work at hand.

(b) Position items to facilitate economy of movement and to minimize movement across the centre of the working area where the handling of cultures will be performed.

(c) Always leave the front and rear vents in the workspace as clear as possible for best airflow.

(d) The cells being cultured (or the tissue for primary culture) should always be the last item transferred into the hood.

(e) Wipe all items, as far as possible, with 70% ethanol before introducing them into the MSC. This will reduce the amount of dust and other particles introduced into the hood, and may kill bacteria that could be present on items.

(f) Incubators – particularly humidified incubators – and water baths are a prime source of microbial contamination. Any items from the incubator or water bath should be dried, then sprayed or wiped with 70% ethanol before they are placed in the MSC.

ii. Pipetting and the prevention of aerosol formation   Pipetting mistakes can be a cause of both microbial contamination and cellular cross-contamination. There are some simple rules to follow for good technique.

(a) Mouth pipetting is *never* permissible under *any* circumstances in the cell culture laboratory.

(b) Pipetting aids should be used throughout the laboratory. If possible, one should be dedicated to each laminar flow hood. These devices can become contaminated, so should be disinfected periodically.

(c) In order to maintain sterility, only plugged pipettes should be used for transferring medium.

(d) Withdraw the pipette carefully from its canister or wrapper. It must not touch any surface that is not sterile. If it accidentally touches a non-sterile surface (including the lip or edge or a bottle, dish, or flask) discard it. Similarly, if you even think that there is an outside chance that you may have contaminated the pipette in any other way, discard it.

(e) **Never use a pipette more than once**. Never enter a culture vessel or reagent bottle with a used pipette.

(f) Never draw so much liquid into the pipette that you wet the plug.

(g) Do not allow fluid to drip from a pipette tip. Never drop fluids so that they splash. Avoid expelling fluid above the mouth of any culture vessel or tube.

(h) Never blow bubbles in media or cell suspensions. This can create aerosols that disperse fluid within the hood. Aerosols can spread contaminating micro-organisms and, importantly, can introduce cells into the air, potentially leading to cellular cross-contamination (see Section 4.2.1.2). Also, at the cellular scale, the bursting of a bubble can release large amounts of energy that can damage or kill cells in the vicinity of the bubble [46].

(i) When using a pipette to disperse cells, keep the tip below the surface of the liquid and within a deep, narrow-mouthed bottle or flask.

(j) Clean up spillages *immediately* with 70% ethanol. Likewise, disinfect the work area at the completion of culture work, or before handling a different cell line.

iii. Additional considerations for good aseptic technique

Wearing gloves is good practice, may be essential for success with antibiotic-free culture and may well be a safety requirement at your institution. Gloves reduce the risk to you of contacting a pathogen. In addition, they protect your cultures, as desquamated skin can be shed from the hands and carry microbes into your work. However, the use of gloves does not entirely eliminate particles from the field and certainly does not allow you to be any less careful. Indeed, the gloves themselves can be a source of particles (e.g. talc). The use of sterile, talc-free gloves reduces this risk, but since some gloves may be sticky and the environment in the hood (and surroundings) is not sterile, it is possible to pick up particulates and carry them over the working area. For this reason some workers periodically rinse their gloves with 70% ethanol during use. In spite of the advantages that the use of gloves provides, many workers who handle supposedly pathogen-free cultures get along well without them (but see Section 4.2.8.2).

Although it is a common practice to pour media and other biologicals from bottles, this is risky and can easily lead to contamination. The mechanism by which contamination most frequently occurs is through the formation of a liquid bridge between the sterile inside of the bottle and the exterior, non-sterile surface of the bottle's lip. To be safe, always assume that the lip of a bottle, flask or tube is not sterile and thus should not be touched by either a sterile liquid or a pipette. For those intent on violating the 'no-pour rule', bottles with drip-resistant lips pour much better than the conventional type. Pouring only once from a bottle (i.e. transferring the entire contents in one pour) also improves the chances of maintaining sterility. Some workers flame the mouth of bottles before and after pouring, but this is not good practice in the MSC since it requires a fairly large open flame that disrupts the hood's air flow. For those who choose to use a burner in the hood, gas burners are available in which the working flame is turned on and off by depressing a pressure plate (e.g. Touch-O-Matic, from Fisher and other distributors).

For procedures where it is necessary to set down a lid, cap or similar item, it is a good idea to keep a sterile dish handy to receive it. Alternatively, prepare a sterile field using several dry, sterile, absorbent wipes, towelling or similar material. This can also be a useful way to protect dissection instruments during the preparation of primary cultures. It is useful to keep a supply of sterile caps and lids handy to replace any that might be accidentally dropped or otherwise contaminated.

Some additional guidelines for good sterile technique are listed below.

(a) As you wish to avoid wetting the mouth or lips of bottles, tubes and flasks, you should also avoid wetting the caps.

(b) Never pour from a culture dish or flask.

(c) Do not overfill culture flasks and dishes since medium is easily splashed over the lip.

(d) Keep a supply of 70% ethanol and sterile swabs or towels available for use in disinfecting spills.

(e) Never invert caps or lids.

(f) Never work directly over an open culture vessel or bottle. Tilt bottles or flasks to allow access from the side.

(g) Unless it is absolutely necessary for the procedure being performed, do not allow more than one person to work in the MSC at any one time. This is essential for work with pathogens and is good practice in routine cell culture. (For applications where two-person operation is essential, longer (e.g. 6-foot wide) MSCs are available.)

4.2.1.2 Prevention of cellular cross-contamination

A major scientific controversy erupted in the mid-1960s when it was discovered that a large number of cell lines, each supposedly derived independently and from different primary tissue sources at different laboratories, and often by well-known and influential cell culture scientists, all exhibited isoenzyme patterns and/or karyotypes characteristic of the HeLa cell line [47, 48]. When investigators tracked the culture history of the suspect cell lines they found that the cultures had once resided in laboratories where HeLa cells were also used. HeLa cells had apparently been inadvertently introduced into these lines and had overgrown them. Cross-contamination was also detected involving cell lines other than HeLa [49].

Despite being first reported more than 40 years ago, cellular cross-contamination remains an ongoing and important – but often ignored – issue in cell culture. In 1999 MacLeod *et al* [50] reported that 18% of 'new' cell lines submitted to the DSMZ (German cell bank) were actually other established cell lines, and despite efforts to increase awareness of the problem, it shows no sign of going away [51–56] and continues to lead to the waste of vast amounts of money and the invalidation of huge amounts of scientific endeavour.

The fact that cells can inadvertently be introduced into other cultures underlines the need for additional precautions in cell handling, and means that IT IS ESSENTIAL THAT YOU PERIODICALLY CHARACTERIZE ANY CELL LINES YOU USE IN THE LABORATORY (see Chapter 9, Section 9.2.6). Increasingly, journals (including *Nature* [54]) are demanding that the identity of all cell lines used in a paper be verified before publication will be considered.

The potential for cellular cross-contamination is, of course, much greater in laboratories that work with more than one continuously culturable line, and is less likely to be a problem when studies involve only cultures with limited *in vitro* lifespan.

Some suggestions for steps to prevent cellular cross-contamination are:

(a) Never work in the MSC with more than one cell line at a time (unless this is an essential for the work being undertaken (e.g. in the generation of hybrids, or the use of feeder cells)). Disinfect the work area after handling each cell line, and allow the hood to run for an additional 5 min before another cell line is transferred to the MSC.

(b) Eliminate aerosols. Aerosols of cell suspensions generated during routine culture maintenance activities are likely sources of cellular cross-contamination.

(c) Never work in the MSC with more than one cell line at a time.

(d) Designate separate bottles of medium for the feeding of different cell lines and permit bottles of media in the MSC only when they are in use.

(e) Never work in the MSC with more than one cell line at a time. (Sorry to keep repeating this, but it is important!).

4.2.1.3 Cell culture at the open bench

The MSC was introduced into general use in the 1960s. Before that time cell culture was performed at the open bench. Many workers still carry out certain procedures outside the MSC, but this must never be done before a full risk assessment is carried out as the operator is not protected from any pathogens that may be present in the culture material (see Section 2.8.1.2). The key to success (i.e. avoiding microbial contamination) is cleanliness and strict attention to aseptic technique. Some suggestions for work outside the hood are as follows.

(a) Work in an isolated or closed portion of the laboratory in order to avoid air turbulence.

(b) Prepare the area to be as free from particulates as possible. A closed glove box or a bench-top dust hood with a partial window can be used to isolate the field further.

(c) Work at arm's length and follow all routine precautions to avoid contaminating critical surfaces.

(d) Some workers use a burner to flame bottles, caps and instruments.

(e) Replace instruments frequently. Keep all instruments covered and pipette canisters closed when not in use.

For further details concerning many of the aspects covered in Section 4.2.1, see reference 57.

4.2.2 General procedures for the cell culture laboratory

Proper preparation of the culture area along with systematic, routine protocols for maintenance of equipment and stock cultures promotes productivity, lessens the risk of contamination and improves safety.

4.2.2.1 Maintenance of the laboratory

i. Laboratory start-up and shut-down   Establish routines for start-up (*Protocol 4.1*) and shut-down (*Protocol 4.2*) of the laboratory that keep it clean and safe, and ensure that both the cultures and the frozen cell stocks remain in the best possible condition.

PROTOCOL 4.1   Laboratory start-up (daily)

Equipment and reagents

- Equipment as mentioned below

- Paper towels

- 70% ethanol solution in spray bottle, and/or other disinfectant(s)

- 0.5% and 0.05% sodium hypochlorite solutions

Method

1 Check the incubators first. Record temperature, humidity (if relevant) and concentration of CO_2 (and of other gases if in use). If there is a problem such as power failure, or the CO_2 concentration or the temperature are outside the set limits, deal with it immediately.

2 Check the water level indicator of any water-jacketed incubator.

3 Check the incubator humidification system (if used). Be sure the water pan on the floor of the chamber is filled. If the incubator has an exterior reservoir check the water level and be sure the tubing line is open.

4 Check gas cylinder pressures and replace any cylinders that are (or are nearly) empty. Check gas regulators to ensure that line pressure is correct for the incubator. Check gas cylinder moorings as a routine precaution.

5 Keep an inventory that locates and identifies incubator contents. This is particularly valuable when multiple users are involved. Refer to this record and scan each tray on each shelf for overtly contaminated (i.e. unusually cloudy or (when using Phenol Red) yellow) cultures. Discard any contaminated cultures. Contaminated cultures can be immersed (10 min) in 0.5% sodium hypochlorite (10% household bleach) or placed directly into an autoclave bag for sterilization.

6 Wipe down all work surfaces in the laboratory with a disinfectant such as 70% ethanol or a commercial antimicrobial solution (e.g. Coldspor, Metrex Corp.). Caution: 70% ethanol is flammable, but is a good alternative to caustic or toxic agents such as bleach (0.5–1%) or phenolics (for further discussion, see Chapter 2, Section 2.2.4.2). Dispense 70% ethanol from a squeeze bottle in modest amounts.

7 Turn on the MSC and allow it to run for 10–15 min before use.

8 If the MSC is equipped with an ultraviolet lamp, be sure that it is off. UV exposure at the hood for even a relatively brief period (many minutes to several hours) can result in painful eye injury, often with some visual impairment (usually short-term).

9 Wipe down the interior of the MSC with 70% ethanol. Any apparatus, equipment, media bottles, cultures, and so on, that go into the MSC throughout the day should (as far as possible) receive the same treatment.

10 If an aspirator is used, hook up a clean vacuum bottle (containing 0.05% sodium hypochlorite (1% household bleach solution)) and line trap. If a peripheral pump and trap system are used, likewise assemble a clean trap and tubing line.

(continued overleaf)

11 If a burner is used in the hood (this is not recommended, as the flame disrupts the laminar air flow pattern) check to see that the tubing to the gas supply is in good repair.

12 Check the function of the automatic pipetting devices.

13 Fill and turn on the water bath used to warm culture medium.

14 If relevant, purge the water purification system (e.g. ion-exchange) and check to see that resistivity meets specifications.

PROTOCOL 4.2   Laboratory shut-down

Equipment and reagent

• Equipment as mentioned below

• Paper towels

• 70% ethanol solution in spray bottle, and/or other disinfectant(s)

Method

When work is completed:

1 Remove all equipment from the hood. Turn off vacuum and gas supply valves. Wipe down the hood interior with 70% ethanol.

2 Discard waste medium. Medium decontaminated with 1% bleach can be washed down the sink. Other waste medium should first be autoclaved. Note: medium from 'high-risk' cultures, especially those of human or primate origin, must be autoclaved before disposal. In the UK, spent medium used in the culture of GMOs must be inactivated by a validated method prior to disposal.

3 Check to see that the lids of cryogenic freezers are securely seated. If left open a freezer can lose a substantial volume of liquid nitrogen (LN) overnight.

4 Check the LN log to see if the level should be measured. Typical 35-l freezers are fairly efficient and may not need replenishment for up to 2 weeks, depending on frequency of use. Excess use of LN may indicate that the freezer is defective. Establish a calendar for LN replenishment (e.g. first, tenth and twentieth days of the month). Never take the risk of running out of LN. The frozen cell inventory may be the laboratory's most valuable resource.

5 Check the incubators and their gas supplies.

ii. Care and maintenance of cell cultures

• *Daily inspection of ongoing cultures*

It is good practice to view ongoing cultures by phase contrast microscopy on a daily basis. In this way one becomes familiar with the growth pattern and morphology of the cell line and may be able to spot irregularities in culture form or individual cell morphology that are indicative of changes in culture characteristics or problems with culture conditions. It may also be possible to spot microbial contamination that is not evident to the naked eye. It is very useful to keep a photographic record of culture morphology. For this purpose it is valuable to have access to an inverted photomicroscope equipped with a digital camera.

- *Maintenance of stock cultures*

Consistency and reproducibility within and between experiments requires that stock cultures be maintained by a routine protocol. Attention to several points will help improve culture performance:

(a) Establish and maintain culture stocks for a given experimental series under identical seeding density, medium and incubation conditions.

(b) Replenish the medium according to a routine schedule and before nutrients are severely depleted or the pH drifts outside acceptable limits. Frequency of feeding will depend on population density. Maintain cultures at a cell population density that permits replacement of the medium at practical intervals. For example, it is common to feed cultures on a set schedule three times weekly.

(c) Passage stock cultures on a routine schedule and when they are at their healthiest (i.e. during logarithmic growth phase, see Section 4.2.5.2).

- *Antibiotic-free stock cultures*

The uninterrupted use of antibiotics in culture media can mask low-level contamination and encourage the development of antibiotic-resistant micro-organisms. Thus some laboratories carry two sets of stocks for each cell line, one with and one without antibiotics/antimycotics. Provided no microbial contamination occurs, the antibiotic-free stock is then used to replenish the working stock. It should be noted that certain antibiotics have significant cytotoxic effects on some cell lines. Thus, for many applications the use of antibiotic-free culture conditions is a necessity. This is also the case when producing therapeutic material from cell cultures as (among other considerations) many potential recipients may be allergic to certain antibiotics.

If the laboratory works exclusively with antibiotic-treated cultures, it is wise to test periodically for the presence of low-level contamination. A simple test for the presence of bacterial or fungal contamination is to plate replicate cultures in which one set is in antibiotic-free medium. These cultures are then followed for 1–2 weeks to see if overt contamination develops. More thorough, definitive testing to detect and identify microbial contaminants should also be performed. A detailed protocol is given in Chapter 9 (*Protocol 9.1*).

Mycoplasma contamination can be more difficult to detect. Mycoplasma (formerly known as pleuropneumonia-like organisms – PPLO) can originate from a variety of sources, the most common being serum, the tissue used to establish primary cultures, and other cell lines. It can also come from individuals who handle the cultures. These organisms can cause numerous problems such as acute cell lysis, but can also lead to alteration in cell structure, function, karyotype, metabolism and growth characteristics that may not be apparent for many cell generations. Mycoplasma are extremely small (300–800 nm diameter) and many strains can readily pass through 0.2-µm filter sterilization membranes. In addition, some strains are unaffected by the antibiotics in common use. For these reasons periodic mycoplasma screening is recommended. Chapter 9 contains protocols for the fluorescent staining technique (*Protocol 9.2*) and the culture isolation method (*Protocol 9.3*), as well as details of other methods in Section 9.2.3.3*iii*. For laboratories that find in-house testing to be problematic, a number of companies offer mycoplasma screening services (e.g. ATCC, Bionique, BioReliance, ECACC, Mycoplasma Experience).

- *Characterization and verification of cell line identity*

The importance of verifying the identity of each and every cell culture with which one works cannot be overstated. Inadvertent cross-contamination and consequent replacement of one cell line by another, either of a different [58] or the same species [47–49], has been well documented since the 1960s. However, it remains a severe problem to this day [50–56] and the amount of time, effort and money wasted as a consequence (not to mention the number of misleading papers published and reputations damaged) is incalculable. Yet it is very easy to ensure that this problem does not become an issue in *your* laboratory. Good practice, as described in this book and elsewhere [22], should reduce the chances of cross-contamination occurring, but this MUST be combined with periodic revalidation of culture identity using one or more established techniques such as short tandem repeat (STR) profiling or isoenzyme analysis (see Chapter 9, Section 9.2.6). This will help alert the laboratory to any problems in cell line integrity. As new cell lines are introduced to, or developed in, the laboratory, they too should be characterized. These assays may be beyond the expertise of the typical laboratory, but are available commercially (e.g. from ATCC and ECACC).

4.2.3 Preparation of culture media

In many laboratories, and for a great deal of routine cell culture, bottles of prepared, sterile, quality-controlled medium are purchased from a commercial supplier. These simply require the aseptic addition of the relevant volumes of any additives (e.g. glutamine, serum, antibiotics) to be ready for use (see also Chapter 5, Section 5.2.1.1.).

However, in some laboratories where large volumes of a particular medium are used, it may be more economical to purchase preformulated medium powder (which has a much longer shelf life than liquid medium) and prepare batches of medium in-house. This medium format and its preparation are discussed in greater detail in Chapter 5, Section 5.2.1.2. When considering the economics of producing medium by reconstitution from powder, the existence, or cost of purchase and maintenance, of relevant facilities (e.g. autoclaves, large balances, preparation vessels, filtration equipment, culture-grade water supply) must be taken into account, along with the staff time required for preparation and the costs of the necessary quality control testing. If preparation from powder still seems a good proposition, it can be performed as described in Chapter 5, *Protocol 5.1*.

4.2.4 The culture of attached cells

Cell culture most commonly involves the growth of cells in stationary incubation on a rigid substrate (usually plastic) (*Protocol 4.3*). Cells are seeded in liquid medium into a culture flask or dish, the cells attach and proliferate and can be subcultured following release from the substrate by brief exposure to a solution typically containing trypsin and EDTA (although, less commonly, mechanical methods such as scraping are used).

PROTOCOL 4.3 Subculture of substrate-adherent cells (LLC-PK1)

Equipment and reagents

- LLC-PK1 (porcine kidney epithelial, ATCC CL101) cells (near-confluent) in serum-supplemented DMEM/F12 (1 : 1) in a 75-cm^2 flask

- Ca^{2+}- and Mg^{2+}-free phosphate-buffered saline, pH 7.4 (CMF-PBS)

- 0.1% Trypsin, 0.05% EDTA in CMF-PBS[a]

- Deoxyribonuclease (DNAse) 2 mg/ml in CMF-PBS[b]

- Benchtop centrifuge

Method

1 Remove culture medium and gently rinse the cell sheet with 12 ml of CMF-PBS.

2 Remove CMF-PBS and add 3 ml of trypsin/EDTA solution to the flask. Tilt the flask to achieve coverage of the whole surface[c,d]

3 Secure the cap and incubate the flask at 37 °C[e]. Be sure that flask lies level on the incubator shelf.

4 Every few minutes, remove the flask from the incubator and examine under an inverted microscope to monitor the progress of the dissociation.

5 When cells start to be released from the substrate, dissociation can be hastened by giving the edge of the flask a sharp rap with the hand[f].

6 Mix 1 ml of DNAse[g] with 35 ml of complete medium[h] in a 50-ml centrifuge tube. Use this medium to rinse cells from the flask, directing the flow from the pipette such that it dislodges adherent cells. Transfer cells to the tube and gently disperse the suspension (by repeatedly pipetting up and down) to break up cell clumps.

7 Collect a 0.5-ml aliquot for a cell count (*Protocol 4.5*). This is used to determine seeding density or culture growth kinetics (see Section 4.2.6).

8 Sediment the cells (100 *g*, 10 min), discard the supernatant medium and resuspend the cells at a known population density (e.g. 1×10^6 cells/ml)[i].

9 For subculture, transfer the desired inoculum (e.g. 1.0×10^6 cells) to a 75-cm^2 flask containing 15 ml of complete medium[j].

(continued overleaf)

10 Loosen the cap and incubate the flask at 37°C in a humidified atmosphere (relative humidity >95%) of 5% CO_2 in air[k].

11 Replenish the medium every 48 h.

Notes

[a]A new innovation that permits the subculturing of attached cells without subjecting them to trypsinization is the UpCell plastic culture surface (Thermo-Fisher). Vessels incorporating this substrate are simply moved from 37 to <32 °C for >15 min, and the attached cells are released due to the change in the surface's physico-chemical properties [59].

[b]The use of DNase reduces cell clumping, but for many cell lines and applications its use is not necessary.

[c]Once the whole surface has been covered, some workers minimize the volume of trypsin/EDTA remaining in the flask by removing all but a thin film of the enzyme solution by aspiration. In this example, 2 ml of the solution could be removed.

[d]Trypsin is normally sourced from pigs. Where it is desirable not to use such animal-sourced material (for example when cell lines are to be used for the production of biopharmaceuticals) other, non-animal-sourced cell detachment solutions are available (e.g. Accutase™ and Accumax™, Millipore; Detachin™, Genlantis). Alternatively, see note [a] above.

[e]With some cell lines, viability and subsequent recovery of dissociated cells may be improved by exposure to trypsin/EDTA at room temperature or 4 °C, but this will extend the time needed to free the cells from the substrate.

[f]This depends on the cell line – with some lines rapping the flask will actually encourage the formation of clumps.

[g]Use of DNAse is optional, but may make it easier to obtain a single-cell suspension.

[h]Mammalian serum has trypsin inhibitor activity, therefore cells are resuspended in complete medium. For serum-free culture, the trypsin can be inactivated by the addition of soybean trypsin inhibitor (0.1 mg/ml).

[i]For many cell lines, this step is unnecessary.

[j]To optimize culture conditions, some workers pre-equilibrate (in the CO_2 incubator) the medium used for subculture and feeding.

[k]Cultures can be incubated with the flask cap tightened, but in this case the flasks should be gassed with 5% CO_2 in air (via a sterile plugged Pasteur pipette) before use. Alternatively, flasks with a gas-permeable filter in the cap are available (see Section 4.2.4.2 i).

4.2.4.1 Routine culture substrates

Anchorage-dependent cells require a physical surface for attachment and growth. Glass, plastic and various types of porous filter membrane are the materials most commonly used as culture substrates, although many other materials can be used for special applications.

i. Glass   For decades, workers grew cells almost exclusively in glass culture vessels. Indeed, the term *in vitro* means on/in glass. Cells adhere well to glass, particularly glass with a high silica content (borosilicate glass). Nowadays, glass is rarely used for mass culture. However, glass coverslips or slides still give the best distortion-free images by light microscopy and are commonly used for DIC microscopy and other demanding microscopy techniques (see Chapter 3). Glass bottomed culture dishes are available (e.g. from MatTek), as are special culture chambers, such as the Sykes-Moore chamber (Bellco) and Leiden dish (Medical Systems), that allow one to use a glass coverslip as the floor of the dish. Also available are pre-assembled, sterile, single- and multi-well culture chambers that use a glass slide as the floor (e.g. Lab-Tek® II Chamber Slide™, Nalge Nunc International). A similar item with a reusable rubber chamber, intended primarily as a tool for cytological analysis, can also be used for cell culture (FlexiPERM® Chamber, Heraeus). Further options and practical details are described in Chapter 3, Section 3.2.3.

Several points should be kept in mind regarding glass as a culture substrate:

(a) Glass coverslips should be cleaned (e.g. with Linbro 7-X detergent, Bellco; or for more demanding applications, with acid – see Chapter 3, *Protocol 3.6*) and rinsed thoroughly before sterilization.

(b) For optimal image quality, the microscope objective and the coverslip thickness must be matched. This may be a most important consideration for some applications [60]. It may also be important to match the refractive index of the coverslip with that of the immersion oil used – see Chapter 3, Section 3.2.1.1.

(c) Photo-etched alphanumeric locator-grid coverslips (e.g. Bellco) may be useful in applications where it is necessary to return to the same area of the culture for sequential study over time.

(d) A useful technique for high-resolution microscopy is to use Silastic adhesive (734RTV, Dow Corning) to mount a coverslip permanently over (or under) a hole drilled in the floor of a plastic dish. Allow the adhesive to cure (60 °C, 48 h), sterilize the dish by immersion in 70% ethanol (15 min), rinse with sterile, de-ionized water, and add cell suspension.

ii. Tissue culture plastic Culture-grade plasticware was introduced in the late 1960s following the development of methods to modify the surface to improve cell adhesion. Polystyrene is most commonly used for cell cultureware, largely because it has good optical clarity. However, virgin (unmodified) polystyrene is rather hydrophobic, and many cell types will not adhere to it, and those that do tend not to spread well. It has been shown that cell shape influences cell proliferation, and that attachment surfaces that increase the extent of cell spreading promote cell proliferation. Thus, treatments that make a surface more hydrophilic generally improve its performance for cell propagation. The commercial preparation of tissue culture plastic commonly involves oxidation of the growth surface by a method such as the gas–plasma discharge technique [61]. Only the growth surface is treated, then the vessel is assembled, packaged and sterilized by gamma-irradiation. Oxygen-plasma treatment increases the O : C and COOH : C ratios of the plastic and imparts an increased negative charge. The magnitude of the charge is important, not its polarity, since cells adhere to both positively and negatively charged, and amphoteric, surfaces [61,62].

Some additional points regarding plastic culture substrates include the following:

(a) Culture flasks made from polyester copolymer (PETG, poly(ethylene) terephthalate glycol) are easier to cut than polystyrene and offer some advantages where improved access to cells is required.

(b) Polystyrene is susceptible to many organic solvents. This should be kept in mind when designing cell constituent extraction systems or when processing for morphological analysis [63].

(c) Thermanox^{TM} plastic coverslips (Nalge Nunc International) are resistant to xylene, acetone and many other organic chemicals.

(d) Plastic coverslips tend to float and can be secured to the floor of a dry dish with a drop of medium or a dab of silicone grease (sterilized by dry heat or autoclaving).

4.2.4.2 Choice of culture vessels

Culture-grade plasticware is available in a variety of sizes (surface areas) and configurations, for example flasks, dishes and multiwell plates. Selection of a specific type of vessel is largely a matter of convenience and the demands of the methods to be used. If direct access to the cells is required, then this is much more easily achieved by using dishes or multiwell plates than flasks. For example, where cells are to be harvested by scraping, dishes are more convenient. Although long-handled scrapers or rakes are available to reach into flasks, this can be laborious and inefficient, especially when harvesting multiple flasks. Direct access to the cells is necessary for some cell cloning procedures (see Chapter 8), and makes processing of cultures for ultrastructural study easier.

i. Vented versus sealed culture vessels   Most dishes and multiwell plates are vented, the lid and base being constructed with a narrow interposing ridge that prevents them from forming a tight seal. This narrow gap allows the atmosphere outside the vessel to equilibrate with that inside the vessel and in contact with the culture medium, and is essential when using a bicarbonate-buffered medium in a CO_2 incubator. By contrast, screw-capped culture flasks can be sealed tightly and consequently are suitable for use with low-bicarbonate media and organic buffers (e.g. HEPES), or alternatively before sealing must be gassed with the appropriate CO_2/air mixture for the medium in use. For vented culture the cap can simply be loosened by about half a turn. Some manufacturers supply flasks with caps incorporating a sterile, gas-permeable filter to improve gas-medium equilibration (e.g. Corning Costar; Nunc; Greiner). Such caps overcome the potential for ingress of contamination posed by the use of a flask with a loose cap. Whether using a loose cap or a filter cap, gas exchange will tend to be better nearer the cap, and this may lead to non-homogeneous growth within the flask.

Depending on the buffering capacity of the medium, one can expect a noticeable increase in pH when a vented dish or flask is removed from the CO_2 atmosphere of the incubator for any period, such as for feeding or viewing with the microscope. This problem can be minimized by incorporating an organic buffer (e.g. 10–20 mM HEPES) in the medium. Taking a culture out of the incubator for a short period of time (minutes) during routine maintenance is usually not a problem. However, changes in pH and temperature can and do affect cell function, so cultures should be disturbed as little as possible especially during timed experiments. Cells at very low population densities, such as those encountered during cloning procedures, are particularly sensitive to such changes, and again should be disturbed as little as possible and special care taken to avoid subjecting the cells to high pH conditions (see Chapter 8).

Some additional points should be taken into account when using vented or filter-cap culture vessels.

(a) Always check the position of the cap before transferring a flask to the incubator.

(b) Handle the vessel carefully to avoid splashing medium into the cap of a flask or between the lip and lid of a dish. This can restrict gas exchange and can lead to the ingress of contamination.

(c) Vented and filter-cap culture vessels allow evaporation from the medium. Evaporation can be substantial if humidity in the incubator is not sufficiently high. Excessive evaporation will increase the osmolality of the medium, which can affect cell function and even cause cell death. It is wise to test for evaporative loss over time by determining weight change in blank dishes (i.e. without cells), and to do this for shelf sites throughout the incubator.

ii. Culture dish Inserts   Culture dish inserts are culture chambers in which the floor is made of a semi-permeable, microporous filter membrane material. They are available in various materials (e.g. nitrocellulose, polyolefin, poly(ethylene) terephthalate (PET), fluorinated ethylene propylene (FEP), ceramic) and different porosities and surface areas from a number of suppliers, and may be translucent or optically clear. These units are designed to be placed within a larger dish, or the well of a multiwell plate. Inserts may be either free-standing, and thus have short feet to hold them off the floor of the outer well, or may hang suspended from the lip of the outer well. Some membrane types are hydrophobic and must be coated with an extracellular matrix (ECM) component or other attachment factor to promote cell adhesion (see *Protocol 4.4*). ECM-pretreated culture dish inserts are also available.

Cells growing within a culture dish insert are in contact with both the fluid that overlies them and that which diffuses upwards through the porous substrate. Importantly, this environment can promote cell differentiation and the expression of differentiated cell functions. Culture in this configuration presents the opportunity to perform studies that cannot be carried out when cells are grown on a non-permeable substrate. For example, when epithelial cells are grown to confluency within a filter insert, the cells form a barrier that separates the culture environment into two compartments, one that bathes the apical cell surface and one that bathes the lateral-basal cell surface. This makes it possible to manipulate selectively the composition of medium in one or other compartment, or to collect media samples selectively, and is useful for studies of vesicular trafficking and transcytosis, transepithelial ion and fluid transport, the sidedness of cell responses to agents or toxins, and the maintenance of barrier function. For certain types of study, cells of a second type can be grown either on the base of the well containing the insert [64] or on the other surface of the membrane. Using such inserts it is also possible to grow cells at the liquid–air interface, and this has facilitated the establishment of cultures of differentiated bronchial epithelial cells [65].

4.2.4.3 Use of attachment factors and biosubstrates

Attachment factors are substances that promote cell adhesion to culture substrates. Many also enhance cell spreading, proliferation and substrate coverage. Attachment factors include basic polymers (e.g. polylysines) used to increase the surface charge of the substrate to encourage cell surface binding, as well as a variety of purified ECM components (e.g. fibronectin, laminin, collagens I and IV) and complex ECM preparations (e.g. Matrigel^{TM} – see below) to which cells bind via specific transmembrane receptors. ECMs and complex ECM preparations are sometimes referred to as biosubstrates. It is well established that ECM preparations can influence the differentiation, structure and function of cultured cells, and indeed *in vitro* studies with ECM components have provided much of our understanding of the biology of cell–ECM interactions.

Many common attachment factors and biosubstrates are commercially available and can be easily reconstituted for use. However, these agents are not always supplied in sterile form, and it may be necessary to sterilize a preparation before or after coating [66]. Also, some ECMs are expensive, and for certain applications may require additional purification. Thus some laboratories find it worthwhile to make their own preparations. Alternatively, commercial firms (e.g. BD Biosciences) market cultureware pre-coated with attachment factors.

A popular biosubstrate is Matrigel^{TM} (BD Biosciences), a solubilized basement membrane preparation extracted from the Englebreth–Holm–Swarm (EHS) tumour. Matrigel^{TM} is supplied as a sterile liquid and can be used to form a thin substrate layer or a three-dimensional gel [67]. Matrigel^{TM} also normally contains a variety of growth factors, but a growth factor-reduced preparation is available along with a number of other forms (e.g. Phenol Red-free). In addition, Matrigel^{TM}-coated culture dish inserts are available.

BD Cell-Tak^{TM} Cell and Tissue Adhesive is a biosubstrate of polyphenolic proteins derived from the marine mussel *Mytilus edulis*. This material, used at 1–5 µg/cm^2, promotes non-specific cell attachment and has also be used to anchor organ explants *in vitro* [67].

When selecting an appropriate attachment factor (or biosubstrate) and method of coating, it will probably be necessary to optimize coating conditions for specific applications. This is because, as described in *Protocol 4.4*, there are numerous methods for preparing ECM-coated substrates, and the components are used over a range of concentrations.

PROTOCOL 4.4 Preparation of attachment factor- and ECM-coated culture substrates

A. Polylysine

Poly-D- and poly-L-lysine are available over a range of molecular weights and can be purchased in sterile form (e.g. from Sigma-Aldrich). The D- and L-isomers of polylysine have a comparable cell adhesion promoting effect. Since L-isomer amino acids are biologically active, poly-D-lysine is usually used.

Equipment and reagents

- Poly-D-lysine
- PBS
- Sterile culture-grade water.

Method

1 Prepare sterile poly-D-lysine (0.1 mg/ml) in PBS.

2 Flood the substrate with poly-D-lysine solution for 5 min at room temperature.

3 Aspirate the excess and rinse substrate with sterile culture-grade water.

4 Allow surface to dry before seeding.

B. Collagen I (thin coating)

Collagen I from rat tail tendon is available as a sterile solution (3.5–5.0 mg/ml) in 0.02 M acetic acid (e.g. from BD Biosciences). Other collagen types (e.g. types III, IV and V) from various species, and recombinant versions, are also available. Suppliers provide suggested coating protocols for each. Pre-coated cultureware, and 3D matrices can also be purchased.

Equipment and reagents

- Collagen solution
- 0.02 M acetic acid (sterile)
- PBS

Method

1 Dilute the collagen solution to 50 µg/ml in sterile 0.02 M acetic acid.

2 Add sufficient solution to the substrate to give a concentration of 5 µg/cm^2. Allow to stand for 1 h at room temperature.

(continued overleaf)

3 Aspirate the solution, and rinse the substrate (and neutralize) with PBS or serum-free medium. Use immediately or allow to air dry before seeding.

C. Gelatin

Gelatin sourced from cows, pigs or fish skin is available as a powder or sterile solution (e.g. 2%, from Sigma-Aldrich).

Equipment and Reagents

- 2% gelatin solution
- Incubator or water bath set at 37 °C
- PBS

Method

1 Allow the 2% gelatin solution to completely liquefy at 37 °C. Prepare a 0.2% solution in PBS. (Gelatin can be autoclaved, but appreciable hydrolysis occurs which may adversely affect the strength of the gel formed subsequently.)

2 Add sufficient solution to the substrate to give a concentration of 100–200 $\mu g/cm^2$. Allow to stand for several hours (or overnight) at 4 °C.

3 Aspirate the excess solution, and allow substrate to dry for at least 2 h before adding medium and cells.

D. Fibronectin

Plasma and cellular fibronectin from both human and animal sources, as well as derived peptides and recombinant versions, are available in sterile form (e.g. from Sigma-Aldrich or BD Biosciences).

Equipment and reagents

- Sterile 0.1% fibronectin solution
- PBS

Method

1 Dilute the fibronectin solution to a concentration of 50 $\mu g/ml$ or less in PBS.

2 Add sufficient solution to the substrate to give a concentration of 1–5 $\mu g/cm^2$. Allow to stand for 1 h (or remove after a few moments and air dry).

3 Aspirate any excess and rinse the dish with medium before seeding.

(continued)

E. Laminin

Equipment and reagents

• 1 mg/ml laminin solution

• PBS

• Incubator set at 37 °C

Method

Commercial laminin is commonly derived from fibroblasts, the placenta, or the EHS mouse tumour and is available in sterile form (e.g. from Sigma-Aldrich or BD Biosciences).

1 Dilute the laminin solution to a concentration of 10–50 μg/ml in PBS.

2 Add sufficient solution to the substrate to give a concentration of 1–2 μg/cm^2.

3 Incubate at 37 °C for 30 min. Aspirate any excess, rinse with medium, then seed.

4.2.4.4 Other culture substrates

Cells have been grown on a variety of substrates other than glass or plastic, including silicone rubber [68,69], and various metals [70–72]. With the burgeoning interest in tissue engineering, three-dimensional matrices are becoming increasingly important and new substrates for this purpose – mostly artificial polymers – are being reported on an almost weekly basis. For this reason the topic will not be covered further here, and readers are encouraged to refer to the current literature for up-to-date information.

Certain other substrates commonly used in large-scale culture applications (e.g. micro-carrier beads) are covered in Chapter 10.

i. Non-adherent substrates for small-scale culture   For some applications such as the formation of histotypic multicellular aggregates [73] and tumour cell spheroids [74], cell adherence to the substrate is undesirable. In such cases, a hydrophobic surface is used. Non-tissue-culture-treated or bacteriological-grade polystyrene may be suitable for some applications, but certain cell types may nevertheless attach to these dishes. Cell adherence can be discouraged by incubating the dishes on a rocker platform or gyratory shaker. A simple, cheap alternative is to coat the dish with a layer of 0.5–1.0% agarose [75]. Another method is to coat the vessel with poly(2-hydroxy-ethyl methacrylate). This technique has been used to inhibit cell attachment and spreading in studies on the role of cell shape as a regulator of cell proliferation [76].

4.2.5 *In vitro* cell growth behaviour

4.2.5.1 Adaptation to culture

When a primary culture is established or a cell line is subcultured, the cells may be subjected to considerable stress. For example, enzymatic dissociation of organ fragments or substrate-dependent cultured cells breaks cell–cell and cell–substrate interactions, commonly causing the cells to change shape (round up), lose phenotypic polarity and show alterations in the distribution of plasma membrane proteins. Not all cells survive such manipulations, and those that do may experience some degree of physical injury. In order to survive and resume growth, or to express its differentiated function, a cell must be able to repair such injuries in a timely manner, adapt to the changes in its environment and (usually) attach to a new patch of substrate. These adaptations take time and are strongly influenced by culture conditions. It has been demonstrated that cells 'condition' their environment, releasing into the medium substances (such as ECMs or growth factors) that promote attachment and growth. Many cell lines adapt and proliferate more quickly or differentiate sooner or more fully if the medium has been 'conditioned' by exposure to actively growing cells. This is particularly true for cells cultured at extremely low population densities, as for instance during cloning (see Chapter 8). The use of mitotically inactivated 'feeder cells' can also be useful when cloning or when attempting to establish primary cultures of certain cell types (see for example Chapter 7, *Protocol 7.6*, and reference 77).

Cells also require time following attachment to re-establish their differentiated functions. A culture seeded at very high density might show rapid attachment and growth, such that it is near confluency within hours, and yet it may be dysfunctional for a feature of interest. This is, for example, known to be the case for some epithelial cell lines, where vectorial ion and fluid transport may not be observed until several days after the attainment of confluency [42].

4.2.5.2 Phases of cell growth

Cells in culture typically show a sigmoidal proliferation pattern, reflecting adaptation to culture, conditioning of the environment, and the availability of the physical substrate and nutrients necessary to support the production of new cells (Figure 4.2).

The phases of cell growth in culture are defined below.

i. Lag phase   Following seeding, cells exhibit a lag phase during which they do not divide. The length of the lag phase is dependent on at least two factors, including the growth phase at the time of subculture and the seeding density. Cultures seeded at lower density need more time to condition the medium and consequently have a longer lag phase. Cultures seeded from actively growing stock (i.e. in log phase) have a shorter lag phase than cells from quiescent stocks.

ii. Logarithmic (log) growth phase   This period is characterized by active proliferation during which the number of cells increases exponentially. In log phase the percentage of cells in the cell cycle may be as high as 90–100%, and at any given time cells will be randomly distributed throughout the phases of the cell cycle. During log phase the cell population is considered to be at its 'healthiest', so it is common to use cells in log phase when assessing cell function. Cell proliferation kinetics during log phase are determined both by the innate characteristics of the cell line and by the conditions under which it is being grown, and it is during this phase that the population doubling time is determined (see Section 4.2.6.2). Factors that influence the length of the log phase will include the seeding density, the rate of cell growth, and the saturation density of the cell line. The

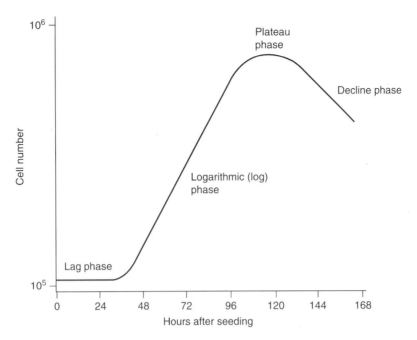

Figure 4.2 Diagrammatic representation of the phases of cell growth. Note that the y-axis is logarithmic, while the x-axis is linear.)

second and third of these will in turn depend on the conditions (medium composition, oxygen availability, temperature, etc.) under which the cells are being grown.

iii. Plateau (or stationary) phase The rate of cell proliferation slows down and levels off as the cell population attains confluency and/or depletes the medium of essential nutrients. This is the plateau phase. Many cell lines in substrate-dependent culture still show some degree of mitotic activity at confluency. Cells may continue to divide and pack tighter, or daughter cells may fail to attach and be released into the overlying medium. During plateau phase, cell division tends to be balanced by cell death, and the percentage of cells in the cell cycle may drop to 0–10%. For some anchorage-dependent cell lines the plateau phase can be extended if the medium is replenished. For most cell lines this is not a stable phase, and cells may be more susceptible to injury.

iv. Decline phase The plateau phase is followed by a period of decline in cell number (decline phase) as the rate of cell death outstrips the rate at which cells are replaced via mitosis. This is not merely a function of nutrient supply.

4.2.6 Determinations of cell growth data

4.2.6.1 Calculation of in vitro age

It is well established that normal cell lines undergo *in vitro* age-related changes in structure and function and have a finite lifespan [26, 78]. Transformed cell lines may also show changes with successive generations. For example, the distribution of morphologically distinct cell types within phenotypically heterogeneous lines, such as Madin–Darby canine kidney (MDCK) cells, changes with increasing subculture. Since cell lines do not remain stable throughout their lifespan it is important to track the *in vitro* age of the line. Studies may then be restricted to cultures within a selected age range as a measure to help achieve consistency.

Two methods are commonly used to track cell age. Many workers simply record the number of times the line has been subcultured (passaged). Passage number simply documents the number of times a culture has been handled, and is a crude estimate of the age of a cell line. It does not take into account the number of population doublings a line may have gone through, which may vary with each passage depending on the conditions and seeding densities employed. A more useful method is to calculate the number of cell generations the line has undergone, to determine the number of cumulative population doublings.

i. Population doubling level (generation number)   The concept of population doubling level (PDL) makes the assumption that cells in culture undergo sequential symmetric divisions such that the population increases in number exponentially (1 to 2 to 4 to 8, etc.). Thus at the end of n generations each cell of the original seeded inoculum will have produced 2^n cells. Hence the total number of cells at a given time following inoculation is given by $N_H = N_I 2^n$, where N_H is the number of cells present (harvested) at that time, and N_I is the number of cells inoculated. The number of generations is then taken as being equivalent to the number of population doublings, and can be expressed using common logarithms (base 10) as:

$$2^n = N_H/N_I \text{ or } n\log 2 = \log\left\{\frac{N_H}{N_I}\right\} = \log N_H - \log N_I$$

$$\text{since } \log 2 = 0.301, 0.301n = \log N_H - \log N_I$$
$$\text{so } n = 3.32\,(\log N_H - \log N_I).$$

Example. Calculation of PDL

1.5×10^5 cells seeded	$N_I = 1.5 \times 10^5$	$\log 1.5 \times 10^5 = 5.176$
8.0×10^6 cells harvested	$N_H = 8.0 \times 10^6$	$\log 8.0 \times 10^6 = 6.903$
$n = 3.32\,(6.903 - 5.176) = 5.73$		

Therefore, to yield 8.0×10^6 cells, 1.5×10^5 cells underwent 5.73 population doublings.

The cumulative PDL is obtained by adding the number of generations during the most recent growth interval (time between seeding and harvest) to the previous PDL number. Cumulative PDL is recorded in the experimental records and on the flask, and similar calculations are performed at each subsequent subculture.

PDL cannot be calculated without an accurate determination of the number of cells in the original inoculum and so is usually not attempted with primary cultures. Thus by convention the primary culture at confluency is designated PDL 1.

The calculation of PDL makes the assumption that no cells die during culture, which is extremely unlikely. Thus in reality the PDL is always an underestimate of the actual number of generations through which the cell population has gone.

4.2.6.2 Multiplication rate and population doubling time

Two additional calculations can give useful information on the growth characteristics of a cell line during log phase. **Multiplication rate** (r) is the number of generations that occur per unit time and is usually expressed as population doublings per 24 hours. **Population doubling time** (PDT) is the time, expressed in hours, that it takes for the cell number to double, and is the reciprocal of the multiplication rate (i.e. $1/r$).

$$PDT = total\ time\ elapsed/number\ of\ generations$$

Example. Calculation of multiplication rate and PDT

Supposing that 1.5×10^5 cells were seeded at time zero (t_1), and 8.0×10^6 cells were harvested at 120 h (t_2).

Multiplication rate (r) $= 3.32 (\log N_H - \log N_1)/(t_2 - t_1)$
$r = 3.32 (6.903 - 5.176)/120 = 0.048$ generations per hour
or 1.15 population doublings/24 h
PDT $= 1/r$
PDT $= 1/(1.15/24)$ h $= 24/1.15$ h $= 20.9$ h per doubling.

Again, no allowance is made in the above calculations for cells dying during culture. Multiplication rate and PDT describe the growth characteristics of the cell population as a whole, and strictly speaking do not characterize the division cycle of individual cells. By definition, PDT (commonly 15–25 h) is not the same as cell generation time (cell cycle time), that is the interval between successive divisions (mitosis to mitosis) for an individual cell. In practice, however, PDT is used as an estimate of cell cycle time.

Several points should be considered in order to obtain meaningful estimates of PDL and PDT.

(a) Accurate estimates of cumulative PDL, and PDT, can only be obtained with consistent handling of cultures.

(b) Cells must be subcultured at regular intervals and when the culture is in the log phase of growth.

(c) Each subculture should be seeded at the same population density, and culture conditions must be consistent, including the same type of substrate, size of flask, volume of medium, and formulation of medium.

(d) The method used to perform cell counts must be consistent and accurate (see Sections 4.2.6.3 and 4.2.6.4, and *Protocol 4.5*).

(e) When combining (e.g. averaging) data from multiple cultures for the estimation of PDT, it is best to calculate the individual multiplication rates first, average these, and finally calculate the PDT, otherwise a single slow-growing flask (giving a long PDT) can disproportionately skew the result.

4.2.6.3 Counting cells in suspension

Many cell culture protocols require an estimate of the number of cells at plating or at harvest. Cells in suspension can either be counted manually, using a haemocytometer and a light microscope, or using any one of a number of electronic, semi-automatic counting devices.

i. Manual counting using a haemocytometer   This is the simplest, most direct and cheapest method of counting cells in suspension. It also allows the percentage of viable (intact) cells to be determined using a dye-exclusion method (see *Protocol 4.5*).

The haemocytometer is a modified microscope slide that bears two polished surfaces each of which displays a precisely ruled, subdivided grid (Figure 4.3). The grid consists of nine primary squares, each measuring 1 mm on a side (area 1 mm^2) and limited by three closely spaced lines (2.5 µm apart). These triple lines are used to determine if cells lie within or outside the grid. Each of the primary squares is further divided to help direct the line of sight during counting. The plane of the grid rests 0.1 mm below two ridges

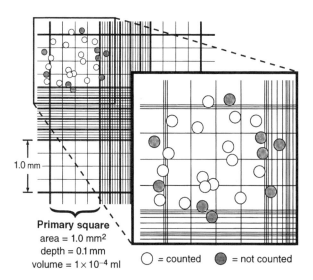

Primary square
area = 1.0 mm^2
depth = 0.1 mm
volume = 1 × 10^{-4} ml

○ = counted   ● = not counted

Figure 4.3 Grid pattern of an improved Neubauer haemocytometer. The inset shows cells (enlarged for clarity) distributed over a primary square. All cells that are within the square and do not touch a boundary, as well as those that touch the left or top boundary, are counted. However, those that are outside the square, and those that touch the bottom or right boundary, are not counted. (Reproduced by permission of Oxford University Press from "Basic Cell Culture: A Practical Approach, 2nd ed." by John Davis (2002).)

that support a sturdy (and ideally, optically flat) coverslip.[1] There is a bevelled depression at the outer edge of each polished surface, where cell suspension is added to be drawn across the grid by capillary action.

A number of points should be borne in mind when performing cell counts with a haemocytometer (see *Protocol 4.5*).

(a) For accuracy and reproducibility, counts must be performed in the same way each time. Standardize the cell dissociation or other preparative procedure for a given cell line (or tissue for primary culture) and use the same counting protocol and conventions (which cells to include or exclude - see Figure 4.3).

(b) When the haemocytometer is properly loaded, the volume of cell suspension that will occupy one primary square is 0.1 mm^3 (1.0 mm$^2 \times 0.1$ mm) or 1.0×10^{-4} ml.

(c) In practice, one should count and total the number of cells within 10 primary squares (five primary squares per chamber), to give the number of cells within 1.0 mm^3 (10×0.1 mm^3) or 1×10^{-3} ml. Total cell concentration in the original suspension (in cells/ml) is then:

$$\text{total count} \times 1000 \times \text{dilution factor}$$

(d) A dilution factor [(volume of sample + volume of diluent)/volume of sample] is needed if the suspension is diluted with buffer, or with a dye used to perform a viable cell count. Any diluent used must be isotonic.

- *Determining the number of viable cells by dye exclusion*

Dye exclusion involves mixing an aliquot of the cell suspension with a volume of buffer or balanced saline containing a water-soluble (i.e. membrane lipid-insoluble) dye – for example 0.4% erythrosin B, or trypan blue – that is visible when it leaks into cells that have damaged plasma membranes. By counting the number of unstained (undamaged) and stained (damaged) cells one can calculate the percent viability (see *Protocol 4.5*).

One should be cautious when interpreting cell 'viability' data obtained by dye exclusion methods. Dye uptake marks cells that have grossly disrupted membranes, and may not detect other forms of injury that may affect cell attachment or lead to cell death. Also, this method does not account for cells that have fully lysed, that is, are no longer identifiable as cells; where an estimate of lysed cells is required, other methods should be employed, for example estimation of lactate dehydrogenase release [79]. In addition, the choice of dye may influence results. Trypan blue has a greater binding affinity for protein in solution than for injured cells, and thus theoretically should only be used in protein-free solution [80]. However, in practice satisfactory results can still generally be obtained in protein-containing solutions. On the other hand, when assessing cell killing

---

[1]Note that counting chambers are also available that have a depth of 0.2 mm, and these may be better for use with some larger cell types. Note therefore that the calculations in points b and c and the whole of *Protocol 4.5* refer to the use of a 0.1-mm-deep chamber.

by hypochlorite (e.g. for validating cell inactivation procedures) trypan blue itself is inactivated by the chemical and an alternative dye such as erythrosin B must be used instead.

PROTOCOL 4.5   Haemocytometer counting for total and per cent viable cells

Equipment and reagents

- Haemocytometer and coverslip

- Upright[a] bright-field microscope, mechanical stage, 20× objective

- Ca^{2+}- and Mg^{2+}-free phosphate-buffered saline (CMF-PBS)

- Multikey manual tally counter

- 0.4% (w/v) erythrosin B in CMF-PBS[b]

Method

1 Prepare a dissociated cell suspension in complete medium (e.g. *Protocol 4.3*).

2 Mix the suspension, then remove a small (>50 μl) aliquot and mix it with an equal volume of erythrosin B solution in an Eppendorf or similar tube. Place the tube on wet ice for 5 min[c].

3 Seat a clean coverslip squarely on top of the haemocytometer. Lightly moisten the edges of the coverslip before pressing it into position on the supporting surfaces of the slide[d]. If the supporting surfaces are polished, Newton's rings (alternating light and dark interference fringes) should appear in both the areas of contact – if not, remove the coverslip, clean both it and the slide, and try again. Note that if the supporting surfaces are ground instead of polished, Newton's rings will not be seen[d].

4 Using a Pasteur or other (e.g. Gilson) small volume pipette, resuspend the cells gently, then draw up a small volume (>10 μl) into the pipette.

5 Load the haemocytometer by placing the tip of the pipette by the edge of the coverslip over the counting chamber, and expelling a small volume of the cell suspension. This should be drawn into the counting chamber by capillary action. Add suspension only until this side of the 0.1 mm deep counting chamber is full[e]. Repeat the procedure to fill the other side of the counting chamber.

6 Using the 20× objective, locate the upper left primary square of one grid and defocus the condenser to improve visibility of the cells.

7 Count the cells in the centre and four corner primary squares of each grid (10 primary squares in total)[f]. Use separate keys of the tally counter to record unstained (viable) and stained (non-viable) cells. Use the following counting conventions:

(a) The middle of the triple lines separating each primary square is the boundary line. Cells that touch the upper or left boundaries are included, those that touch the lower or right boundaries are excluded (Figure 4.3, inset).

(continued)

(b) If greater than 10% of particles are clusters of cells, attempt to disperse the original cell suspension more completely, and collect another sample. Some workers assign all clusters containing more than five cells a value of five.

Example. Total and percent viable cell count

Aliquot (0.5 ml) collected from 40 ml of cell suspension and mixed with 0.5 ml erythrosin B solution.

Stained/unstained cells per primary square

Grid A

| 2/28 | 4/35 | 1/25 | 6/32 | 4/44 | = 17/164 |

Grid B

| 1/30 | 3/32 | 0/26 | 4/40 | 4/29 | = 12/157 |

Total in 10 primary squares = 29/321

Total cell count	350
Percent viable cells	$(321/350) \times 100 = 91.7\%$
Dilution factor	$(0.5 + 0.5)/0.5 = 2$
Total cells/ml	$350 \times 1000 \times 2 = 7.0 \times 10^5$
Total no. of cells	$(7.0 \times 10^5$ cells/ml$) \times$ 40 ml $= 2.8 \times 10^7$
Viable cells/ml	$7.0 \times 10^5 \times 91.7\% = 6.4 \times 10^5$

Notes

[a] An inverted microscope can also be used, provided the working distance of the objective is sufficient to allow visualization of the counting grid.

[b] This protocol can equally be used with trypan blue at a concentration of 0.1–0.4% (w/v).

[c] For viable cell counts, allow a set time (e.g. 4–5 min) for dye to diffuse into the cells, as too rapid a count may underestimate the number of damaged cells.

[d] Further confirmation that the coverslip is correctly pressed into position on the slide can be obtained by turning the slide and coverslip into a vertical position, such that the coverslip might be expected to slide off (place hand underneath to catch it in case it does!). A correctly positioned moistened coverslip will instead be held in position and retained on the slide by surface tension.

[e] The deeper areas around the counting chambers are 'drains' to bear away excess cell suspension. Although a little overflow into these is acceptable, they must not be filled. Rather, the aim should be to fill the flat 0.1 mm deep counting chamber but to stop adding suspension before it starts flowing into these 'drains'.

[f] If necessary, dilute original suspension such that when mixed with the erythrosin B solution it gives c. 25–50 cells/primary square. This will yield around 250–500 cells total in the count, which should yield an acceptable degree of accuracy for most purposes. The suspension should be uniform and, ideally, monodisperse (i.e. without cell clusters). Clumping can be minimized by diluting the aliquot in CMF-PBS.

ii. Semi-automatic counting   A number of devices are now on the market that will count cells semi-automatically, that is, one still has to manually prepare a suitable sample of cell suspension, but once this is loaded onto the device the counting process is automated (to a greater or lesser extent). As with counting using a haemocytometer, the quality of the cell suspension is critical, and accurate, reproducible counts require the use of a consistent method of preparation.

Some of these devices will enumerate viable and dead cells as stained by trypan blue, using the techniques of microscopy and image analysis (e.g. Vi-CELL, BeckmanCoulter; Cellometer, Nexcelom Bioscience; Cedex, Roche Innovatis AG).

The NucleoCounter from ChemoMetec uses a different principle, employing propidium iodide (PI) – which is excluded from intact cells by the plasma membrane – to stain cell nuclei in untreated and in lysed cell preparations. In the former, only dead cells are detected, whereas in the latter all cells are detected. The instrument uses fluorescence microscopy and image analysis to detect PI-stained nuclei.

Other counters, such as those of the Coulter Counter® series (Beckman Coulter) and the CASY® series (Roche Innovatis AG) use the electronic impedance of particles (cells) suspended in an electrolyte to give an estimate of cell size, with dead cells appearing smaller (having a lower impedance) than large cells. Whilst these instruments may give other information, such as the cell size distribution in a population, they may need calibrating for individual cell lines in order to give meaningful estimates of viability.

Recently, Millipore have introduced a hand-held counter (the Scepter) that works on the same principle as the Coulter counter. It is similar in size and format to a Gelman or similar micropipette, and the counting is performed in a disposable flow cell that fits like a pipette tip to the handpiece, which acts both as a pipette to draw the sample into the flow cell and as a housing for the data collection electronics. Data can be downloaded to a computer.

As a recent addition to the range of available cell counters, the advantages and disadvantages of the Scepter have yet to be fully identified. However, some of its apparent advantages include:

- its small size allows it to be used easily within an MSC without restricting the workspace

- it can easily be moved within and between laboratories

- its disposable tip means it can be used with biohazardous materials

- its cheapness relative to a normal Coulter-type counter will make it attractive for many laboratories.

With all of these instruments (except possibly the Scepter), the high purchase price, the cost of consumables and the time required to maintain such instruments is liable to restrict their use to laboratories regularly performing large numbers of viable cell counts, or having need of specific information that cannot be obtained using simpler equipment.

4.2.6.4 Counting/quantifying cells adherent to a substrate

For some applications an estimate of the number of cells adherent to the substrate may be required. A number of methods have been used.

i. Visual counting of cells or nuclei   The number of living, or fixed and stained, cells can be counted within fields of known area under the microscope. This is time-consuming and would normally be performed on electronic images (see Chapter 3). There are a number of sources of potential error in this technique, and the following issues must be addressed.

(a) Sampling within a field or culture must be representative, and in order to avoid bias images must be collected from a number of positions on the slide using a systematic method.

(b) Criteria for the identification of individual cells must established. These must be unambiguous, and incorporate conventions to deal with instances where cell boundaries may not be visible.

(c) Counting must be performed using appropriate conventions to include or exclude cells that straddle the boundaries of the field (see, for example, those employed when using a haemocytometer – Figure 4.3).

(d) Counting nuclei, frequently performed using fluorescence microscopy and DNA dyes such as Hoechst 33342 or propidium iodide, is an alternative, but multinucleate cells (if present) can lead to errors.

Laboratories frequently using such techniques might consider the purchase of a dedicated instrument (such as the Cellavista, Roche Innovatis AG) for the purpose. Although expensive, this would free up a great deal of staff time and also remove some of the problems of human subjectivity inherent in these techniques.

ii. Densitometry of fixed and stained cultures   Densitometry has been used to quantify the density of cell growth within a culture vessel [81]. The cells are fixed then stained (e.g. with 0.5% crystal violet) and the absorbance of the specimen at the relevant wavelength is determined. The method is most useful for detecting relative differences in culture density or substrate occupancy within the same experiment and has been applied, for example, to growth factor response studies [82].

iii. DNA assays   DNA content can be useful as a surrogate marker for cell number, for example when establishing growth curves or assessing culture responses to added growth factors. Several techniques are available, including those based on the binding of DAPI [83], Hoechst 33258 [84] and picoGreen ([85]; kit from Invitrogen) to DNA. However, it must be borne in mind that the DNA content of an individual cell increases as it progresses though the S-phase of the cell cycle. Moreover, many cell cultures contain bi- or multinucleate cells in addition to mononucleate ones. Consequently, DNA content may only poorly reflect the actual cell number. In addition, DNA assays do not discriminate between viable and dead cells.

iv. Colorimetric assays   Many methods have been developed to estimate cell number based on the cellular content of a specific enzyme or substrate, or the uptake and subsequent quantitative extraction of a dye. Perhaps the commonest of these employs the conversion of a tetrazolium salt (substrate) to a coloured formazan (product) by the cellular mitochondrial dehydrogenase activity present in living cells [86]. MTT (3-[4, 5-dimethylthiazol-2-yl]-2,5-diphenyl-tetrazolium bromide) is the most widely used of these substrates, but related molecules with slightly different properties can also be used, including XTT; WTS-1, -3, -4, and -5; and MTS+PMS [87] (see Abbreviations section for full chemical names). Since the level of this mitochondrial enzyme is relatively consistent among cells of a specific type (if grown under standard culture conditions), the amount of formazan produced under set incubation conditions is proportional to live cell number. There is, however, variation between different cell types. This assay has found widespread use for investigating and quantifying the cytotoxic or cytostatic effects of substances such as drugs, toxins and antitoxins, but can also be used in the assessment of growth-promoting substances. An example of how this assay is performed is given in *Protocol 4.6.*

PROTOCOL 4.6   MTT assay

As presented here, the assay is performed in a 96-well plate format to assess antitoxin activity. However, it can easily be adapted for other purposes and formats.

Equipment and reagents

- 96-well flat-bottomed tissue culture plates

- Adhesive plate-sealing film

- Class II MSC

- Assorted Gilson or similar micropipettes and tips

- 37 °C CO_2 incubator

- Indicator cells: these must grow attached to the substrate under the conditions used, and be susceptible to the cytotoxic effects of the toxin in use. (e.g. Vero cells and diphtheria toxin)

- Standard toxin preparation[a]

- Antitoxin test preparation[a]

- Complete medium with penicillin and streptomycin (CM+PS) – for example Eagle's Minimal Essential Medium + 15 mM HEPES + 5% heat-inactivated newborn calf serum + 100 U/ml penicillin + 100 µg/ml streptomycin (Gentamycin has a wider anti-microbial spectrum and can be used instead of or in combination with these if required.)

- Sterile solution of 5 mg/ml MTT in PBS[b]

(continued)

- pH adjustment solution: 8.75 ml distilled water + 1.25 ml 1 M HCl + 40 ml of glacial acetic acid

- Extraction buffer: mix 200 ml of dimethyl formamide (DMF) with 200 ml of distilled water, then add 50 g of sodium dodecyl sulphate. Stir until dissolved. Adjust to pH 4.7 with pH adjustment solution. Bring the volume to 500 ml by the addition of a 1:1 mixture of distilled water and DMF.

- 96-well plate reader set for 565 nm

Method

1 Design the plate layout. This must include relevant positive and negative controls for the major components under test (toxin, antitoxin) as well as the cells employed, and a dilution series of the toxin standard without antitoxin. The remainder of the plate can be used for serial dilutions of the antitoxin in the presence of a standard quantity of toxin known from previous experiments to be a little more than sufficient to kill all the cells.

2 Prepare relevant dilutions of the toxin and antitoxin preparations in CM+PS.

3 Add antitoxin to the appropriate wells on the plate, making serial dilutions in CM+PS as required. The volume of solution in each well at the end of this stage should be 50 μl.

4 Add 50 μl of diluted toxin to the appropriate wells, using a clean pipette tip for each well to avoid contamination of the toxin solution, or other wells, with antitoxin.

5 Incubate the plate (e.g. for 1 h) to allow the antitoxin to react with the toxin. Conditions should be used that will not result in either the medium in the wells turning alkaline (purple, if the medium contains Phenol Red), or significant evaporation of water from the medium in the wells.

6 During the incubation in step 5, prepare (trypsinize – *Protocol 4.3*) and count (*Protocol 4.5*) the indicator cells. There must be sufficient cells for all the wells (except the cell negative controls) in all the plates being used. Actual numbers will depend on the cell line/toxin/incubation period being used, but 2×10^4 viable cells per well might be a suitable starting point. As these cells will be added in a volume of 50 μl, this corresponds to 4×10^5 cells/ml in CM+PS.

7 Once incubation of the plate is complete, place it in the Class II MSC and check all the relevant controls have been added.

8 Add 50 μl of cell suspension to every well except the cell negative control, which should receive 50 μl of CM+PS instead.

9 Incubate the plates in the CO_2 incubator for a period (e.g. 60 min) to allow the pH of the medium to equilibrate, then remove the lid and seal each plate with an adhesive sealing film. The lids can be replaced over the sealing film.

10 Incubate the plates for an appropriate period that will allow the cells to die in the presence of active toxin, or grow to confluence in the absence of active toxin. This period will depend on the toxin/cell type/cell number used, but would typically be several days.

(*continued overleaf*)

11 When the incubation period is complete, remove the plates from the incubator. Examine visually – if using Phenol Red-buffered medium, there should be both red and yellow wells on each plate. The wells can also be examined briefly under the microscope; red wells should have few if any live calls, whilst the yellow wells should contain many live (attached) cells.

12 Remove the lid and adhesive sealer from the top of each plate, being careful not to jolt the plate or splash liquid between the wells.

13 Add 10 µl of the 5 mg/ml MTT solution to each well, using a fresh pipette tip for each well. Replace the lid on each plate and return to the 37 °C CO_2 incubator for 4 h[c].

14 Using a Gilson pipette, carefully remove the liquid from every well, being careful not to disturb the attached cells[d].

15 Add 100 µl of extraction buffer to each well. Reseal the plates and return to the incubator overnight.

16 Remove the plate sealer and read the absorbance of each well at 565nm. Interpret the results using standard procedures.

Notes

[a] Employ appropriate precautions to deal with any potential biohazard posed by these reagents, and contain possible spillages.

[b] Although MTT is not classified as dangerous according to European Directive 67/548/EEC, its toxicological properties have not been thoroughly investigated and it may be an irritant and mutagen, so always wear gloves and take other appropriate precautions when handling this chemical. Also note that, although this chemical has the alternative name thiazolyl blue tetrazolium bromide, it is in fact yellow.

[c] When setting up the assay, this incubation period may be adjusted as required for the cells and test system in use; however, 4 h has been found to be satisfactory with many different cell types. Once the length of this incubation has been decided upon, it must subsequently be adhered to closely in order for the assay to yield reproducible results.

[d] This step is performed using a Gilson-type pipette, as frequently the cells used in this technique are not attached strongly enough to the plastic substrate to permit the medium to be flicked out of the wells manually.

v. Dye uptake methods   The binding of certain dyes to cells can be employed to estimate cell numbers. Methylene blue (1%), for example, binds to negatively charged groups such as nucleic acids in fixed cells and, after the removal of excess dye, can be extracted by washing with 0.1 M HCl [88]. Janus green (like trypan blue) stains only cells that have damaged membranes. However, if it is added to ethanol-fixed cells at a concentration of 1mg/ml, all the cells will stain. After washing, the dye can then be extracted with ethanol and quantified spectrophotometrically [89].

4.2.6.5 Expressions of culture 'efficiency'

The ability of cells to attach to a substrate, and the capacity of cells – whether attached or not – to form colonies, can be quantified. Such measurements can be useful for assessing various aspects of cell behaviour, including survival following cryopreservation, cell–matrix interactions and the effects of cell ageing. These assays can also be used in toxicity testing, cell culture quality control testing (see Chapter 9), and for the screening of sera and reagents. Attachment efficiency is discussed below, and colony-forming efficiency (CFE) is covered in Chapter 8, Section 8.2.2 and *Protocol 8.1.*

An alternative (for attached cells) to the approach to CFE assessment described in Chapter 8 is to seed cells at a known low population density, allow them to grow for several days, then fix the culture and stain with Crystal Violet (or other suitable dye). The number of colonies in fields of known area (chosen systematically to avoid bias) can then be counted using low-magnification microscopy, and the number in the whole vessel calculated from this. CFE is then derived as in Chapter 8, Section 8.2.2.

i. Attachment efficiency   This is expressed as the percentage of cells that attach to the culture substrate within a given period of time. A known number of cells is seeded, and after incubation under defined conditions for a set period, the unattached cells are collected and counted. The attachment efficiency can then be calculated.

Attachment efficiency (%) = [(number of cells seeded − number of cells recovered)/
(number of cells seeded)] × 100

4.2.7 Transportation of cells

4.2.7.1 Transporting frozen cells

Frozen cells in vials or ampoules can be packaged in dry ice ($-78\,°C$) and shipped by overnight courier – note that some postal services, such as the UK Royal Mail, do not accept packages containing dry ice. Special refrigerator-type containers (dry shippers) are also available (e.g. CryoPak, Taylor-Wharton; Arctic Express, Barnstead-Thermolyne) for transport of multiple vials in LN vapour. These units are modified Dewar vessels contain a LN-absorbent material surrounding the specimen canister. The absorbent material is saturated with LN, any non-absorbed liquid poured off, and then vials are placed within the canister for shipment (*Protocol 4.7*). Dry shippers are well insulated and have a long holding time (i.e. many days), and there is no danger of spilling LN during transit. They are approved by IATA as safe for carriage on aircraft.

PROTOCOL 4.7   Transportation of cell cultures in LN vapour using a dry shipper

Equipment and reagents

- Dry shipper
- LN
- Personal protective equipment for handling LN (see Section 4.2.8.1)

(continued overleaf)

- Vials of frozen cells

- Aluminium canes to which vials can be attached

- Outer shipping case for the dry shipper[a]

Method

Filling with LN

1 Remove from the dry shipper the cap/neck tube assembly and the fitted canister that will hold the canes carrying the cell vials.

2 CAREFULLY AND SLOWLY (particularly if the shipper has warmed internally to room temperature) and taking all the precautions for handling LN outlined in Section 4.2.8.1, introduce LN from a storage vessel[b]. Fill to near the top, and replace cap/neck tube.

3 Leave for 10 min. Remove cap/neck tube. Repeat step 2.

4 Leave for 30 min. Repeat step 2.

5 Leave for 2 h. Repeat step 2.

6 Leave overnight. Next morning repeat step 2 once more.

7 If there is still LN visible in the vessel at the end of the day, the absorbent material is saturated and the dry shipper is ready for use. If not, fill up again and repeat as required.

Preparation for shipping

1 Check that the recipient is ready to receive the cells

2 Arrange collection time/date with carrier (e.g. FedEx)

3 Remove cap/neck tube assembly

4 CAREFULLY pour off **ALL** (= excess) LN, using all normal safety equipment and precautions (see Section 4.2.8.1).

5 Place the canister (removed in step 1 of the previous section) back in the vessel.

6 Retrieve the relevant vials of cells from LN storage. Check labels and tightness of caps.

7 Attach and secure to aluminium cane(s), and place these inside the canister.

8 Replace cap/neck tube. Secure cap to prevent tampering.

9 Place dry shipper in padded and insulated shipping case, add a covering letter[c], and secure lid.

10 Label 'DRY SHIPPER, DELICATE UNIT, NON-REGULATED, NON-HAZARDOUS, SHIP UPRIGHT' along with anything else required by the carrier. Label clearly with both the recipient's address and the sender's (i.e. your) address.

11 Dispatch with carrier. Inform the recipient that the cells have been dispatched.[d]

Notes

[a] Dedicated shipping containers can generally be supplied by the manufacturer of the dry shipper. These will contain insulating padding made to fit the shape of the dry shipper, a lid that can

(continued)

be secured in place, a hard shell, and (usually) lifting handles. Any container in which a dry shipper is placed must NOT be completely sealed, as nitrogen gas is given off over time as the LN absorbs heat from the environment.

[b] Remember that LN is one of the most hazardous substances used in the cell culture laboratory (see Section 4.2.8.1).

[c] This should include all necessary information, including (as a minimum) the identity of the cells and instructions for their culture following receipt. Also ensure that any biohazard implications have been adequately addressed before dispatch, as this may affect (among other things) the packaging of the container, its labelling and whether a particular delivery service will agree to undertake its transportation.

[d] On receipt, vials can be transferred directly to a LN cell storage refrigerator, from which they can be thawed using *Protocol 6.5* of Chapter 6.

4.2.7.2 Transporting growing cells at ambient temperatures

Many cultured cell lines will survive at room temperature (15–25 °C) for several days without evident adverse effects if they are then returned to normal incubation temperature. Thus in principle growing cultures can be transported under ambient conditions without the need for elaborate equipment (*Protocol 4.8*). However, 'ambient conditions' will mean different things in different locations and at different times of the year. So the average temperature in London varies during the year from 3 to 16 °C, the average temperature in Helsinki in January is −6 °C, and the highest temperature recorded in Los Angeles in September is 43 °C. Thus transportation at ambient temperature should only be considered if the transit period is no longer than around 2 days, and there is reasonable assurance that the cells will not be subjected to extremes of temperature and that the cells will be returned to their normal incubation temperature immediately on receipt.

i. Preparation and transportation   This is performed as described in *Protocol 4.8*.

PROTOCOL 4.8   Transportation of cell cultures at ambient temperature

A. Non-substrate-dependent cells

Equipment and reagents

- Healthy, actively growing culture of the relevant cell line

- Complete medium (including FBS and/or other growth factors as required)

(continued overleaf)

- Robust, sterile, screw-capped tubes
- Parafilm
- Insulated shipping container
- Cushioning material
- Absorbent material (e.g. paper towelling)

Method

1 Contact the receiving laboratory and ensure that they are ready to receive the cells, and that they have the correct medium, serum and so on. to grow the cells upon receipt.

2 Transfer an aliquot of an actively growing (log phase) cell suspension to a sterile, screw-capped tube.

3 Fill the tube with medium, cap it securely, and wrap the cap with Parafilm[a].

4 Place the tube in an insulated container, along with cushioning material (e.g. bubble-wrap) and enough absorbent material to absorb all the liquid in the tube if it should break. Add a covering letter[b] (in a waterproof bag), seal the container and ship by overnight courier[c,d].

5 Advise the receiving laboratory that the cells have been dispatched.

B. Substrate-dependent cells

i Growing in a screw-capped flask

Equipment and reagents

- Healthy, actively growing culture of the relevant cell line in a culture flask having a screw cap with no integral filter
- Complete medium (including FBS and/or other growth factors as required)
- Parafilm
- Insulated shipping container
- Cushioning material
- Absorbent material (e.g. paper towelling)

Method

1 Proceed as in Section A above.

2 In place of step 2, select a healthy culture in mid-logarithmic growth phase and 20–50% confluent. Fill the flask to the neck with gas-equilibrated (optimal pH) complete medium. Cap the flask securely, and seal it with Parafilm.

3 Proceed from step 4 of Section A above, with the flask in place of the tube.

(continued)

ii Growing on microcarrier beads

Equipment and reagents

- Healthy, actively growing culture of the relevant cell line in a screw-capped culture flask
- Complete medium (including FBS and/or other growth factors as required)
- Microcarrier beads
- Phase contrast microscope
- Robust, sterile, screw-capped tubes
- Parafilm
- Insulated shipping container
- Cushioning material
- Absorbent material (e.g. paper towelling)

Method

1 Carry out step 1 of Section A (above).

2 Suspend hydrated microcarrier beads[e], prepared according to the manufacturer's instructions, in equilibrated culture medium and transfer to a flask containing the actively growing culture for transport. As a guideline, use 10 mg dry weight of beads per cm^2 of flask area. Incubate in stationary culture for 24–48 h to allow the cells to migrate onto the beads. Using a phase contrast microscope, confirm that this has happened.

3 Resuspend the beads and remove them from the flask. Concentrate them by centrifugation, resuspend them in fresh, equilibrated complete medium, and transfer to a sterile screw-capped tube.

4 Proceed from step 3 of Section A above.

Notes

[a] The cells should now be at 20–50% of their maximum population density.

[b] This should include all necessary information, including (as a minimum) the identity of the cells and instructions for their culture following receipt.

[c] Any biohazard implications must be adequately addressed before dispatch, as this may affect (among other things) the nature of the container in which the cells are packaged, its labelling and whether a particular delivery service will agree to undertake its transportation.

[d] If shipping to a warm-climate destination it may be a good idea to include a cold pack in the shipping box to reduce the chance of the cells getting too warm. In general, cells are far more tolerant of temperatures below their optimal incubation temperature (so long as they are not frozen) than they are of temperatures above it.

[e] For further details, see Chapter 10, Section 10.2.1.3.

Further advice on the transportation of cells can be found in reference 21. The regulations covering transportation of cells are forever changing, and vary from one country to another. Any biohazard associated with the cells will clearly be important in this respect, but less obvious factors can also be pivotal. For example, the source of any FBS present can be vitally important if sending cells to the US. If FBS must be used, it is best to use US source material. Some information relevant to the UK can be found at www.royalmail.com and www.hse.gov.uk/cdg/index.htm, and some EU/UN regulations can be found at www.unece.org.

ii. Reinitiating incubation following transportation   Whenever a cell line is brought into the laboratory from an external source, steps should be taken to keep it separate from ongoing cultures (i.e. to quarantine it), in order to minimize the chances of introducing microbial contaminants to the laboratory (see Chapter 9, Section 9.2.1.2). It should also be characterized (Chapter 9, Section 9.2.6) to check that it is in fact the required cell line (even cell culture collections have been known to make mistakes) and that it retains the desired characteristics and has not been adversely affected by the trauma of transportation.

On receipt of a culture that has been transported at ambient temperature, incubation at normal growth temperature should be reinitiated immediately as described in *Protocol 4.9*. Frozen cells can be stored in LN or LN vapour until convenient.

PROTOCOL 4.9   Reinitiation of incubation of cell cultures transported at ambient temperature

Equipment and reagents

- Complete medium
- Incubator
- Sterile paper towel (or similar)
- Centrifuge and tubes
- Culture flasks
- Inverted phase contrast microscope
- Sterile swabs
- 70% ethanol solution

Method

1 Prepare the relevant complete medium for the cell line, and warm it to 37 °C.

2 Unpack the flask or tube and inspect for breaks or leakage. If it is broken, discard immediately into an appropriate biohazard waste receptacle, or autoclave. A small amount of leakage at the cap may not be a problem, but extra care must be taken when opening the vessel. In this case, carefully remove any Parafilm, then use the edge of a sterile absorbent paper towel to blot medium from the interface between the cap and neck. Remove the cap and blot any additional medium from the outside surface of the neck.

(continued)

3 Cells that normally grow attached may have been dislodged from the substrate during transportation. If this is the case, they can be sedimented and replated. The same procedure must be used for cells that grow in suspension. Use a pipette to transfer medium from the flask to sterile centrifuge tubes. As in the routine handling of culture flasks, assume that the lip is contaminated and avoid touching it with the pipette. Centrifuge the medium at 150 g for 10 min, resuspend any cells (or cells on microcarriers) in fresh medium, and transfer to a new flask (or the original flask) in an appropriate volume.

4 If the cells were not transported in a tube, examine the cells in the original flask by phase-contrast microscopy. If a substantial number of cells remain adherent to the substrate, retain the flask; if not, discard it. If the flask is retained and there was any evidence of leakage of medium at the cap, wipe the outside of the neck with a sterile swab soaked in 70% ethanol, and replace the cap. If it was not necessary to carry out step 3, the volume of medium in the flask should be reduced to half of that normally used, and the volume made up to the normal level with fresh pre-warmed complete medium.

5 Return the flask to the incubator. If using a CO_2-buffered medium, the cap of the flask should be left loose until the atmosphere in the flask has had time to equilibrate with the CO_2-containing atmosphere in the incubator, and the pH of the medium has stabilized. Bear in mind that the flask is more susceptible to the ingress of contamination while the cap is loose.

4.2.8 Safety in the cell culture laboratory

In the cell culture laboratory, all the general risks common to all laboratories are present, for example from chemicals, pressurized systems, equipment and so on, and these must be assessed and handled according to both the relevant legislation and the institutional policies in force at your location. That being said, however, the general cell culture laboratory, that is, one in which activities are limited to work involving well characterized cell lines or cells derived from pathogen-free subprimate species, is a relatively safe place to work. In this setting, whilst the breakage of glass pipettes – usually when inserting them into a pipetting aid – can pose a significant threat (which can largely be eliminated by holding the pipette close to the point of insertion, and/or using plastic pipettes), the principal safety concern is probably the handling of LN.

4.2.8.1 Liquid nitrogen

If LN is not handled correctly, there is the potential for injury (frostbite) or, in extreme cases, death from either asphyxiation [90] or the explosion of a poorly designed pressurized vessel [91]. Risks can be minimized by following the correct procedures, wearing the correct personal protective equipment, and only using the right equipment that is specifically designed to the relevant standards, intended for use with LN, and properly maintained (for example, in the UK *pressurized* vessels used for LN storage must be checked regularly in compliance with the Pressure Systems Safety Regulations 2000). Advice can be sought from LN suppliers, but some of the important points are listed below.

- Only store vessels containing LN in a well-ventilated area.

- Similarly, only handle LN in a well-ventilated area.

- Immediately before handling significant volumes, make sure you inform someone not involved that you are about to handle LN, and where, and make sure you inform them when you have finished.

- When handling ANY volume of LN, ALWAYS wear the correct protection:

  - laboratory coat with long sleeves (and ideally without pockets, where LN could pool)

  - visor (your eyesight is incredibly precious)

  - insulated gloves (ideally, loose fitting so that they can be removed quickly, but with elasticated wrists)

  - shoes that do *not* have open toes

  - some people/institutions also favour the wearing and use of personal oxygen depletion monitors to warn when oxygen levels are getting low (usually <19%).

- Take extra care when filling warm vessels as the boil-off can be extremely fierce, generating large volumes of nitrogen vapour and possibly spraying LN significant distances in all directions.

For further discussion of safety in relation to the use of LN in the cell culture laboratory, see Appendix 1 of reference 22.

4.2.8.2 Pathogens

The other main risk in the cell culture laboratory is the presence of pathogens in the cultures or reagents in use.

The greatest risk of infection with a pathogen probably comes from viruses, but certain bacteria, fungi, mycoplasma and parasites are also potential pathogens. With regard to the potential for virus infection, Caputo [92] reported in 1988 that there were no known incidents of laboratory-acquired infection among workers exclusively handling cell lines considered to be free from infectious virus, and to the best of this author's knowledge that is still the case. Nevertheless, the potential exists for cell lines to carry latent viruses (or viruses that have yet to be identified) and for transformed lines to spontaneously produce viruses with oncogenic potential in man [93]. Thus it is essential to carry out a risk assessment before starting to culture any cell line [94–98]. Unfortunately, such an assessment requires knowledge of the history of the cell population, and the history of many cell lines that were isolated decades ago is unclear. In addition, due to the expense involved few cell lines have been extensively tested to determine if they harbour a potential pathogen. If the work involves animal or human tissues or primary cultures, it is important to know the pathogen status of the donor. Thus cell culture can be performed at the lowest level of containment (Containment Level 1) only where the work solely involves well characterized or authenticated cell lines with a low risk of endogenous infection with a biological agent, that present no apparent risk to laboratory workers and which have been tested for the most serious pathogens [94]. Otherwise, Level 2 should be employed unless there are known risks that require use of a higher level [96]. Note that even rodent cells can harbour viruses potentially pathogenic in man (see Chapter 9, Table 9.3).

When the work of the laboratory deliberately involves pathogens carried by cell lines, by the animals used to establish primary cultures, or possibly in other materials or samples used in the laboratory, or where there is a known risk of the presence of such agents, all work must be performed under the appropriate containment conditions. In the UK the Advisory Committee on Dangerous Pathogens specifies four categories of agent and the conditions under which they must be handled [96], and the US [97] and WHO [98] categories are essentially the same. Guidance specific to cell culture has also been published by the UK Health and Safety Executive [94].

It should be realized that the risks are not merely theoretical. There have been reports of worker infection (and subsequent deaths) associated with the primary culture of cells from virus-infected animals [99, 100]. Note that the fact that the animals were virus-infected may not have been known at the time the primary cells were cultured.

i. Awareness of the increased risk associated with human-sourced materials

Because human cells and bodily fluids can carry agents that are infectious to humans, such as (but not limited to) HIV, HTLV, Herpes and hepatitis viruses, human cells clearly pose a greater potential risk to health than those from sub-primate species. It is important to remember that:

(a) HIV has been isolated from human cells and tissues, cell extracts, whole blood (and blood products), and a wide range of other bodily fluids and secretions [97].

(b) All work with human cells must be carried out under the assumption that the specimen may carry an infectious agent. There is also the risk that certain human or primate cell lines, if introduced to the body, could have oncogenic potential [101].

(c) If human-sourced reagents are used, the donors should have been screened for the usual range of blood-borne viruses (see above). However, a negative test does not guarantee the absence of these (or other) pathogens, so such reagents must always be handled under the assumption that they may carry an infectious agent.

ii. Precautions in handling pathogenic organisms and human cells

All workers in the cell culture laboratory must be instructed in the relevant good practices and proper techniques for handling pathogens. Several useful guidelines have been published that focus on the cell culture laboratory [92, 94, 102]

Protocols must be conducted with the understanding that infection only takes one exposure and that, depending on the pathogen involved, the result could be serious illness or death. Wherever the potential exists that pathogens may be present in the laboratory, the practice of **universal precaution** should be adopted, that is, all specimens should be handled as if they present a real risk of infection [97].

Some practical precautions include:

(a) Vaccines are available against a range of pathogenic agents, and (if relevant to the potential risks in the laboratory) workers should be vaccinated before starting work in the laboratory.

(b) Health surveillance should be carried out on all workers at risk.

(c) All work involving the handling of potential pathogens or cell cultures containing pathogens must be performed in a Class II or other relevant class of MSC (see Chapter 1, Section 1.2.3.1 and references 93, 94, 96–98, 103).

(d) Mouth pipetting is **NEVER** permitted.

(e) Hands must be washed before and after handling cells.

(f) Gloves must be worn, and replaced if torn or punctured.

(g) Avoid touching unprotected body surfaces (e.g. eyes and mouth) with gloves or unwashed hands.

(h) Decontaminate all surfaces and equipment that might have come into contact with a pathogen.

(i) Laboratory coats or gowns must be worn, and removed before leaving the laboratory.

(j) Use of sharps such as needles and scalpel blades should be avoided as far as possible. When sharps have to be used they must be disposed of properly into a leak-proof, rigid container clearly marked 'Biohazard', which itself must be disposed of by approved procedures.

(k) Once used, pipettes and instruments should be discarded into a stainless steel pan (with lid) containing distilled water, within the MSC. This container must be covered when it is removed from the hood for autoclaving.

(l) All reusable items must be autoclaved before they are cleaned for re-use. All disposable contaminated items must be autoclaved or incinerated. Transportation of materials from the cell culture laboratory to the autoclave or incinerator must be by institutionally approved procedures. In many cases, autoclave bags may be inadequate for this: suitable rigid containers (e.g. disposable 'Biohazard' bins) may be more appropriate.

(m) Contaminated media must be autoclaved. or treated with suitable chemical disinfectants using a validated inactivation protocol. This approach is best applied to all cell cultures/used media.

(n) Cultures that may harbour pathogens must be clearly labelled, and a separate incubator designated specifically for such cultures.

4.3 Troubleshooting

Artefacts in cell growth pattern sometimes occur but are seldom due to defects in the culture substrate [104]. Irregular growth patterns can usually be traced to errors in routine handling or problems with the incubation conditions. Some of the more common problems encountered in routine cell culture include.

• *Effect of volume of medium* – Improper media volume may affect cell distribution and survival. If the initial volume of cell suspension is too low, the liquid layer may be too thin, and the meniscus effect will pull cells to the edge of the dish or flask, leading to reduced cell attachment near the centre. For the same reason, if used medium is removed from a growing culture and replaced with an insufficient volume of fresh medium, this will pool at the edges and can lead to cell injury and death towards the centre of the dish. If the incubator shelves are not level the depth of medium across the dish will vary, and this too may create an irregular growth pattern.

- *Uneven distribution of cells* – The pattern of cell attachment can be influenced by a number of factors. Inadequate mixing at the time of seeding can lead to non-uniform cell distribution. Also, even with a well-mixed cell suspension of adequate volume, cells tend to settle unevenly, often concentrated toward the centre of the vessel. This is particularly evident when using round dishes and can be minimized by gently tilting freshly seeded cultures side to side then forwards and backwards to disperse the cells more evenly prior to static incubation.

  Creating bubbles at seeding can cause 'spotting'. If the medium is shallow, a bubble may reach the culture surface and prevent cells from attaching at that site, leading to vacant patches within an otherwise uniform or confluent culture. Thus one should avoid creating bubbles, not only to prevent attachment artefacts, but also to minimize the potential for creating aerosols which can lead to microbial and cellular contamination (see Chapter 9). In addition, excessive bubbling can cause cell injury [46].

- *Peeling of the cell sheet* – This can be caused by several factors. Cell lines differ in how tightly they adhere to the culture substrate; some types may be prone to peeling or sloughing, especially if handled too roughly. Confluent cultures are often more likely to detach or peel, particularly if the cell sheet is scratched, for example by a pipette, or if medium is squirted directly onto the cell sheet during feeding. Some fluid-transporting epithelial cell lines that are avid dome-formers [105] may separate from the substrate over fairly large areas as fluid accumulates between the base of the cells and the plastic, and consequently may require particularly gentle handling.

- *Effect of vibration* – Vibration can disrupt or delay cell attachment and may cause cells to aggregate within specific areas of the culture vessel. Cells plated at very low population densities may be particularly sensitive to this problem.

- *Effect of temperature variation* – The attachment and/or proliferation of some cell lines may be particularly sensitive to temperature. An example has been reported where the cell growth pattern reflected the position of perforations in the shelving upon which the culture vessel rested during incubation [104].

A more common temperature-dependent artefact occurs when the temperature within an incubator is not uniform, causing vessel-to-vessel variability in cell growth. This may be a problem in incubators that are opened frequently or have a poorly insulated door, a leaky door gasket, or an inadequate stopper in the rear access port. Condensation on vessels and trays may also be an indicator of temperature gradients within the incubator. This should be investigated without delay as condensation also promotes the growth of microorganisms.

The reader may also find it useful to refer to the troubleshooting section of Chapter 7 (i.e. Section 7.3), which, although written with primary cultures in mind, is also almost entirely applicable to the culture of cell lines.

Acknowledgements

Many thanks to Peter Roberts, Rosalyn Masterton, John Clarke and Dorothy Bennett for critical reading of all or part of the manuscript, and for their numerous helpful suggestions.

References

1. *Methods in Cell Science* (formerly *Journal of Tissue Culture Methods* (Kluwer) and since January 2005 published as a section within Cytotechnology (Springer)) – *This journal publishes protocols in cell culture, tissue and organ culture for application to biotechnology, cellular and molecular toxicology, cell biology, cellular pathology, developmental biology, growth-differentiation-senescence, genetics, immunology, infectious disease, neurobiology, plant biology and virology.*

2. Kruse, P.F., Jr and Patterson, M.K., Jr (ed.) (1973) *Tissue Culture Methods and Applications*, Academic Press, New York.

3. Jakoby, W.B. and Pastan, I.H. (eds) (1979) *Methods in Enzymology*, vol. **58**, Academic Press, New York.

★★★ 4. Freshney, R.I. (2010) *Culture of Animal Cells: A Manual of Basic Technique and Specialized Applications*, 6th edn, John Wiley & Sons, Inc., Hoboken, New Jersey, USA. – *Excellent book on general cell culture. Essential reading for any scientist using cell culture, and an important reference book that should be in every cell culture laboratory.*

5. Freshney, R.I. and Freshney, M.G. (2002) *Culture of Epithelial Cells*, Wiley-Liss, New York.

6. Butler, M. and Dawson, M. (ed.) (1992) *Cell Culture Labfax*, BIOS Scientific Publishers, Oxford.

7. Pollard, J.W. and Walker, J.M. (ed.) (1990) *Animal Cell Culture. Methods in Molecular Biology*, vol. **5**, John Wiley & Sons, Ltd, Chichester.

8. Doyle, A. and Griffiths, J.B. (2000) *Cell and Tissue Culture for Medical Research*, John Wiley & Sons Ltd, Chichester, UK.

9. Butler, M. (2004) *Animal Cell Culture and Technology*, 2nd edn, BIOS Scientific Publishers, Oxford, UK.

10. Barnes, D.W., Sirbasku, D.A. and Sato, G.H. (ed.) (1984) *Methods for Preparation of Media Supplements, and Substrata for Serum-Free Animal Cell Culture. Cell Culture Methods for Molecular and Cell Biology*, vol. **1**, Alan R. Liss, New York.

11. Doyle, A., Griffiths, J.B. and Newell, D.G. (ed.) (1995) *Cell and Tissue Culture: Laboratory Procedures*. John Wiley & Sons, Ltd, Chichester, UK.

12. Spier, R.L. (ed.) (2000) *Encyclopedia of Cell Technology*, John Wiley & Sons, Inc., New York.

13. Celis, J.E. (ed.) (2006) *Cell Biology: a Laboratory Handbook*, Elsevier, Amsterdam, Holland.

14. Spector, D.L., Goldman, R.D. and Leinwand, L.A. (1997) *Cells: a Laboratory Manual*, Cold Spring Harbor Laboratory Press, Plainview, NY.

15. Harrison, M.A. and Rae, I.F. (1997) *General Techniques of Cell Culture*, Cambridge University Press, Cambridge, UK.

16. Mather, J.P. and Barnes, D. (1998) *Methods in Cell Biology, Vol. 57, Animal Cell Culture Methods*, Academic Press, London.

17. Pollard, J.W. and Walker, J.M. (ed.) (1997) *Methods in Molecular Biology, vol. 75. Basic Cell Culture Protocols*, The Humana Press Inc., Totowa, NJ.

18. Poertner, R. (ed.) (2007) *Animal Cell Biotechnology; Methods and Protocols*, 2nd edn, Humana Press, Totowa, NJ.

19. Stacey, GN. and Davis, J. (2007) *Medicines from Animal Cell Culture*, John Wiley & Sons, Ltd, Chichester, UK.

20. Flickinger, M.C. (ed.) (2010) *The Encyclopedia of Industrial Biotechnology: Bioprocess, Bioseparation and Cell Technology*, John Wiley & Sons, Inc., New York, NY.

21. Masters, J., Twentyman, P., Arlett, C. *et al.* (1999) *UKCCCR Guidelines for the Use of Cell Lines in Cancer Research*, 1st edn, UKCCCR, London, UK. [Also published in *British Journal of Cancer*, **82**, 1495–1509 (2000)]

★★★22. Coecke, S., Balls, M., Bowe, G. *et al.* (2005) Guidance on good cell culture practice. A report of the second ECVAM task force on good cell culture practice. *Altern. Lab. Anim.*, **33**, 261–287. [This can also be downloaded from http://www.esactuk.org.uk/. Click on the Best Practice (GCCP) tab.] - *Essential reading for anyone involved with cell culture.*

23. Schaeffer, W.I. (1990) Terminology associated with cell, tissue and organ culture, molecular biology and molecular genetics. *In Vitro Cell. Dev. Biol.*, **26**, 97–101.

24. McPherson, I. (1969) Agar suspension culture for quantitation of transformed cells, in *Fundamental Technique in Virology* (ed. I. McPherson, K. Habel, and N.P. Salzeman), Academic Press, New York. pp. 214–219.

25. Davis, J.M. (1986) A single-step technique for selecting and cloning hybridomas for monoclonal antibody production. *Methods Enzymol.*, **121**, 307–322.

26. Hayflick, L. (1985) The cell biology of aging. *Clin. Geritiatr. Med.*, **1**, 15–27.

27. Lee, K.M., Choi, K.H. and Ouellette, M.M. (2004) Use of exogenous hTERT to immortalize primary human cells. *Cytotechnology*, **45**, 33–38.

28. Milo, G.E., Casto, B.C. and Shuler, C.F. (ed.) (1992) *Transformation of Human Epithelial Cells: Molecular and Oncogenic Mechanisms*, CRC Press, Boca Raton, FL.

29. Fridman, A.L. and Tainsky, M.A. (2008) Critical pathways in cellular senescence and immortalization revealed by gene expression profiling. *Oncogene*, **27**, 5975–5987.

30. Degerman, S., Siwicki, J.K., Osterman, P. *et al.* (2010) Telomerase upregulation is a postcrisis event during senescence bypass and immortalization of two Nijmegen breakage syndrome T cell cultures. *Aging Cell*, **9**, 220–235.

31. Bettger, W. and Ham, R. (1982) The nutrient requirements of cultured mammalian cells. *Adv. Nutr. Res.*, **4**, 249–286.

32. Leibowitz, A. (1963) The growth and maintenance of tissue-cell cultures in free gas exchange with the atmosphere. *Am. J. Epidemiol.*, **78**, 173–180.

33. Reitzer, L.J., Wice, B.M. and Kennell, D. (1979) Evidence that glutamine, not sugar, is the major energy source for cultured HeLa cells. *J. Biol. Chem*, **254**, 2669–2676.

34. Butler, M. and Christie, A. (1994) Adaptation of mammalian cells to non-ammoniagenic media. *Cytotechnology*, **15**, 87–94.

35. McKeehan, W.L. (1982) Glycolysis, glutaminolysis and cell proliferation. *Cell. Biol.* Int. Rep., **6**, 635–650.

36. Ham, R.G. and McKeehan, W.L. (1979) Media and growth requirements. *Methods Enzymol.*, **58**, 44–93.

37. Iscove, N.N. and Melchers, F. (1978) Complete replacement of serum by albumin, transferrin, and soybean lipid in cultures of lipopolysaccharide-reactive B lymphocytes. *J. Exp. Med.*, **147**, 923–933.

★★★38. Festen, R. (2007) Understanding animal sera: Considerations for use in the production of biological therapeutics, in *Medicines from Animal Cell Culture* (eds G.N. Stacey and J.M. Davis), John Wiley & Sons, Ltd, Chichester, UK, pp. 45–58. – *Excellent review on animal sera for use in cell culture.*

39. European Union (2004) Note for guidance on minimising the risk of transmitting animal spongiform encephalopathy agents via human and veterinary medicinal products. *Official Journal of the European Union*, **28.1.2004**, C24/6–C24/19. This can be downloaded at http://www.ema.europa.eu/docs/en_GB/document_library/Scientific_guideline/2009/09/WC500003 700.pdf. (Accessed November 2010).

40. Hodgson, J. (1993) Fetal bovine serum revisited. *Bio/Technology*, **11**, 49–53.

41. Cartwright, T. and Shah, G.P. (2002) Culture media, in *Basic Cell Culture: A Practical Approach*, 2nd edn (ed. J.M. Davis), Oxford University Press, Oxford, UK, pp. 69–106.

42. Perkins, F.M. and Handler, J.S. (1981) Transport properties of toad kidney epithelia in culture. *Am. J. Physiol.*, **241**, C154.

43. Nardelli, L. and Panina, P.F. (1976) 10-years experience with a 28,800 roller bottle plant for FMD vaccine production. *Develop. Biol. Stand.*, **37**, 133–138.

44. Panina, G.F. (1985) Monolayer growth systems: multiple processes, in *Animal Cell Biotechnology* (ed. R.E. Spier, and J.B. Griffiths), Academic Press, London, pp. 211–242.

45. Lubiniecki, A.S. (ed.) (1990) *Large-Scale Mammalian Cell Culture Technology. Bioprocess Technology*, vol. **10**, Marcel Dekker, New York.

46. Handa, A., Emery, A.N. and Spier, R.E. (1987) On the evaluation of gas-liquid interfacial effects on hybridoma viability in bubble column bioreactors. *Dev. Biol. Stand.*, **66**, 241–253.

47. Gartler, S.M. (1967) Genetic markers as tracers in cell culture. *Natl. Cancer Inst. Monogr.*, **26**, 167–195.

48. Nelson-Rees, W.A. and Flandermeyer, R.R. (1976) HeLa cultures defined. *Science*, **191**, 96–98.

49. Nelson-Rees, W.A., Daniels, D.W. and Flandermeyer, R.R. (1981) Cross-contamination of cells in culture. *Science*, **212**, 446–452.

50. MacLeod, R.A., Dirks, W.G., Matsuo, Y. *et al.* (1999) Widespread intraspecies cross-contamination of human tumor cell lines arising at source. *Int. J. Cancer*, **83**, 555–563.

51. Masters, J.R. (2002) False cell lines: The problem and a solution. *Cytotechnology*, **39**, 69–74.

52. Chattergee, R. (2007) Cases of mistaken identity. *Science*, **315**, 928–931.

53. Hughes, P., Marshall, D., Reid, Y. *et al.* (2007) The costs of using unauthenticated, over-passaged cell lines: how much more data do we need? *BioTechniques*, **43** (5), 575–583.

54. Editorial (2009) Identity crisis. *Nature*, **257**, 935–936.

55. Lucey, B.P., Nelson-Rees, W.A. and Hutchins, G.M. (2009) Henrietta Lacks, HeLa cells, and cell culture contamination. *Arch. Pathol. Lab. Med.*, **133**, 1463–1467.

56. Boonstra, J.J., van Marion, R., Beer, D.G. *et al.* (2010) Verification and unmasking of widely used human esophageal adenocarcinoma cell lines. *J. Natl Cancer Inst.*, **102**, 271–274.

57. Davis, J.M. and Shade, K.L. (2010) Aseptic techniques in cell culture, in *Encyclopedia of Industrial Biotechnology: Bioprocess, Bioseparation and Cell Technology* (ed. M. Flickinger), John Wiley & Sons, Inc., New York, pp. 396–415.

58. Coriell, L.L. (1962) Detection and elimination of contaminating organisms. *Natl. Cancer Inst. Monogr.*, **7**, 33–53.

59. Cell harvesting by temperature reduction. http://www.nuncbrand.com/en/frame.aspx?ID =11867 (Accessed December 2010).

60. Delly, J.G. (1988) *Photography Through the Microscope*, Eastman Kodak, Rochester, NY.

61. Ramsey, W.S., Hertl, W., Nowlan, E.D. and Binkowski, N.J. (1984) Surface treatments and cell attachment. *In Vitro*, **20**, 802–808.

62. Davis JM (2007) Systems for cell culture scale-up, in *Medicines from Animal Cell Culture* (eds G.N. Stacey and J.M. Davis), John Wiley & Sons, Ltd, Chichester, UK, pp. 145–171.

63. Dougherty, G.S., McAteer, J.A. and Evan, A.P. (1986) Simultaneous processing of substrate-dependent monolayer cultures for scanning and transmission electron microscopy. *J. Tiss. Cult. Methods*, **10**, 239–244.

64. Spiers, V., Ray, K.I. and Freshney, R.I. (1991) Paracrine control of differentiation in the alveolar carcinoma, A549, by human foetal lung fibroblasts. *Br. J. Cancer*, **64**, 693–699.

65. Schwab, U.E., Fulcher, M.L., Randell, S.H. *et al.* (2009) Equine bronchial epithelial cells differentiate into ciliated and mucus producing cells in vitro. *In Vitro Cell. Dev. Biol. Anim.*, **46**, 102–106.

66. Reid, L.M. and Rojkind, M. (1979) New techniques for culturing differentiated cells: Reconstituted basement membrane rafts, in *Methods in Enzymology*, vol. **58** (eds W.B. Jakoby and I.H. Pastan), Academic Press, New York, pp. 263–278.

67. Elliget, K.A. and Trump, B.F. (1991) Primary cultures of normal rat kidney proximal tubule epithelial cells for studies of renal cell injury. *In Vitro Cell. Dev. Biol.*, **27A**, 739–748.

68. Harris, A.K., Jr (1984) Tissue culture cells on deformable substrata: biomechanical implications. *J. Biomech. Eng.*, **106**, 19–24.

69. Terracio, L., Miller, B.J. and Borg, T.K. (1988) Effects of mechanical stimulation of the cellular components of the heart: In vitro. *In Vitro*, **24**, 53–58.

70. Birnie, G.D. and Simmons, P.J. (1967) The incorporation of 3H-thymidine and 3H-uridine into chick and mouse embryo cells cultured on stainless steel. *Exp. Cell Res.*, **46**, 355–366.

71. Litwin, J. (1973) Titanium disks, in *Tissue Culture Methods and Applications* (ed. P.F. Kruse, Jr. and M.K. Patterson, Jr.), Academic Press, New York, pp. 383–387.

72. Westermark, B. (1978) Growth control in miniclones of human glial cells. *Exp. Cell Res.*, **111**, 295–299.

73. Moscona, A.A. (1961) Rotation-mediated histogenetic aggregation of dissociated cells. A quantifiable approach to cell interactions in vitro. *Exp. Cell. Res.*, **22**, 455–475.

74. Bjerkvig, R. (ed.) (1992) *Spheroid Culture in Cancer Research*, CRC Press, Boca Raton, FL.

75. Nitsch, L. and Wollman, S.H. (1980) Suspension culture of separated follicles consisting of differentiated thyroid epithelial cells. *Proc. Natl Acad. Sci. USA*, **77**, 472–476.

76. Folkman, J. and Moscona, A. (1978) Role of cell shape in growth control. *Nature*, **273**, 345–349.

77. Rheinwald, J.G. (1989) Methods for clonal growth and serial cultivation of normal human epidermal keratinocytes and mesothelial cells, in *Cell Growth and Division: a Practical Approach* (ed. R. Baserga), IRL Press, Oxford, pp. 81–94.

★ 78. Hayflick, L. and Moorhead, P.S. (1961) The serial cultivation of human diploid cell strains. *Exp. Cell Res.*, **25**, 585–621. - *Original description of the limited proliferation capacity of normal cells.*

79. Racher, A.J., Looby, D. and Griffiths, J.B. (1990) Use of lactate dehydrogenase release to assess changes in culture viability. *Cytotechnology*, **3**, 301–307.

80. Phillips, H.J. (1973) Dye exclusion tests for cell viability, in *Tissue culture methods and applications* (ed. P.F. Kruse Jr. and M.K. Patterson Jr.), Academic Press, New York, pp. 406–408.

81. Terracio, L. and Douglas, W.H.J. (1982) A densitometer for the evaluation of cell growth in primary cultures: Construction and operation. *J. Tiss. Cult. Methods*, **7**, 5–8.

82. Terracio, L. and Douglas, W.H.J. (1982) Densitometric and morphometric evaluation of growth in primary cultures of rat ventral prostate epithelial cells. *Prostate*, **3**, 183–191.

83. Brunk, C.F., Jones, K.C. and James, T.W. (1979) Assay for nanogram quantities of DNA in cellular homogenates. *Anal. Biochem.*, **15**, 497–500.

84. Labarca, C. and Paigen, K. (1980) A simple, rapid, and sensitive DNA assay procedure. *Anal. Biochem.*, **102**, 344–352.

85. Blaheta, R.A., Kronenberger, B., Woitaschek, D. *et al.* (1998) Development of an ultrasensitive in vitro assay to monitor growth of primary cell cultures with reduced mitotic activity. *J. Immunol. Meth.*, **211**, 159–169.

86. Mosmann, J. (1983) Rapid colorimetric assay for cellular growth and survival: application to proliferation and cytotoxicity assays. *J. Immunol. Methods*, **65**, 55–63.

87. Cory, A.H., Owen, T.C., Barltrop, J.A. and Cory, J.G. (1991) Use of an aqueous soluble tetrazolium/formazan assay for cell growth assays in culture. *Cancer Commun.*, **3**, 207–212.

88. Oliver, M.H., Harrison, N.K., Bishop, J.E. *et al.* (1989) A rapid and convenient assay for counting cells cultured in microwell plates: application for assessment of growth factors. *J. Cell Sci.*, **92**, 513–518.

89. Rieck, P., Peters, D., Hartman, C. and Courtois, Y. (1993) A new, rapid colorimetric assay for quantitative determination of cellular proliferation, growth inhibition, and viability. *J. Tiss. Cult. Methods*, **15**, 37–41.

90. Safety problems led to lab death. http://news.bbc.co.uk/1/hi/scotland/798925.stm (Accessed November 2010).

91. Rupture of a Liquid Nitrogen Storage Tank, Japan, 28th August 1992. http://www. hse.gov.uk/comah/sragtech/caseliqnitro92.htm (Accessed November 2010).

92. Caputo, J.L. (1988) Biosafety procedures in cell culture. *J. Tiss. Cult. Methods*, **11**, 223- 227.

93. Weiss, R.A. (1978) Why cell biologists should be aware of genetically transmitted viruses. *Natl. Cancer Inst. Monogr.*, **48**, 183–189.

94. Health and Safety Executive (2005) Biological agents: Managing the risks in laboratory and healthcare premises. http://www.hse.gov.uk/biosafety/biologagents.pdf, pp. 68–70. (Accessed November 2010).

95. Stacey, G.N. (2007) Risk assessment of cell culture procedures, in *Medicines from Animal Cell Culture* (eds G.N. Stacey and J.M. Davis) John Wiley & Sons, Ltd, Chichester. UK, pp. 569–588.

96. Advisory Committee on Dangerous Pathogens (2004) *The Approved List of Biological Agents*, HMSO, Norwich, UK. This can be viewed/downloaded at http://www.hse.gov.uk/ pubns/misc208.pdf. (Accessed November 2010).

97. Chosewood, L.C. and Wilson, D.E. (ed.) (2009) *Biosafety in Microbiological and Biomedical Laboratories*, 5th edn, Department of Health and Human Services, Washington, DC, USA. This can be viewed/downloaded at http://www.cdc.gov/biosafety/ publications/bmbl5/BMBL.pdf. (Accessed November 2010).

98. World Health Organisation (2004) *Laboratory Biosafety Manual*, 3rd edn, WHO, Geneva, Switzerland. This can be viewed/downloaded at http://www.who.int/csr/resources/ publications/biosafety/WHO_CDS_CSR_LYO_2004_11/en/ (Accessed November 2010).

99. Hummeller, K. Davidson, W.L. Henle, W. *et al.* (1959) Encephalomyelitis due to infection with Herpesvirus simiae (herpes B virus); a report of two fatal, laboratory-acquired cases. *N Engl. J. Med.*, **261**, 64–68.

100. Barkley, W.E. (1979) Safety considerations in the cell culture laboratory, in *Methods in Enzymology*, vol. **58** (ed. W.B. Jakoby and I.H. Pastan), Academic Press, New York, pp. 36–44.

101. Gugel, E.A. and Sanders, M.E. (1986) Needle-stick transmission of human colonic adenocarcinoma. *N Engl. J. Med.*, **315**, 1487.

102. Grizzle, W.E. and Polt, S.S. (1988) Guidelines to avoid personnel contamination by infective agents in research laboratories that use human tissues. *J. Tiss. Cult. Methods*, **11**, 191–199.

103. Chosewood, L.C. and Wilson, D.E. (ed.) (2009) Appendix A - Primary Containment for Biohazards: Selection, Installation and Use of Biological Safety Cabinets, in *Biosafety in Microbiological and Biomedical Laboratories*, 5th edn, Department of Health and Human Services, Washington, DC, USA. This can be viewed/downloaded at http://www.cdc.gov/biosafety/publications/bmbl5/BMBL.pdf (Accessed November 2010).

104. Ryan, J.A. (2008) *Corning Guide for Identifying and Correcting Common Cell Growth Problems*, Corning Inc., New York, NY.

105. Lever, J.E. (1985) Inducers of dome formation in epithelial cell cultures including agents that cause differentiation, in *Tissue Culture of Epithelial Cells* (ed. M. Taub), Plenum Press, New York, NY, pp. 3–22.

5

Development and Optimization of Serum- and Protein-free Culture Media

Stephen F. Gorfien[1] and David W. Jayme[2]

[1]*Cell Systems Division, Life Technologies Corporation, Grand Island, USA*
[2] *Dept of Biochemistry and Physical Sciences, Brigham Young University - Hawaii, Laie, Hawaii, USA*

5.1 Introduction

The evolution in quality, consistency and biochemical definition of exogenous nutrient formulations for cell culture has paralleled the evolution over the past century in breadth and complexity of cell types, bioreactors, feeding regimens and target applications. This chapter is intended to illustrate fundamental concepts in the development, optimization and troubleshooting of nutrient media for existing and emerging eukaryotic cell culture applications.

The earliest successful attempts at cultivation of higher eukaryotes utilized tissue explants and cells derived from the circulation and from organ infusions placed in isosmotic, buffered salt solutions, supplemented with selected organic nutrients or crude animal-derived fractions intended to mimic the native *in vivo* environment [1]. Beyond providing a useful window to probe normal and abnormal cellular processes, such culture systems eventually produced effective vaccines against polio and other human ailments.

Although initial eukaryotic cell culture applications generally used differentiated cells maintained in primary cultures, or targeted the longer term cultivation of cells immortalized by chemical or viral transformation, contemporary applications require genetically modified or progenitor cell types derived from characterized cell banks (see Chapter 6)

Animal Cell Culture: Essential Methods, First Edition. Edited by John M. Davis.
© 2011 John Wiley & Sons, Ltd. Published 2011 by John Wiley & Sons, Ltd.

and maintained in lengthy campaigns lasting weeks to months to produce biomolecules (e.g. recombinant proteins, monoclonal antibodies) for exacting human diagnostic and therapeutic applications [1, 2]. Cultivation vessels have evolved from glass Petri plates and roller bottles to disposable plasticware and computer-controlled process bioreactors.

These technological advances in eukaryotic culture have necessitated alternative feeding regimens, evolving from classical batch introduction of nutrient fluids to fed-batch and perfusion feeding techniques that optimize biomass expansion to approximate tissue densities, extend bioreactor longevity and productivity over prolonged campaigns, and facilitate downstream harvest and purification of target biomolecules. This evolution has resulted in both quantitative and qualitative modifications in exogenous nutrients to comply with technical, commercial and regulatory demands [3, 4].

A fundamental step in the evolution of nutrient media involved the successful replacement of crude animal-derived supplements (typically bovine sera) with medium constituents that fulfilled the manifold roles of the serum additive but with greater biochemical definition and reduced inherent variability [5, 6]. Initial attempts at the development of serum-free formulations were often characterized by limited performance relative to serum-supplemented media or by inadequate stability and robustness for large-scale processes [7–10]. However, engineering improvements in bioreactor design and process monitoring permitted the elucidation of the nutrient consumption patterns of production cell types and facilitated elimination of unnecessary or inhibitory constituents and augmentation of bioproduction-limiting nutrients [11–14].

5.1.1 Nomenclature definition

To ensure understanding, we now define several critical terminologies associated with development of next-generation nutrient formulations [2, 9, 15]. Basal nutrient formulations, originally developed decades ago [1, 9, 16–20] and containing a metabolically balanced mixture of carbohydrates, amino acids, vitamins and nucleosides within a buffered inorganic salt solution, will maintain eukaryotic cells with normal cellular function for less than a few hours. Such media typically required supplementation by serum or other blood fractions or organ extracts to extend the culture period and to promote cell proliferation. Animal-derived sera and/or organ extracts are commercially available from a variety of vendors and often exhibit lot-to-lot variability, have fluctuating availability and/or price, can mask detection of mediators and may harbour adventitious agents.

Serum-free media (SFM) are typically based upon an enriched basal nutrient formulation, augmented by constituents designed to substitute for the metabolic and biophysical contributions of the serum additive, including growth and attachment factors, carrier proteins, supplemental buffers, intermediary metabolites and protective agents. However diverse in their biochemical composition, the unifying element of SFM is that they do not require supplemental serum to sustain cell proliferation [7]. Although most new bioprocessing applications have moved towards protein-free and chemically defined systems, other culture applications like stem cell culture are just beginning to make use of serum-free methods. Elimination of serum supplementation makes it possible to better define critical nutrients and factors impacting growth and/or differentiation.

Protein-free media (PFM) refined the diversity of the SFM from which they were derived by eliminating serum albumins as carriers of lipids and other nutrients/mediators, and by replacing other protein components with lower molecular weight alternatives [21–23]. Still lacking biochemical definition, most PFM contained protein hydrolysates and lipid fractions as critical components. PFM have proven most useful in suspension culture systems like those used with recombinant Chinese hamster ovary (CHO) and myeloma cells. Cells that require adherence to a substrate often have difficulty adapting to PFM and require pre-treatment of the culture surface and/or supplementation of the culture medium with an attachment protein like fibronectin or collagen.

Chemically defined media (CDM) represented the pinnacle in reproducibility and regulatory quality assurance, requiring that every constituent be biochemically defined [24, 25]. By some definitions, CDM allow for the inclusion of highly purified proteins in the formulation (i.e. those in which the chemical structure is known and impurities are absent), but others define CDM as also being protein free, so if this distinction is important for your application, it is important to be able to review the formulation or specifically ask the medium provider whether proteins are present.

In the midst of the evolution of this increasing sequence of biochemical definition of nutrient additives emerged the awareness of potential human illnesses that might result from the utilization of protein components or intermediary metabolites derived from animal sources that might harbour adventitious viruses or prion agents. Although primarily of concern in biopharmaceutical production and related processes, there was also (at least in theory) a risk to workers in other types of laboratories. Consequently, nutrient media were redeveloped to eliminate peptides, extracts and nutrient ingredients obtained originally from animal sources and even extrapolated to recombinant factors produced in prokaryotic fermentation media containing additives of animal origin or purified by uncontrolled methods or systems that might expose them to adventitious contamination [26–28].

While initially laborious and time-consuming to develop, these refined nutrient formulations offered substantial benefits to eukaryotic cell culture. Previous serum-supplemented formulations tended to vary in performance based upon the lot-to-lot variability of the serum additive. Serum also intrinsically contained elements that prevented undifferentiated maintenance of progenitor species and that preferentially promoted fibroblast overgrowth in mixed cell cultures. Other serum-associated problems included neutralizing antibodies that reduced virus titres in cultures designed for virus production, and intrinsic enzymatic activities (e.g. protease, glycosidase) that accelerated degradation of secreted products [29–32].

However, the ultimate driver for elimination of serum became the regulatory obstacles rather than the technological impediments. To ensure public safety from viral and prion contaminants, regulatory agencies demanded such extensive documentation and validation of process removal of potential adventitious contaminants, and the projected product and civil liabilities that might result from a contaminated product became so prohibitive, that it became cost-effective for the biotechnology industry to develop and validate processes that minimized risk through the adoption of animal-origin-free SFM [26, 28].

5.1.2 Eukaryotic cell culture applications

Superior nutrient formulations have permitted expansion in the breadth of current cell culture applications. Basic research in signalling mechanisms that govern normal and aberrant cell functions associated with the regulation of proliferation, differentiation and senescence has been vastly accelerated by the ability to cultivate indicator cells under more defined exogenous environments. Defined media enabled the cultivation of epithelial cultures exhibiting normal physiological morphology and function, and facilitated elucidation of the differentiation lineages of progenitors [2, 7].

In addition to resolving regulatory concerns for bioproduction applications, in many situations the emergence of SFM also accelerated proliferation, elevated biomass, enhanced total and specific biological productivity, and facilitated downstream product recovery and purification. Overcoming concerns regarding adequate availability of qualified serum for large-scale bioproduction applications, adoption of SFM has generally permitted more accurate projection of production costs and qualified inventory of critical nutrient materials.

Technical, regulatory and ethical issues remain to the expansion of cell culture applications for the promising remedial medicine applications associated with cell and gene therapy. However, increasing biochemical definition through serum elimination, enhancing the reproducibility and robustness of these cell-based techniques, and ensuring an acceptably low risk of introducing adventitious contaminants through developing effective SFM represent fundamental enabling steps towards therapeutic implementation [10].

5.1.3 Impact of culture process and nutrient medium format

Another contributor to cell culture nutrient medium complexity is the variety of bioreactors and culture processes. Some of the quantitative and qualitative adjustments necessitated by transitioning from batch culture to fed-batch environments were predictable, whereas others were only discovered through spent medium analysis and evaluation of consumption kinetics of critical nutrients under high-density cultivation conditions [12, 14]. The adoption of perfusion bioreactors required both technical and economic consideration of nutrient partitioning between the extracellular and perfusion phase components [4]. The complexity of the nutrient medium is also dramatically influenced by the propensity of cells to proliferate and synthesize biological product in suspension versus adherent culture.

A final contributor worthy of mention in this chapter is the delivery format for the nutrient medium to optimize process cost and the associated capability of sterilizing the medium components within that delivery format to minimize risk of culture or product contamination. Ready-to-use liquid medium provides convenience and high sterility and quality assurance, but at the expense of higher purchase and freight costs, increased storage logistics and diminished biochemical stability. Dry medium formats [33] produced by ball-milling partially overcome the purchase and storage cost challenges, but settling of constituents during the particle attrition process and degradation of thermo-sensitive nutrients can lead to variable and diminished culture performance [34]. Hybrid benefits may be derived by production of bulk basal nutrient formulations as a dry-format component and supplying sensitive, costly or trace components in stable concentrated liquid format, particularly if this concentrated component may be sterilized by flash pasteurization or other validatable process [35,36].

A novel format for dry-form nutrient medium delivery that is receiving increasing interest for bioproduction applications involves fluidized bed granulation [32,37–39]. The granulated medium exhibits homogeneity superior to ball-milled formulations of comparable composition, while offering many of the cost, storage and stability benefits of conventional powders and many of the convenience and performance-enhancing features of the liquid medium format.

Having briefly reviewed some of the contributory factors to the design complexity of nutrient media designed for animal cell culture applications, we will now focus upon specific issues relating to preparation, optimization and performance screening of these nutrient fluids.

5.2 Methods and approaches

5.2.1 Preparation of medium

As noted previously, the complex nutrient mixtures commonly utilized to promote proliferation and biological production of mammalian and other higher eukaryotic cells may be formulated in multiple formats. Several basal nutrient formulations [16–20] (details of formulations also can be found in most commercial media supplier catalogues) developed over the past few decades have been widely used and variably supplemented to produce the enriched formulations required to support specialized cell types under serum-free culture conditions.

Comparison of these common basal formulations indicates qualitative adjustments in nutrients corresponding to intended use. The prototypical Minimal Essential Medium (MEM) formulation of Eagle [16] was differentially modified by subsequent investigators. DMEM [17] augmented levels of amino acids and vitamins and provided additional sources of metabolic energy and buffering capacity consistent with high-density cultivation. By contrast, Ham and co-workers were attempting to isolate cells at clonal densities in serumless medium, and consequently required lower levels of bulk nutrients but needed supplemental lipids and trace metals. Investigators wanting to initiate cultures at low population density but expand them to higher density exploited the beneficial growth properties of both nutrient formulations by developing volumetric admixtures, for example DMEM–Ham's F12. The principle of evaluating various ratios of suboptimal formulations to develop a superior mixture remains beneficial today, facilitated by automated fluid mixing technologies and computer-assisted statistical design methodologies. Cultivation of non-adherent cell types, such as human lymphocytes, required further qualitative and quantitative modification, as evidenced in the development of RPMI 1640 medium [20].

5.2.1.1 Use of liquid medium and supplemental additives

The most convenient format comes directly from commercial suppliers as a pre-sterilized, ready-to-use liquid medium. Such formulations may require no further supplementation or processing, although an initial user should refer to the product literature or label directions to avoid confusion. Basal nutrient formulations typically require volumetric supplementation with animal sera, with organ fractions, with tissue homogenates, with conditioned culture supernatants or with biochemical factors to promote normal cell proliferation and biological function. Owing to the spontaneous deamidation of glutamine in aqueous solution over a relatively short time period, many commercial manufacturers supply liquid nutrient medium either glutamine-free or substituted with a glutaminyl dipeptide that is resistant to spontaneous breakdown under refrigerated storage conditions. For many applications, glutamine-free formulations must be supplemented with appropriate levels of this nutrient to achieve specified performance.

Many formulations contain supplemental nutrients and other factors previously supplied by the serum supplement and are designed for serum-free, protein-free or chemically defined cell culture applications. Such formulations may be challenging to produce as milled powders or may yield inconsistent results initially when reconstituted from dry format because of factors described in the troubleshooting section of this chapter (Section 5.3). Initial examination of a nutrient formulation may be most effectively performed by evaluation of a single-strength liquid medium to minimize process variables.

It is often desirable to add concentrated nutrient feeds to liquid medium. Such nutrient concentrates may be added directly to the batch formulation vessel and processed as an integral component of the complete medium. Alternatively, it may be desirable to add nutrient supplements either to a fed-batch bioreactor or in the course of a perfusion feeding regimen, to replenish depleted nutrients or to alter the metabolic poise of the biomass to redirect it from proliferation to biological production.

5.2.1.2 Reconstitution of dry-format media

Less convenient, but offering significant cost benefits for large-scale production applications, are the various dry formats. Historically first were the ball-milled powders which placed chemical and biochemical constituents, either all at once or following a sequential regimen, within a ceramic-lined mill along with a high-density ceramic grinding material. As the mill rotated within a humidity-controlled environment, the medium constituents were pulverized by the grinding material, resulting in a fine, freely flowing powder. Scrupulous sanitization of the mills and grinding material was required to prevent cross-contamination between batches.

These ball-milling processes continue to be used to manufacture the majority of dry-format basal nutrient formulations. However, the mechanical shearing and the heat accumulation resulting from the pulverization process can adversely affect certain critical constituents of more novel formulations designed for serumless cultivation applications. Alternative milling processes that reduced the residence time diminished the degradation of thermo-labile and other sensitive constituents and resulted in high-quality powdered media. However, all of these milling approaches were limited by the reality that the milling process was designed for particle attrition, rather than homogeneous mixing. Milled powders tended to settle and stratify by particle size during shipping and storage, and analytical technologies were unable to assure that trace ingredients were homogeneously distributed throughout a milled batch. To ensure a greater degree of biochemical homogeneity, multiple milled batches were frequently blended together.

An alternative approach, based upon the fluidized-bed granulation technology common to the pharmaceutical and other industries, also found application in nutrient medium manufacture, incorporating biochemical constituents into homogeneous granules. Trace components or elements difficult to solubilize or sensitive to milling were sprayed onto levitated milled components to produce granulated particles. These granules demonstrated a greater degree of biochemical homogeneity than comparable formulations prepared as milled powders. Other features and practical benefits of this granulated format have been described elsewhere [37–39].

With either dry format, the initial step in the preparation of fully formulated liquid medium (*Protocol 5.1*) is hydration with high-quality water. Milled powders typically require the addition of sodium bicarbonate to the formulation vessel, along with additional supplements inconsistent with the milling production process, and acid or base to adjust the formulation to the desired pH. Granulated media minimize these supplemental additions since all or most medium components may be homogeneously incorporated into the granules and the formulation may be designed to equilibrate automatically to a desired target pH.

Following complete solubilization of nutrient constituents (see Section 5.3) from the dry-format intermediate precursor (as determined by visual inspection or by analysis of medium conductivity or osmolality) liquid medium must be processed by membrane filtration or an alternative method to eliminate adventitious microbial contaminants.

PROTOCOL 5.1   Preparation of media from milled powder

Equipment and reagents

- Analytical balance

- Graduated cylinder (or, for larger volumes, a suitable balance)

- Mixing vessel of appropriate volume

- Tissue culture-grade water

- Appropriate quantity of dry-format medium

- 1 N sodium hydroxide and 1 N hydrochloric acid as required

- Magnetic stirrer unit (with magnetic stir bar) or motorized overhead mixing unit

- Pre-weighed additives, including sodium bicarbonate and other required supplements for the formulation

- Properly calibrated pH meter

- Sterile 0.2-μm or, if required, 0.1-μm filter(s) (0.45-μm pre-filter(s) may be necessary for high protein-containing formulations)

(*continued overleaf*)

- Sterile pre-labelled bottles (or other suitable vessels, such as sterile bags) into which to dispense the filtered medium
- Class II microbiological safety cabinet
- Osmometer
- 70% isopropanol or 70% ethanol in spray bottle

Method

1 Using the graduated cylinder (or for larger volumes, the balance), measure out 90% of the required volume of tissue culture-grade water and dispense it into the mixing vessel. The temperature of the water should be approximately 20–30 °C.

2 Slowly add the appropriate amount of powdered medium to the water while gently stirring. Rinse the powdered medium container with tissue culture-grade water and add to the mixing vessel. Stir until all the powder has dissolved.

3 Add the required amount of additives and sodium bicarbonate.

4 Adjust the pH to 0.2–0.3 pH units below the desired pH using 1 N NaOH or 1 N HCl. (With bicarbonate-buffered media the pH will normally rise 0.1–0.3 pH units during filtration.)

5 Dilute the medium to the desired volume with tissue culture-grade water.

6 Assemble the filtration system to deliver filtered medium into the microbiological safety cabinet. Place the sterile medium bottles or other vessels in the cabinet and spray with 70% isopropanol or 70% ethanol. Leave to air dry.

7 Sterilize the medium by filtration using a membrane of porosity of 0.2 μm or less (a positive pressure system is recommended – see Chapter 2, Section 2.2.5 and Chapter 2, *Protocol 2.7*).

8 Using aseptic technique, fill the filtered medium into sterile medium bottles or other vessels in the microbiological safety cabinet. Tighten bottle caps/closures.

9 Store the medium at 2–8 °C in the dark.

5.2.2 Nutrient optimization – determination of basal medium formulation

There are many commercially available culture media for a wide range of applications. Additionally, extensive published literature exist describing media formulations and culture methods for numerous types of cells. For many research applications, existing media from one of these sources is adequate. It is relatively easy to assemble a panel of four or five media and screen the cells in each for the desired biological effect over one or more passages. For some research applications and for many clinical or bioproduction applications, it is often necessary to further optimize an existing medium or develop a new formulation. Several strategies may be successfully employed, often in concert, to achieve improvements in biological performance. Commonly used methods include analytically based approaches, profiling (which may include 'omics'-based) approaches and DOE (design of experiment) approaches.

5.2.2.1 Analytical approaches to medium optimization

Analytical approaches rely on the comparison of concentrations of nutrients and/or metabolites in spent culture media obtained at various time points against fresh culture media. Many media components can be quantitatively analysed by a variety of different technologies. Although it is possible to identify academic and/or industrial laboratories with chromatographic equipment, they may not have methods developed that are able to eliminate interferences from other components of cell culture media. For this reason, it is important to utilize an analytical testing laboratory with experience in cell culture media analysis. Most methods for analysis of media components are chromatographic in nature, but biosensors are also used, particularly for at-line analyses. Examples of methods commonly used for certain cell culture media components are as follows.

Amino acids can be measured using reverse-phase HPLC with fluorescence detection. Fluorescent derivatives are prepared by reacting the amino acids with 6-aminoquinolyl-*N*-hydroxysuccinimidyl carbamate [40–42]. The use of fluorescent derivatives increases the sensitivity of the analysis and minimizes potential interferences from other media components. The excitation and emission frequencies of the detector are optimized for the fluorescent label. Separation of the amino acid derivatives is performed on a high-efficiency NovaPack C18 column using an acetate buffer/acetonitrile gradient. This assay can measure 20 of the 21 amino acids typically used in cell culture media, and if the sample has been collected and stored properly (i.e. in polypropylene tubes with minimal headspace) ammonia can also be determined. Limitations to the assay include the inability to measure tryptophan owing to internal quenching of the fluorescence, and the inability to distinguish between L-cysteine and L-cystine since they co-elute. Additional interferences from primary or secondary amines are possible.

An assay for certain *water-soluble vitamins* employs gradient reverse-phase ion-pair chromatography. Multiple wavelengths are used to optimize sensitivity of the assay, since the concentration of vitamins approaches the limit of detection of the method. Separation is achieved using a Partisil C18 column with a phosphate buffer/acetonitrile gradient. Folic acid, niacinamide, riboflavin, thiamine HCl and vitamin B12 can be quantitated using this method. In addition, L-tryptophan and Phenol red can be determined by this assay.

Glucose and L-lactate can be determined using a Yellow Springs Instruments (YSI) Model 2700 SELECT biochemistry analyser. The YSI analyser employs biosensor technology and is specific for glucose and L-lactate. Specificity is achieved by placing a membrane containing an immobilized enzyme between the sample and an electrochemical probe. As the sample diffuses across the membrane, the immobilized enzyme oxidizes the sample and hydrogen peroxide is generated. Oxidation of the hydrogen peroxide by a silver–platinum electrochemical probe produces a current that is proportional to the concentration of hydrogen peroxide generated.

Gas chromatography can be used to measure the concentration of *cholesterol and fatty acids* in cell culture media. The concentrations of these lipids in culture media are typically near or below the detection limit of gas chromatography; therefore, a concentration step is required for sample preparation. Concentration of the lipids is accomplished using solid-phase extraction (SPE). SPE has the additional benefit of removing potential interfering components. In the case of cholesterol, the non-polar nature of the tetracyclic hydrocarbon skeleton allows it to be retained on a C18 SPE matrix. Chromatography is performed on a 5% diphenyl/95% dimethyl polysiloxane column. Analysis by gas chromatography eliminates potential interferences from other media components, since only volatile components will pass through the gas chromatograph. Isolation and concentration of the fatty acids from cell culture medium is accomplished using a modified procedure of Kaluzny *et al.* [43]. At the normal pH range for mammalian cell culture media, the fatty acids are fully ionized and will be retained by an anion exchange SPE packing. Following the solid phase extraction, the fatty acids are converted to their methyl esters to increase the volatility for the gas chromatographic analysis. Separation of the fatty acids is carried out on a 50% cyanopropylphenyl/50% dimethyl polysiloxane column.

Observed depletion or accumulation of various compounds may suggest a need for supplementation or reduction of individual medium components. It is important to consider the metabolic pathways involved at various stages of the culture and to sample at multiple time points to obtain useful information. Sampling only at the end of the culture will show depletion of many components; samples taken at various time points during growth and expression phases will provide a more balanced indication of metabolic processes. Further, supplementing the culture with nutrients that appear to have been depleted may result in inefficient metabolism and generation of toxic waste products like ammonium. Alternative strategies include use of pyruvate, glutamate or glutaminyl dipeptides instead of L-glutamine, or an alternative carbon source like galactose in place of glucose. Note that when using glutaminyl dipeptides, it may be useful to maintain some L-glutamine in the system to minimize the lag in growth that can occur while the enzyme required to cleave the dipeptide is upregulated.

5.2.2.2 Metabolic profiling and statistical analysis as optimization tools

Profiling technologies have become useful tools to identify specific biomarkers and metabolic pathways, leading to a more complete appreciation of the complexity of the cell. Microarrays exist for analysis of the transcriptome of several species, including human and murine. A review of available gene expression microarray platforms is found in Kuo *et al.* [44]. Of interest to biopharmaceutical scientists is the CHO gene array [45]. These DNA microarrays allow identification of genes activated or deactivated under the culture conditions from which the sample was taken. Wong *et al.* [46] used CHO cDNA microarrays to profile apoptotic pathways in batch and fed-batch cultures and then generated cell lines overexpressing early apoptosis signalling genes resulting in improved growth, and product expression and sialylation [47]. Interpretation of DNA microarray data may be challenging since the effectors of cellular function are proteins rather than mRNA. Integration of genomic and proteomic profiling techniques is becoming important for understanding the regulation of cell function with applications for cell line engineering (for reviews, see Korke *et al.* [48] and Kuystermans *et al.* [49]), but practical application to cell culture media and process development is only just starting. Metabolomic analysis may ultimately prove to be more useful for process optimization since many of the analytical tools already exist, and have been applied to the study of bodily fluids and tissue extracts [50].

DOE approaches have been used successfully to develop and optimize media. The complexity of modern SFM, which may contain more than 90 individual components, may discourage researchers from attempting DOE approaches, but a relatively simple, generic approach can be described as follows:

Media mixtures strategies can be used to take up to four different basal media and combine them in varying proportions to achieve multiple new formulations [51]. It is advisable to choose media for the mixtures experiment that differ significantly in composition (e.g. at least one of the formulations should be a relatively minimal formulation and at least one should be fairly complex with respect to components or component concentrations). In this way, a great diversity of intermediate formulation will be formed, and can be tested in a design matrix; an example is shown in Table 5.1. Comparison of the resulting media can be done by using one or more biological assay systems, for example comparing growth rate, doubling time, peak viable cell density, integral viable cell density or productivity. Commercially available statistical design programs like Design Expert® DOE software (Stat-Ease, Inc., Minneapolis, MN) can be used to analyse the results of a mixtures experiment and determine the best basal formulation. Figure 5.1 (Plate 1) shows an example of a contour plot generated by Design Expert® software from the hypothetical results of the mixtures matrix in Table 5.1. The predicted optimal combination is ~60% Medium 1 and ~40% Medium 2 (with no contribution by Medium 3 or Medium 4).

Once a suitable basal medium has been determined, it can be further optimized by a two-step process (each step may be repeated more than once). First, a two-level factorial design in which additives are tested either at a high concentration and a low concentration or as present or absent is employed. This highlights components that have positive (or negative) effects on some measurable biological endpoint. Table 5.2 and Figure 5.2 (Plate 2) illustrate a two-level factorial design and the resulting effect on cell growth from a hypothetical experiment. Components B (vitamins) and E (hydrolysate) were shown to have positive effects while the interaction of D (trace metal salts) and E (hydrolysate) had a negative effect. Since the vitamins and hydrolysate were the only components shown to have a positive significant effect, they were used in a central composite design (CCD) experiment to optimize their concentrations (Table 5.3 and Figure 5.3 (Plate 3)). In a CCD experiment, individual media components (or process parameters) can be tested at varying levels to determine the optimum concentration. Table 5.3 and Figure 5.3 show the CCD matrix and resulting improvement in biological performance due to optimization of the concentrations of hydrolysate and vitamin supplements.

5.2.3 Combination of basal medium, nutrient feeds and other process parameters

Cell culture media must be considered as one part of the overall culture process [52, 53]. Just as different types of cells and different clones of the same cell type often have

Table 5.1 Media mixtures matrix.

Std	Run	Component 1 Medium 1	Component 2 Medium 2	Component 3 Medium 3	Component 4 Medium 4	Response 1 IVCD day 5	Response 2 IVCD day 6	Response 3 IVCD day 7
1	15	1	0	0	0	2.48E+07	3.68E+07	5.33E+07
2	9	0.5	0.5	0	0	3.08E+07	4.58E+07	6.65E+07
3	10	0.5	0	0.5	0	2.20E+07	3.26E+07	4.73E+07
4	8	0.5	0	0	0.5	2.13E+07	3.15E+07	4.56E+07
5	2	0	1	0	0	2.83E+07	4.20E+07	6.10E+07
6	3	0	0.5	0.5	0	2.33E+07	3.45E+07	5.00E+07
7	12	0	0.5	0	0.5	2.20E+07	3.26E+07	4.73E+07
8	1	0	0	1	0	1.70E+07	2.51E+07	3.63E+07
9	13	0	0	0.5	0.5	1.33E+07	1.95E+07	2.81E+07
10	11	0	0	0	1	1.58E+07	2.33E+07	3.36E+07
11	6	0.625	0.125	0.125	0.125	2.08E+07	3.08E+07	4.46E+07
12	14	0.125	0.625	0.125	0.125	2.48E+07	3.68E+07	5.33E+07
13	4	0.125	0.125	0.625	0.125	1.95E+07	2.89E+07	4.18E+07
14	7	0.125	0.125	0.125	0.625	1.83E+07	2.70E+07	3.91E+07
15	5	0.25	0.25	0.25	0.25	2.08E+07	3.08E+07	4.46E+07

Sample matrix using four different basal nutrient media combined in varying volumetric proportions to achieve multiple new formulations. Hypothetical responses on days 5, 6 and 7 are shown as integral viable cell density (IVCD).

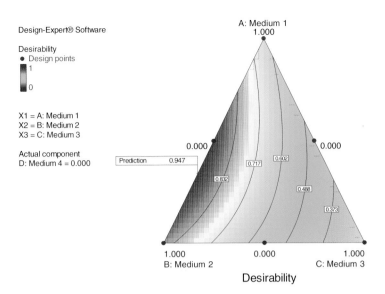

Plate 1 Figure 5.1 Example of a contour plot generated by Design Expert® software from the hypothetical results of the mixtures matrix in Table 5.1. Note that this is a two-dimensional representation of the contribution from four factors with the fourth factor (Medium 4) held constant at 0% (shown in the figure as 'Actual component D; Medium 4 = 0.000'); Media 1–3 are represented by each vertex of the two-dimensional triangle shown. The predicted optimal combination (Shown in the Figure as 'Prediction 0.947') is ~60% Medium 1 and ~40% Medium 2 (with no contribution by Medium 3 or Medium 4).

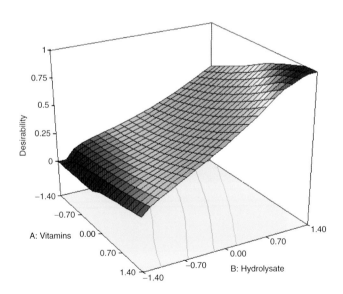

Plate 2 Figure 5.3 Response surface plot showing results of central composite design experiments (data from Table 5.3) with improvement in biological performance due to increasing concentrations of hydrolysate and vitamin supplements. The plateau as both x- and y-axes approach 1.3 suggests further increases in the concentrations of these components would have minimal positive effect.

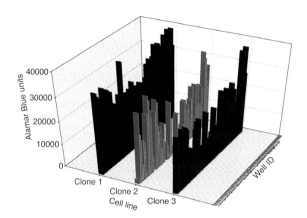

Plate 3 Figure 5.4 Screening of media against three CHO clones. Panel of 60 media (developed using the Hamilton STARplus liquid handler (Hamilton Company, Reno, NV)) by combining a group of known media components resulting in 55 unique blends and five controls (as indicated by 'Well ID') used to screen growth of three different CHO clones (Clones 2 and 3 were derived from a common parental line and expression vector system, and Clone 1 was from a different parental line and expression system). All three clones expressed IgG. The media components were chosen so that there would be no overt deficiencies. Concentrations were selected that spanned appropriate cell culture ranges. Cell metabolic activity was estimated by Alamar Blue Units [60], in which viable cells reduce resazurin to resorufin, which can be quantified by fluorescence detection methods.

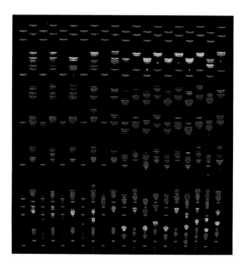

Plate 4 Figure 9.8 STR-PCR profile of a human cell line microsatellite DNA, as generated by the Applied Biosystems AmpFSTR® Identifiler® Plus PCR Amplification Kit. The colour coded banding pattern translates into digital code (see Figure 9.9), which is stored on a DNA profile database.

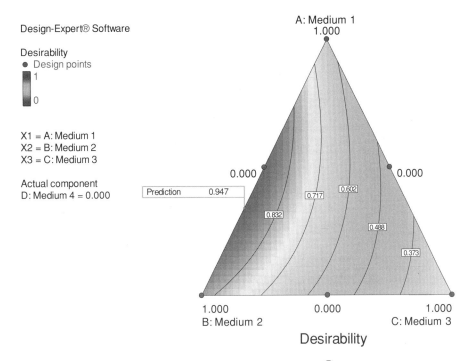

Figure 5.1 Example of a contour plot generated by Design Expert® software from the hypothetical results of the mixtures matrix in Table 5.1. Note that this is a two-dimensional representation of the contribution from four factors with the fourth factor (Medium 4) held constant at 0% (shown in the figure as 'Actual component D; Medium 4 = 0.000'); Media 1–3 are represented by each vertex of the two-dimensional triangle shown. The predicted optimal combination (Shown in the Figure as 'Prediction 0.947') is ~60% Medium 1 and ~40% Medium 2 (with no contribution by Medium 3 or Medium 4). See Plate 1 for the colour figure.

different nutritional requirements, the process by which the cells are grown impacts the media composition. Very rich culture media are often employed for batch-type cultures in which the cells are allowed to reach peak density and are then harvested upon reaching a predetermined minimum viability. A limitation of this method is that waste products like ammonium ions or lactic acid may accumulate in the medium, both of which can have inhibitory effects on cell growth. Oxygen limitations may also be reached in batch cultures, but often it is either total consumption of a key nutrient or build up of toxic metabolites that limit further cell growth. Fed-batch culture has become the process of choice for many biomanufacturing facilities. Increasing the peak cell density and/or extending the viability through addition of nutrients can generally be expected to improve the yield of the expressed product.

Determining which components should go into the basal medium versus the feed may be challenging. Considerations include the solubility of the individual components, stability, toxicity potential and the efficiency with which the cells utilize the nutrient. For example, most mammalian cells can tolerate high levels of glucose in culture media, but

Table 5.2 Example of two-level factorial design matrix and the resulting effect on cell growth from a hypothetical experiment.

Std	Run	A: amino acids	B: vitamins	C: lipids	D: trace metal salts	E: hydrolysate	Integral viable cell density day 5	Integral viable cell density day 6	Integral viable cell density day 7
12	1	1	1	−1	1	−1	2.0E+07	2.8E+07	3.8E+07
5	2	−1	−1	1	−1	−1	1.6E+07	2.2E+07	3.0E+07
10	3	1	−1	−1	1	1	2.2E+07	3.1E+07	4.3E+07
3	4	−1	1	−1	−1	−1	1.7E+07	2.4E+07	3.3E+07
16	5	1	1	1	1	1	2.5E+07	3.5E+07	4.7E+07
6	6	1	−1	1	−1	1	2.3E+07	3.3E+07	4.5E+07
2	7	1	−1	−1	−1	−1	1.3E+07	1.9E+07	2.5E+07
8	8	1	1	1	−1	−1	1.7E+07	2.3E+07	3.2E+07
14	9	1	−1	1	1	−1	1.7E+07	2.4E+07	3.2E+07
11	10	−1	1	−1	1	1	2.6E+07	3.7E+07	5.0E+07
4	11	1	1	−1	−1	1	2.8E+07	4.0E+07	5.5E+07
9	12	−1	−1	−1	1	−1	1.8E+07	2.5E+07	3.4E+07
13	13	−1	−1	1	1	1	2.3E+07	3.2E+07	4.4E+07
1	14	−1	−1	−1	−1	1	2.3E+07	3.3E+07	4.5E+07
15	15	−1	1	1	1	−1	1.7E+07	2.5E+07	3.3E+07
7	16	−1	1	1	−1	1	2.7E+07	3.9E+07	5.2E+07

Table 5.2 (*continued*)

Constraints

Name	Goal	Lower Limit	Upper Limit	Lower Weight	Upper Weight	Importance
Amino acids	is in range	−1	1	1	1	3
Vitamins	is in range	−1	1	1	1	3
Lipids	is in range	−1	1	1	1	3
Trace metal salts	is in range	−1	1	1	1	3
Hydrolysate	is in range	−1	1	1	1	3
IVCD day 5	maximize	1.33E+07	2.83E+07	1	1	3
IVCD day 6	maximize	1.88E+07	4.04E+07	1	1	4
IVCD day 7	maximize	254E+07	549E+07	1	1	5

Solutions

Number	Amino acids	Vitamins	Lipids	Trace metal salts	Hydrolysate	IVCD day 5	IVCD day 6	IVCD day 7	Desirability	Selected
1	−1	1	−1	−1	1	2.75E+07	3.86E+07	5.20E+07	0.920	Selected
2	−1	1	−1	−1	1	2.75E+07	3.86E+07	5.20E+07	0.918	
3	0	1	−1	−1	1	2.75E+07	3.85E+07	5.20E+07	0.918	
4	−1	1	−1	−1	1	2.73E+07	3.85E+07	5.20E+07	0.916	
5	0	1	−1	−1	1	2.74E+07	3.85E+07	5.19E+07	0.916	

Various component groups (amino acids, vitamins, lipids, trace metal salts, and hydrolysate) were tested in different combinations at two concentrations each. The amino acids, vitamins, trace metal salts and hydrolysate were tested at high and low concentrations while the presence or absence of lipids was evaluated. IVCD values for days 5–7 are shown. The highest importance was weighted to IVCD on day 7 using Design Expert® software.

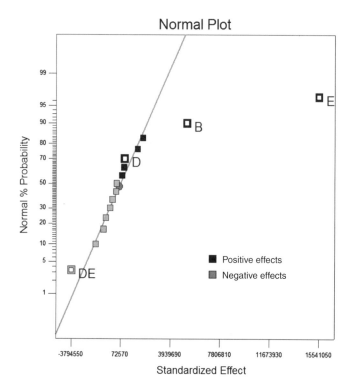

Figure 5.2   Normal plot of two-level factorial design (data from Table 5.2). In this hypothetical modelling experiment, components B (vitamins) and E (hydrolysate) were shown to have positive effects, while the interaction of D (trace metal salts) and E (hydrolysate) had a negative effect on the day 7 IVCD.

will utilize this nutrient inefficiently, rapidly generating lactic acid, which is not actually directly toxic (and can actually be converted to pyruvate and utilized in the tricarboxylic acid cycle by many cells), but may lead to indirect toxicity by causing a decrease in pH of the culture. Some components only have limited solubility (e.g. cysteine, lipids) [54, 55] or may have directly toxic effects if added at high levels (e.g. certain iron chelates). In such cases, it is usually better to include these components in the feed.

A common approach for fed-batch culture has been to add a rich feed supplement to a rich basal medium. Although this strategy may be successful in providing nutrients that extend cell population density, viability and/or productivity, there can be a downside to this approach. Inexperienced process development scientists often attempt to reconstitute a complete dry powder medium in less volume, but this may cause some of the components to be at concentrations above the limits of solubility, leading to the formation of a precipitate (which may pull other components out of solution) that will ultimately be removed upon filtration. This makes it difficult to develop a scalable media preparation process and can lead to variable performance results at large scale. Further, the high salt content of such a feed supplement can increase the culture osmolality to levels to which the cells cannot rapidly adapt, and nutrients which may normally prove beneficial can

Table 5.3 Example of a Central composite design (CCD) matrix in which individual media components can be tested at varying levels to determine the optimal concentration.

Std	Run	A: vitamins	B: hydrolysate	Integral viable cell density day 5	Integral viable cell density day 6	Integral viable cell density day 7
4	1	1	1	1.95E+07	2.89E+07	3.96E+07
3	2	−1	1	1.83E+07	2.67E+07	3.60E+07
6	3	1.4	0	1.73E+07	2.48E+07	3.30E+07
10	4	0	0	1.58E+07	2.21E+07	2.83E+07
9	5	0	0	1.63E+07	2.28E+07	2.92E+07
5	6	−1.4	0	1.45E+07	2.01E+07	2.56E+07
2	7	1	−1	1.73E+07	2.43E+07	3.16E+07
7	8	0	−1.4	1.43E+07	1.99E+07	2.56E+07
8	9	0	1.4	1.83E+07	2.81E+07	4.00E+07
1	10	1	−1	1.73E+07	2.43E+07	3.16E+07

Constraints

Name	Goal	Lower Limit	Upper Limit	Lower Weight	Upper Weight	Importance
Vitamins	is in range	−1.4	1.4	1	1	3
Hydrolysate	is in range	−1.4	1.4	1	1	3
Integral viable cell density day 5	maximize	1.43E+07	1.95E+07	1	1	3
Integral viable cell density day 6	maximize	1.99E+07	2.89E+07	1	1	4
Integral viable cell density day 6	maximize	2.61E+07	4.57E+07	1	1	5

Solutions

Number	Vitamins	Hydrolysate	Integral viable	Integral viable	Integral viable	Desirability	Selected
1	1.28	1.33	1.97E+07	3.17E+07	4.57E+07	1.00	
2	1.3s6	1.30	197E+07	31SE+07	4.58E+07	1.00	
3	1.21	137	1.97E+07	3.17E+07	4.57E+07	1.00	

In this hypothetical example, vitamins and hydrolysate were tested at five concentrations (1.4, 1, 0, −1 and −1.4). IVCD values for days 5–7 are shown and the data were analysed with highest importance weighted to day 7 IVCD. Predictions of the most desirable component concentrations using the Design Expert® software are shown as solutions no. 1–3.

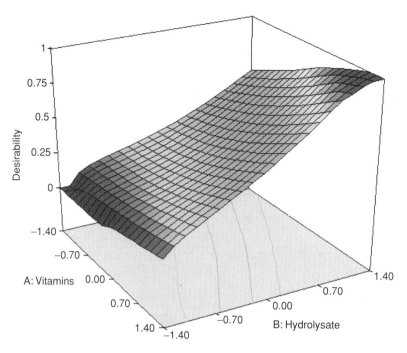

Figure 5.3   Response surface plot showing results of central composite design experiments (data from Table 5.3) with improvement in biological performance due to increasing concentrations of hydrolysate and vitamin supplements. The plateau as both x- and y-axes approach 1.3 suggests further increases in the concentrations of these components would have minimal positive effect. See Plate 2 for the colour figure.

become inhibitory at high concentrations [56]. Starting with a low osmolality basal medium (~270 mOsm/kg) can make addition of high salt-containing supplements more tolerable, but fails to address potential toxicity issues from other media components.

Recently, complex feeds have been developed in which the major salts have been removed and concentrations of low-solubility nutrients have been reduced in order to formulate a stable, 5× liquid supplement. When added to cultures grown in rich basal media, a two- to fourfold improvement in volumetric productivity was achieved [57]. This approach, while ignoring the possibility of toxic metabolite accumulation, was able to eliminate the potential for precipitation of components and scalability problems as well as minimize the osmolality increase normally associated with rich feed supplements. It is possible to quickly determine whether such a strategy is useful, a clear benefit to process development groups with tight timelines. Other possibilities include combining a complex, stable, concentrated supplement with a more minimal basal medium and combining a customized, minimal supplement with a minimal basal medium.

Both approaches are time-intensive, but can lead to more efficient metabolism while still improving productivity several fold [56]. In addition to the composition of the basal media and feed supplements, process parameters routinely investigated include dissolved oxygen (DO), agitation speed, temperature and pH. Combining a feed strategy with a temperature and/or pH shift often provides improved performance of the culture [57].

Temperature and pH shift (often downward for both) is a common way of effecting a metabolic shift from growth to production. Detailed analytical evaluation of spent media to compare nutrient consumption, growth and production kinetics may prove useful in implicating depletion of key nutrients or accumulation of metabolites in metabolic shift.

An important consideration in the development of basal media and feed supplements is the scale-down model used to conduct the experiments. Shake flasks are often used for initial screening studies before moving into bench-top bioreactors, often at 2- to 10-l scale. The disadvantage of shake flasks is the inability to control process parameters, other than temperature. Small bioreactors are preferred to shake flasks since process parameters like dissolved oxygen and pH can be controlled. The disadvantage to small bioreactors is the limited ability to test multiple conditions in parallel with limited space and/or human resources.

Several companies now supply small bench-top systems with multiple vessels (e.g. DASGIP, SixFors). These systems generally employ small vessels (0.1–1.0 l), but have the ability to adjust pH, DO and temperature as in larger systems, although the number of vessels in each system is limited to eight or less. A different approach has been taken by the manufacturers of the SimCell™ cluster tool. This automated microfluidic device employs reactor chambers in a microbioreactor array (MBA) that can hold between 500 and 750 µl of culture volume. There are six chambers per MBA and one incubator can hold up to 42 MBAs on a rotisserie apparatus that keeps the cells in suspension by means of a moving air bubble in the chamber with the cells and culture medium. The MBAs have dyes in hydrogels on the surface of the chamber membrane that can be interrogated with lasers to measure pH or DO. Cell population density is determined by optical means. All cell and fluid dispensing and sampling is done in an automated fashion by means of a programmed robotic arm that can transfer MBAs between the incubator and the filling, sensing and sampling stations. The SimCell™ makes it possible to conduct experiments in which multiple conditions can be varied, including media/feed composition and process parameters (DO, temperature, pH, feed frequency). Many more conditions or replicates are possible than by using other standard bioreactor systems.

5.2.4 Biological performance screening assays

Few discussions regarding the quality assurance of cell culture media, particularly for highly regulated biopharmaceutical production applications, can arouse greater polarity of opinion than the topic of biological performance screening.

One perspective suggests that a biochemically defined nutrient formulation should have thoroughly characterized nutrient components with defined identity and purity criteria, and that formulation of the complete nutrient mixture should have defined manufacturing tolerances for each component, with validated procedures for assembly of each subcomponent and for the final complete mixture. Further, the sanitation processes for all equipment contact surfaces must be thoroughly validated and the quality of formulation water must be carefully specified and monitored. Good Manufacturing Practices (GMPs) specify that a master production record should receive appropriate review and that derived batch-specific documentation must be completed and checked by manufacturing personnel. Given these stringent specifications for a biochemical 'footprint', biological performance screening should be unnecessary.

A passionate howl immediately emerges from the opposing perspective, carefully documenting a situation where virtually every known detail was carefully specified and every procedure was performed according to direction, yet the cell proliferation rate, biomass density, specific productivity or culture longevity was drastically diminished. At considerable expense in terms of lost production time and manpower commitment, the primary cause was ultimately traced to an unknown or undervalued specification, often with a previously perceived non-critical constituent. Consequently, despite all of the documentation in the world, this perspective will passionately advocate the necessity of stringent biological performance evaluation.

Given that excessive scrutiny also has its price, prudence demands effective compromise between these polarized opinions. GMP specifications and procedures are axiomatic for biopharmaceutical production and wise even for non-regulated applications to ensure accuracy and reproducibility. Yet trace impurities and unknown factors can play havoc with biological systems in the absence of some effective type of performance screening assay. We suggest five desirable factors for consideration for biological screening assays of cell culture media: reproducibility, scalability, relevance, sensitivity and robustness.

5.2.4.1 Reproducibility

Perhaps obvious, but experience would suggest worthy of statement, is that accurate interpretation of assay results requires a thorough understanding of the inherent variability of the biological performance assay. Items to consider include inherent variation of replicate samples within a batch; comparison of multiple batches within a single assay; and equipment or technician variability. Prior to establishing clear performance specifications for a particular formulation, prudence dictates that multiple samples from a single batch should be examined to determine the expected variation inherently associated with the assay technique, and the acceptable performance variation due to inevitable nutrient distribution heterogeneity within a large medium production batch. Multiple production batches, preferably a minimum of three separate batches, using (to the extent feasible) different raw material constituents will authenticate the reasonable variation in overall medium performance that might be expected at production scale. Finally, the scientist who developed the assay, who has a thorough knowledge of influential factors and significant experience with the assay (and occasionally using more sophisticated analytical instrumentation) will often demonstrate significantly greater precision than the individual who will ultimately perform final quality release of the medium formulation or final product. All of these factors taken together suggest that provisional specifications, established early, should remain flexible until a sufficient number of batches have been formulated and tested prior to submitting regulatory paperwork or finalizing specifications, in order to avoid inadvertently 'painting oneself into a corner' with quality specifications that are unrealistically tight.

5.2.4.2 Scalability

There are various rules of thumb regarding scaling down of performance assays to permit accurate yet conservative extrapolation to the target parameter. Such details are outside of the scope of this chapter. Suffice it to state that performance assays must mirror (to the extent possible) the physical and biochemical environment of the target production vessel. To the extent that culture vessel dimensions, shapes and contact surfaces, cell cultivation parameters, nutrient and dissolved gas fluxes, and monitoring probes and control features, and so on conform between the test and pilot or production systems, unpredictable artefacts can be minimized. Failure to observe such parameters, or excessive scale-up, could result in either false-positive or false-negative results.

5.2.4.3 Relevance

The critical biological performance screening assay should monitor parameters that correlate with the target application. Wherever possible, positive control and negative control (consciously misformulated) medium samples should be demonstrated to yield the predicted results with the cell types and culture conditions used for *in vitro* assays and (where relevant and practical) should correlate with success or failure in meeting quantitative or qualitative clinical product parameters. Differential criticality may be applied to various assay parameters: for example, if the critical acceptance element for a given formulation is specific productivity of the desired biological, it may be appropriate to allow a wider acceptance range for proliferation rate or maximal cell density, and a narrower range of acceptability for the critical criterion.

5.2.4.4 Sensitivity

Traditional biological systems generally fail to exhibit similar sensitivity to nutrient variations as do their biochemical counterparts. Wide fluctuations in concentration of non-critical nutrients may go unobserved by typical performance assays that analyse the population doubling time of a culture established using a relatively high-density cell inoculum. The sensitivity of such screening assays, either for medium optimization or batch qualification, may be increased by titrating down the seeding density to clonal or intermediate densities. Such limiting dilution assays (see Chapter 8) reduce or eliminate effects due to the masking of a quiescent subpopulation or paracrine supplementation of the exogenous nutrient environment. Sensitivity may also be enhanced through multiple sequential passages in the test formulation to eliminate carryover effects from the former culture medium. Finally, rather than monitoring solely the increase in viable cell count or synthesis of target biological product, effective synergy of molecular biology techniques with cell culture through polymerase chain reaction (PCR) and *in silico* monitoring of transcriptional events by chip hybridization can yield rapid warning of adverse events. However, implementation of such techniques leads us to the final category – robustness.

5.2.4.5 Robustness

A frequent pitfall, both in nutrient medium qualification and in final biological product characterization, is to attempt to utilize identical assay techniques in development and in manufacturing. The labour-intensive techniques that may be effectively utilized to analyse small numbers of samples within the development laboratory often create havoc when transferred directly to the quality control laboratory, because instrumentation is non-identical, because the technician is unable to devote the same level of attention to sample preparation detail due to sheer number of samples to be analysed, or because the development scientist was unaware of the inherent amount of troubleshooting or adjustment required to ensure accurate sample evaluation.

5.3 Troubleshooting

5.3.1 Media Preparation – What Can Go Wrong?

For this discussion, we will assume that the formulation has been optimized, that comprehensive raw material specifications (including shelf life) have been documented and followed for all constituents and that the complex nutrient formulation has been correctly manufactured. For individuals utilizing a commercial medium supplier, ensure that GMPs are strictly followed by thorough periodic audit of the supplier by qualified professionals who know what questions to ask and who know where to look. For individuals who choose to formulate nutrient medium internally from raw materials, conduct as thorough an audit on your in-house medium manufacturing kitchen as you would of a commercial supplier, to ensure quality specifications of constituent biochemicals and formulation water; existence of batch production records and adherence to formulation protocols; documented consistency of filtration media qualification and use; and effective release qualification of formulated product.

Even if these procedures are in place, other factors can influence the stability and performance of nutrient media. Principal among these factors are proper storage, thorough solubilization prior to subsequent processing, and method of sterilization.

5.3.1.1 Storage

Depending upon the format of the medium (e.g. single-strength liquid medium, liquid concentrate, milled or granulated dry format) and storage temperature, nutrient deterioration may be differentially affected. Despite the convenience of single-strength liquid medium, its constituents deteriorate the most rapidly. Certain antibiotics and antioxidants have already undergone substantial degradation by the time of receipt of qualified product. Certain vitamins and amino acids (principally glutamine) may spontaneously degrade to benign or cytotoxic by-products. Elevated storage temperature or sustained exposure to illumination will accelerate such nutrient degradation [58, 59]. Liquid concentrates tend to exhibit accelerated degradation unless specifically formulated to retard spontaneous breakdown [34, 35].

By contrast, dry-format formulations generally exhibit substantially longer constituent and overall performance shelf life. These formats occupy less storage space and may not require refrigerated storage, offering practical benefits. Contrasting features and benefits of milled versus granulated formats have been described elsewhere [37, 39] and may be minimal or exaggerated, depending upon the nutrient composition of the target formulation.

Well documented, but less widely publicized, is the reality that quarantined storage of qualified raw materials and intermediate nutrient medium formats may prevent inadvertent contamination. Obtaining formulation raw materials in batch-driven packaging, so that an entire package may be committed to a particular formulation batch, avoids potential inadvertent contamination resulting from opening and resealing. Maintenance of medium and raw materials in containers, locations and storage conditions that eliminate the potential for introduction of adventitious contaminants from arthropods and rodents prevents frustrating and costly problems.

5.3.1.2 Solubilization

Careful analysis of the diverse constituents of a complex nutrient formulation reveals a broad range of solubilities in aqueous solution. The differential solubility of nutrients remains largely unaltered when each individual biochemical component is combined into a complex mixture. Polar amino acids tend to dissolve more rapidly than non-polar amino acids. Relatively insoluble inorganic salts provided as trace constituents remain relatively insoluble. Lipid constituents retain their inherit hydrophobicity. Given the urgency to complete dissolution and initiate filtration of a medium batch, a natural tendency is to assume that, once the solution within the formulation vessel has clarified, or once it has reached a particular conductivity or osmolality, it is ready for filtration. Remembering that the bulk of the dry constituents is made up of sodium chloride and other similarly soluble inorganic salts should suggest that a prudent additional time should be allowed for thorough mixing and to permit the dissolution of less soluble constituents whose incomplete solubilization would go undetected by visual or conductivity monitoring. Failure to do so might result in increased batch-to-batch performance variability due to capture of critical unsolubilized nutrients by the filtration media.

The temperature of the formulation water and agitation during mixing may also affect medium performance. Although the solubilities of most constituents are enhanced by temperature elevation, this observation is not uniformly true. Further, there may also be significant diversity in nutrient thermostability, so that temperature elevation may substantially reduce its biological potency. Depending upon processing equipment and techniques, elevated temperature may also increase endotoxin levels, creating variable culture effects and necessitating additional downstream removal steps from a potential therapeutic product. Similarly, enhanced vessel agitation will accelerate solubilization of many nutrient components; however, excessive agitation may create foaming that may denature or remove critical nutrients or accelerate deterioration of constituents prone to oxidation.

5.3.1.3 Sterilization

Given the relatively slow proliferative rate of eukaryotic cells (compared with contaminating microorganisms) and owing to the highly regulated nature of culture-derived biologicals, some form of nutrient medium sterilization is required. Steam sterilization and gamma irradiation of complete medium have been demonstrated to destroy certain biologically active constituents, whereas other sterilization practices common to other industries produce cytotoxic or other undesirable residues. While some formulations may be autoclaved, either in whole or in part, such processing is generally not commercially practical for large-volume applications.

Passage of formulated liquid medium through filtration media composed of pharmaceutically approved matrices that exhibit well-characterized porosity is the most widely used method to ensure removal of microorganisms, including water-borne contaminants – see Chapter 2, Section 2.2.5. To ensure consistency of formulated nutrient medium batches, however, careful documentation of filter types is required. Different filter matrices differentially adsorb different medium constituents. Some filter matrices may exhibit a high retention level for a constituent that is performance-critical in your system. There may be qualitative and quantitative differences in clearance of individual medium constituents from different types of filters or even from different porosities of the same filter matrix. No single filter matrix is optimal for every nutrient formulation, and the identical nominal filter matrix from different manufacturers may exhibit different binding properties. Once it has been determined that a particular filter delivers a quality nutrient medium, ensure that it is consistently prepared, scaled and utilized to produce your target nutrient formulation. When a new batch of filters, even from the same supplier, is procured, evaluate the new batch to ensure that medium performance is equivalent to that obtained when using the previously qualified filter lot. Filters made from different matrices or having different porosities, or supplied by new suppliers, must be carefully evaluated to ensure that they will deliver equivalent filtered product.

Since membrane filtration is not a terminal sterilization process, biopharmaceutical manufacturers often require the additional security of a terminal step that will inactivate potential adventitious contaminants, particularly from perceived high risk constituents. Depending upon the relative stability of these biologically-active components, the medium may be subjected to flash pasteurization or other high temperature/short term exposures, gamma ray or electron beam irradiation, or other terminal sterilization processes.

Even meticulous sterilization of nutrient constituents may be inadequate if sufficient process controls for sanitization of process vessels, filter housings, formulation lines, formulation water storage tanks and delivery loops, steam-sterilizable connections and final storage containers are not in place.

5.3.2 Generality versus specificity – Why cannot a single SFM work for every cell application?

The variety of different cell types grown *in vitro* suggests that no one medium formulation would meet the nutritional needs of all. In the early days of cell culture, a relatively small number of different basal media met the needs of the cells cultured at that time, mainly because of the addition of serum and other supplements. As the diversity of cultured cells increased and the drive to eliminate serum intensified, it became apparent that primary cells can have different nutritional requirements from transformed cells, clonal differences in phenotype may translate to differences in nutrient needs, and recombinant cells may have vastly different culture requirements depending on the expression vector system employed and the location of foreign gene integration into the native DNA. Therefore, no single universal cell culture medium exists, although many workflows involving SFM are based on a 'platform' approach in which a common 'backbone' medium is used and one or more supplements are added for each new clone or segment of the process (e.g. different supplements for transfection, cloning, high-density growth and production phases). When developing a culture medium, it is important to understand the cell type and intended application. NS0 myelomas and many of their derivatives are sterol auxotrophs and must have a sterol source when grown in SFM. Cells grown in agitated suspension culture in SFM need a shear protectant like Pluronic® F-68, whereas anchorage-dependent cells may require an attachment factor like fibronectin in order to attach and spread in the absence of serum. Concentrations of calcium and magnesium may need to be altered depending on whether the cells are to be cultured in suspension culture (which may require low calcium/magnesium concentrations) or in anchorage-dependent culture (which may require higher calcium/magnesium levels). Yet even this decision may not be simple, as the concentration of calcium, for example, can determine the ability of a medium to support the growth of particular cell types.

Having determined a good basal medium with or without feed for a particular cell line or clone, the natural inclination is to adopt the same formulation for each new cell line or clone. However, as is evident from Figure 5.4 (Plate 4), different clones of a single cell line may vary in their response to a particular formulation or panel of formulations. We have also observed variability in nutrient consumption in different cell culture systems (e.g. shake flasks vs stirred tank bioreactors) for a given cell line, so two points to consider regarding universality of media formulations include the following.

- If possible, screen multiple media and/or feeds with each new cell line or clone.

- When changing culture vessel design, scale, or other process parameters, it is advisable to perform spent media analysis to assess the impact of the change on nutrient consumption.

5.3.3 Sensitivity to environmental fluctuations – Why are my cells more sensitive when cultured in SFM?

In the absence of the effects of serum (shear protection, buffering, nutrition and signal transduction), cells cultured in serum-free, protein-free or chemically defined media may be more sensitive to a wide variety of chemical and environmental factors. The basal medium components (e.g. HEPES buffer) or supplements (e.g. antibiotics) may inhibit growth of the cells in serum-free systems if used at concentrations commonly used in serum-supplemented media. Similarly, cells may be more sensitive to extremes of pH, osmolality or temperature in the absence of serum. Physical parameters like resistance to shear and centrifugal forces may be diminished and may contribute to poor cell health in

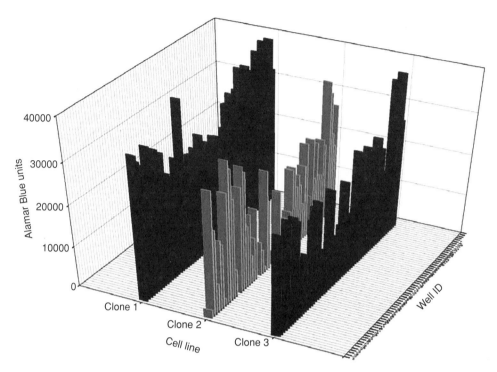

Figure 5.4 Screening of media against three CHO clones. Panel of 60 media (developed using the Hamilton STARplus liquid handler (Hamilton Company, Reno, NV)) by combining a group of known media components resulting in 55 unique blends and five controls (as indicated by 'well ID') used to screen growth of three different CHO clones (Clones 2 and 3 were derived from a common parental line and expression vector system, and Clone 1 was from a different parental line and expression system). All three clones expressed IgG. The media components were chosen so that there would be no overt deficiencies. Concentrations were selected that spanned appropriate cell culture ranges. Cell metabolic activity was estimated by Alamar Blue Units [60], in which viable cells reduce resazurin to resorufin, which can be quantified by fluorescence detection methods. See Plate 3 for the colour figure.

SFM. It is important to titrate any new component added to a serum-free culture system to determine the optimal range of concentration.

5.3.4 Adaptation of cells to SFM - How should I do this, and how and when should banking occur to minimize variability?

There are two different schools of thought on the best way to proceed when moving cells from one set of culture conditions to another. The first allows for the fact that drastic changes in the culture environment, especially the medium composition, will stimulate an adaptation process which, over multiple passages, allows the cells to acclimatize to the new conditions. A lag in growth is expected to occur until the cells have fully adapted to the new environment. However, whether this is really an acclimatization or a selection process in which a subpopulation of cells better able to grow in the new conditions overgrows the culture is often not known. This approach may take an extended time, sometimes weeks to months, and can be attempted either by directly subculturing the cells into the new medium, or by employing a sequential adaptation method as described in *Protocol 5. 2*.

PROTOCOL 5.2 Sequential adaptation to a new medium

Equipment and reagents

- Culture of the cells to be adapted in a standard culture flask (either in suspension or anchorage-dependent culture)

- Class II microbiological safety cabinet

- Electronic cell counter, or microscope and haemocytometer

- New sterile culture flasks

- Adequate volumes of stock medium and new medium

- Appropriate number of sterile pipettes at the appropriate volumes

- Incubator with appropriate, CO_2 concentration, temperature and humidity capabilities (with shaker or spinner platform if needed)

Method

1 From a stock culture with high viability and in mid-log phase (see Chapter 4, Section 4.2.5.2), subculture at twice the normal seeding density into a mixture of 50% new medium + 50% stock medium.

2 Monitor viable cell concentration daily, and when the cells have at least doubled in concentration, subculture again into a mixture of 75% new medium + 25% stock medium. (Keep a back-up culture of cells in the 50:50 mix in case the cells do not adapt to the new mixture)[a].

3 Monitor viable cell concentration daily, and when the cells have at least doubled in concentration, subculture again into 100% new medium. (Keep a back-up culture of cells in the 75:25 mix in case the cells do not adapt to the new mixture)[a].

4 Once the cells are able to reproducibly grow to the same peak viable cell concentration with the same population doubling time, they are considered to be adapted to the new medium and can be subcultured at normal seeding density for subsequent passages.

Note

[a] It may be necessary to subculture at a given level of new : stock medium several times until the cells acclimate.

The second approach attempts to minimize adaptation and instead seeks to identify a set of culture conditions (including media formulation) which minimizes growth lag and facilitates the expected doubling time and peak viable cell concentration. A well-balanced medium formulation is a key component of this strategy and represents an alternative paradigm to the adaptation strategy. Identification of such a formulation may be facilitated by DOE methods, and reduces the chance that a subpopulation of cells that grows well in the new medium (but may not express a recombinant product to high levels) has been selected by the adaptation process. A well-balanced base formulation allows single passage screening, but there is still a need for a multiple passage test once a panel of suitable candidate media has been identified, in order to ensure that any carryover effect of stock media components has been eliminated.

Once the final medium formulation has been determined, it is usually necessary to cryopreserve a new master cell bank and working cell bank(s) in the new formulation. If it can be demonstrated that cells cryopreserved in one formulation can be recovered and grown in a different formulation, and still exhibit all critical characteristics (e.g. karyotype, isozymes, growth/production indicators and quality of expressed product), new cell banks may not be necessary, but either route may be time consuming and costly. Cryopreservation and cell banking is described and discussed in detail in Chapter 6, but three tips are specifically relevant to the cryopreservation of cells growing in SFM (including PFM and CDM).

Success is facilitated by the use of conditioned medium (sterile-filtered supernatant from an actively growing high viability culture) in the solution in which the cells are frozen. It should be combined with fresh growth medium in a 1:1 (v/v) ratio. In addition, if the cells are relatively hardy but the medium into which they are being thawed is still suboptimal, it can sometimes be advantageous to supplement this medium with conditioned medium, up to a maximum concentration of 50%.

- Some cells grown in SFM (including PFM and CDM) can be particularly sensitive to the adverse effects of DMSO, and upon recovery from cryopreservation may require gentle centrifugation (e.g. at $100\,g$ for 5 min) to remove the freezing medium and DMSO.

- After thawing, seed the cells at twice the normal cell concentration.

5.4 Conclusion

The science of eukaryotic nutrient medium optimization has evolved significantly since its origins, with the enhanced complexity of host cells, bioreactors and target biological applications, greater sophistication of analytical methods for monitoring and controlling delivery of nutrients, and statistical design approaches to accelerate and simplify developmental options. Given those advances, however, successful eukaryotic cell culture must rely upon an appropriate blend of bioprocessing technology, basic understanding of intermediary cellular metabolism and application of conventional cell culture wisdom to develop an exogenous environment to promote optimal biomass expansion and biological yield.

Acknowledgments

The authors wish to thank David Rice and Joe Zdanowicz for analytical methods descriptions; David Zhao, Bob Kenerson, Steve Peppers, Rich Hassett, Scott Jacobia and Soverin Karmiol for DOE method descriptions and media panel data; and Bill Paul, Delia Fernandez, Mary Lynn Tilkins, Joyce Dzimian, Barb Dadey, and Jennifer Walowitz for their excellent technical assistance in developing many of the media and processes described in this chapter.

Pluronic® is a registered trademark of BASF Corporation.

References

1. Jayme, D.W. and Blackman, K.E. (1985) Review of culture media for propagation of mammalian cells, viruses and other biologicals, in *Advances in Biotechnological Processes*, vol. **5** (eds A. Mizrahi and A.L. van Wezel), John Wiley & Sons, Ltd, Chichester, UK, pp. 1–30.

★★ 2. Freshney, R.I. (2010) *Culture of Animal Cells: A Manual of Basic Technique and Specialized Applications*, 6th edn, John Wiley & Sons, Inc., Hoboken, New Jersey, USA. – *Excellent reference for beginning cell culture.*

3. Kadouri, A. and Spier, R.E. (1997) Some myths and messages concerning the batch and continuous culture of animal cells. *Cytotechnology*, **24**, 89–98.

4. Vogel, J.H., Prtischet, M., Wolfgang, J. *et al.* (2001) Continuous isolation of rFVIII from mammalian cell culture, in *Animal Cell Technology: From Target to Market* (eds E. Lindner-Olsson, N. Chatzissavidou and E. Lüllau), Kluwer, Dordrecht, Holland, pp. 313–317.

5. Altman, P.L. and Dittmer, D.S. (eds) (1961) *Blood and Other Body Fluids*, Federation of American Societies for Experimental Biology, Bethesda, MD.

★★ 6. Ham, R.G. and McKeehan, W.L. (1979) Media and growth requirements. *Methods Enzymol.*, **58**, 44–93. – *Very good review by some of the pioneers of cell culture media development.*

★★ 7. Barnes, D.W. Sirbasku, D.A. and Sato, G.H. (eds) (1984) *Cell Culture Methods for Molecular and Cell Biology*, vols. **1–4**, A.R. Liss, New York. – *Very good review by some of the pioneers of cell culture media development.*

★★ 8. Waymouth C (1972) Construction of tissue culture media, in *Growth, Nutrition, and Metabolism of Cells in Culture*, vol. **1** (eds G.H. Rothblat and V.J. Cristofalo), Academic Press, New York, pp. 11–47. – *Good methods for those new to cell culture.*

9. Jayme, D.W. (2000) *Encyclopedia of Life Sciences*, article 2558, MacMillan Reference Ltd, London, UK.

10. Jayme, D.W. (2007) Development and optimization of serum-free and protein-free media in *Medicines from Animal Cell Culture* (eds G.N. Stacey and J. Davis), John Wiley & Sons, Ltd, Chichester, UK, pp. 29–44.

11. Jayme, D.W. (1991) Nutrient optimization for high density biological production applications. *Cytotechnology*, **5**, 15–30.

12. Fike, R., Kubiak, J., Price, P. and Jayme, D. (1993) Feeding Strategies for Enhanced Hybridoma Productivity: Automated Concentrate Supplementation. *BioPharm.*, **6** (8), 49–54.

13. Jayme, D.W. and Gruber, D.F. (1998) Development of Serum-Free Media and Methods for Optimization of Nutrient Composition, in *Cell Biology: A Laboratory Handbook*, vol. **1** (ed. J.E. Celis), Academic Press, San Diego, CA, pp. 19–26.

14. Zielke, H.R., Ozand, P.T., Tildon, J.T. *et al.* (1978) Reciprocal regulation of glucose and glutamine utilization by cultured human diploid fibroblasts. *J. Cell Physiol.*, **95**, 41–48.

15. Gruber, D.F. and Jayme, D.W. (1998) Commentary on the evolution of cell culture as a science and updated terminology, in *Cell Biology: A Laboratory Handbook*, vol. **1** (ed. J. E. Celis), Academic Press, San Diego, CA, pp. 547–551.

★ 16. Eagle, H. (1959) Amino acid metabolism in mammalian cell cultures. *Science*, **130**, 432–437. *– Early description of cellular nutrient requirements.*

★ 17. Dulbecco, R. and Freeman, G. (1959) Plaque formation by the polyoma virus. *Virology*, **8**, 396–397. *– Dulbecco's modifications of MEM.*

★ 18. Ham, R. (1965) Clonal growth of mammalian cells in a chemically defined synthetic medium. *Proc. Natl Acad. Sci. USA*, **53**, 288–293. *– Early chemically defined formulation for CHO cells.*

★ 19. Barnes, W.D. and Sato, G. (1980) Methods for growth of cultured cells in serum-free medium. *Anal. Biochem.*, **102**, 255–270. *– Early serum-free formulations and methods.*

20. Moore, G.E., Gerner, R.E. and Franklin, H.A. (1967) Culture of normal human leukocytes. *J. Amer. Med. Assoc.*, **199**, 519–524.

21. Fike, R.M., Pfohl, J.L., Epstein, D.A. *et al.* (1991) Hybridoma growth and monoclonal antibody production in protein-free hybridoma medium. *Biopharm* **4** (3), 26–29.

22. Lobo-Alfonso, J., Price, P. and Jayme, D. (2010) Benefits and limitations of protein hydrolysates as components of serum-free media for animal cell culture applications, in *Protein Hydrolysates in Nutrition and Biotechnology* (ed. V.K. Pasupuleti), Kluwer, Dordrecht, Holland, pp. 217–219.

23. Taylor, W.G. and Parshad, R. (1977) Peptones as serum substitutes for mammalian cells in culture. *Methods Cell Biol.*, **15**, 421–434.

24. Gorfien, S.F., Paul, B., Walowitz, J. *et al.* (2000) Growth of NS0 Cells in protein-free, chemically-defined medium. *Biotechnol. Prog.*, **16**, 682–687.

25. Jacobia, S.J., Kenerson, R.W., Tescione, L.D. *et al.* (2006) Trace element optimization enhances performance and reproducibility of serum-free medium, in *Animal Cell Technology: Basic & Applied Aspects*, vol. **14** (eds S. Iijima and K. Nishijima), Springer, Dordrecht, Holland, pp. 193–199.

26. Jayme, D.W. (1999) An animal origin perspective of common constituents of serum-free medium formulations. *Dev. Biol. Stand.*, **99**, 181–187.

27. Jayme, D.W., Smith, S.R. and Plavsic, M. (1999) Reducing risks of animal origin contaminants in cell culture, in *Animal Cell Technology: Challenges for the 21st Century* (eds K. Ikura, M. Nagao, S. Masuda and R. Sasaki), Kluwer, Dordrecht, Holland, pp. 221–225.

28. Jayme, D.W. and Smith, S.R. (2000) Media formulation options and manufacturing process controls to safeguard against introduction of animal origin contaminants in animal cell culture. *Cytotechnology*, **33**, 27–35.

29. Battista, P.J., Tilkins, M.L., Judd, D.A. *et al.* (1993) CHO cell growth and recombinant protein production in serum-free media, in *Animal Cell Technology: Basic and Applied Aspects*,

vol. **5** (eds S. Kaminogawa, A. Ametani and S. Hachimura), Kluwer, Dordrecht, Holland, pp. 251–257.

30. Price, P.J., Samrock, R.L., Lobo, J.O. and Jayme, D.W. (1993) Sustained inducibility of cytochrome P450 activity in rat hepatocytes cultured in serum-free medium, in *Animal Cell Technology: Basic and Applied Aspects*, vol. **5** (eds S. Kaminogawa, A. Ametani and S. Hachimura), Kluwer, Dordrecht, Holland, pp. 195–201.

31. Jayme, D.W., Price, P.J., Plavsic, M.Z. and Epstein, D.A. (2000) Low serum and serum-free cultivation of mammalian cells used for virus production applications, in *Animal Cell Technology: Products from Cells, Cells as Products* (eds A. Bernard, B. Griffiths, W. Noe and F. Wurm), Kluwer, Dordrecht, Holland, pp. 459–461.

32. Jayme, D.W. and Gruber, D.F. (2006) Development of serum-free media: optimization of nutrient composition and delivery format, in *Cell Biology: A Laboratory Handbook*, 3rd edn, Chapter 5, (ed. J.E. Celis), Elsevier Science, Amsterdam, Holland, pp. 33–41.

33. Young, F.B., Sharon, W.S. and Long, R.B. (1966) Preparation and use of dry powder tissue culture media. *Ann N.Y. Acad. Sci.*, **139**, 108–110.

34. Jayme, D.W., DiSorbo, D.M., Kubiak, J.M. and Fike, R.M. (1992) Use of nutrient medium concentrates to improve bioreactor productivity, in *Animal Cell Technology: Basic & Applied Aspects*, vol. **4** (eds H. Murakami, S. Shirahata and H. Tachibana), Kluwer, Dordrecht, Holland, pp. 143–148.

35. Jayme, D.W., Fike, R.M., Kubiak, J.M. *et al.* (1993) Use of liquid medium concentrates to enhance biological productivity, in *Animal Cell Technology: Basic and Applied Aspects*, vol. **5** (eds S. Kaminogawa, A. Ametani and S. Hachimura), Kluwer, Dordrecht, Holland, pp. 215–222.

36. Jayme, D.W., Fike, R.M., Kubiak, J.M. and Price, P.J. (1994) Improved bioreactor productivity and manufacturing efficiency using liquid medium concentrates, in *Animal Cell Technology: Products of Today, Prospects for Tomorrow* (eds R.E. Spier, J.B. Griffiths and W. Berthold), Butterworth-Heinemann, Oxford, UK, pp. 105–109.

37. Jayme, D., Fike, R., Radominski, R. *et al.* (2002) A novel application of granulation technology to improve physical properties and biological performance of powdered serum-free culture media, in *Animal Cell Technology: Basic & Applied Aspects*, vol. **11** (eds S. Shirahata, K. Teruya and Y. Katajura), Kluwer, Dordrecht, Holland, pp. 155–159.

★ 38. Fike, R., Dadey, B., Hassett, R. *et al.* (2001) Advanced granulation technology (AGTTM): an alternate format for serum-free, chemically-defined and protein-free cell culture media. *Cytotechnology*, **36**, 33–39. – *Novel dry medium format*

39. Radominski, R., Hassett, R., Dadey, B. *et al.* (2001) Production-scale qualification of a novel cell culture medium format. *BioPharm*, **14** (7), 34–39.

40. Cohen, S.A. and Michaud, D.P. (1993) Synthesis of a fluorescent derivatizing reagent, 6-aminoquinolyl-N-hydroxysuccinimidyl carbamate, and its applications for the analysis of hydrolysate amino acids via high-performance liquid chromatography. *Anal. Biochem.*, **211**, 279–287.

41. Cohen, S.A. and De Antonis, K.M. (1994) Applications of amino acid derivatization with 6-aminoquinolyl-N-hydroxysuccinimidyl carbamate: Analysis of feed grains, intravenous solutions, and glycoproteins. *J. Chromatog. A*, **661**, 25–34.

42. Cohen, S. and Wandelen, C. (1997) Using quaternary high-performance liquid chromatography eluent systems for separating 6-aminoquinolyl-N-hydroxsuccinimdyl carbamate-derivatized amino acid mixtures. *J. Chromatog. A*, **763**, 11–22.

43. Kaluzny, M.A., Duncan, L.A., Merritt, M.V. and Epps, D.E. (1985) Rapid separation of lipid classes in high yield and purity using bonded phase columns. *J. Lipid Res.*, **26**, 135–140.

44. Kuo, W.P., Liu, F., Trimarchi, J. *et al.* (2006) A sequence-oriented comparison of gene expression measurements across different hybridization-based technologies. *Nature Biotechnol.*, **24**, 832–840.

45. Wlaschin, K.F., Nissom, P.M., Gatti Mde, L. *et al.* (2005) EST Sequencing for gene discovery in Chinese hamster ovary cells. *Biotechnol. Bioeng.*, **91**, 592–606.

46. Wong, D.C.F., Wong, K.T.K., Lee, Y.Y. *et al.* (2006) Transcriptional profiling of apoptotic pathways in batch and fed-batch CHO cell cultures. *Biotechnol. Bioeng.*, **94**, 373–382.

47. Wong, D.C.F., Wong, K.T.K., Lee, Y.Y. *et al.* (2006) Targeting early apoptotic genes in batch and fed-batch CHO cell cultures. *Biotechnol. Bioeng.*, **95**, 350–361.

48. Korke, R.R.A., Seow, T.K., Chung, M.C. *et al.* (2002) Genomic and proteomic perspectives in cell culture engineering. *J. Biotechnol.*, **94**, 73–92.

49. Kuystermans, D., Krampe, B., Swiderek, H. and Al-Rubeai, M. (2007) Using cell engineering and omic tools for the improvement of cell culture processes. *Cytotechnology*, **53**, 3–22.

50. Lenz, E.M. and Wilson, I.D. (2007) Analytical strategies in metabonomics. *J. Proteome Res.*, **6**, 443–458.

51. Jayme, D.W., Watanabe, T. and Shimada, T. (1997) Basal medium development for serum-free culture: a historical perspective. *Cytotechnology*, **23**, 95–101.

52. Mather, J.P. (1998) Making informed choices: medium, serum, and serum-free medium, in *Animal Cell Culture Methods* (eds J.P. Mather and D. Barnes), Academic Press, San Diego, pp. 20–30.

53. Jayme, D.W. and Greenwald, D.J. (1991) Media selection and design: wise choices and common mistakes. *Bio/Technology*, **9**, 716–721.

54. Walowitz, J., Tescione, L., Paul, W. *et al.* (2005) Optimized feeding strategy in NS0 cells, in *Animal Cell Technology Meets Genomics* (eds R. Godia and M. Fussenegger), Springer, New York, pp. 711–714.

55. Walowitz, J.L., Fike, R.M. and Jayme, D.W. (2003) Efficient lipid delivery to hybridoma culture by use of cyclodextrin in a novel granulated dry-form medium technology. *Biotechnol. Prog.*, **19**, 64–68.

56. Gorfien, S.F., Paul, W., Judd, D. *et al.* (2003) Optimized nutrient additives for fed-batch culture. *Biopharm Intl.*, **16**, 34–38.

57. Fernandez, D., Walowitz, J., Paul, W. *et al.* (2007) Bioreactor process strategies for rNS0 cells, in *Cell Technology for Cell Products* (ed. R. Smith), Proceedings of the 19th ESACT Meeting, Harrogate, UK, 2005, Springer, Dordrecht, Holland.

★★★ 58. Wang, R.J. (1976) Effect of room fluorescent light on the deterioration of tissue culture medium. *In Vitro*, **12**, 19–22. – *Reasons why culture media should be stored in the dark.*

★★★ 59. Taylor, W.G. (1984) Toxicity and hazards to successful culture: cellular responses to damaged induced by light, oxygen and heavy metals. *In Vitro*, **20**, 58–70. – *Reasons why culture media should be stored in the dark.*

60. Lancaster, M.V. and Fields, R.D. (1996) Antibiotic and cytotoxic drug susceptibility assays using resazurin and poising agents. U.S. Patent No. 5,501,959.

6

Cryopreservation and Banking of Cell Lines

Glyn N. Stacey, Ross Hawkins and Roland A. Fleck

National Institute for Biological Standards and Control, South Mimms, Hertfordshire, UK

6.1 Introduction

Cryopreservation of fully functional animal cells was first reported in the scientific literature 1949 by Polge *et al.* [1], who discovered that glycerol added to the culture medium enabled cock sperm to regain motility after being frozen. Cryopreservation techniques permit indefinite storage of cells in 'suspended animation' at ultra-low temperatures, and theoretically there is no limit to the time over which cryopreserved material can be maintained and recovered. A cryopreserved seed stock or 'bank' of cells, comprising a large number of aliquots of cells derived from the same original pooled culture, provides the researcher or manufacturer with a reproducible and reliable supply of cells, sustainable over many decades. Today this is a significant enabling factor in the standardization of any work involving cell cultures, permitting samples of cells which are exactly the same to be used at different sites and at different times. Biomedical products derived from cells (e.g. monoclonal antibodies, recombinant therapeutic proteins, vaccines) take many years to develop, test and license, and throughout this period and the subsequent years of use, identical material must remain available. In addition, patent applications involving cell cultures may require storage of the cells in a stable state for a minimum of 30 years in order to be available in the event of a challenge to the patent. The ability to preserve human tissues and cell cultures also means that they can be thoroughly safety tested prior to use in

Animal Cell Culture: Essential Methods, First Edition. Edited by John M. Davis.
© 2011 John Wiley & Sons, Ltd. Published 2011 by John Wiley & Sons, Ltd.

humans for therapeutic applications. Furthermore, in the case of primary cultures derived directly from animal tissues, cryopreservation has the added advantage of providing an opportunity to reduce the need for routine sacrifice of animals for supplies of fresh cells for research and testing purposes.

The most commonly used technique for preservation of cell lines involves treatment of the cells with a cryoprotective agent (CPA) that depresses the freezing point (typically DMSO), and then subjecting them to a cooling process until the sample is completely frozen. As ice begins to form in the extracellular space, water is progressively removed from the cells in suspension by osmosis. For most nucleated mammalian cells, the cell membrane permeability to water and the changes of this parameter during cooling dictate cooling rates. The cell water content remains effectively in equilibrium with the increasing osmotic potential of the extracellular ice–water matrix. Therefore, the cells become dehydrated and there is little remaining intracellular water to participate in ice formation. The increasing concentration of the cell contents inhibits intracellular ice formation, and the resultant suspension of dehydrated cells encased in ice can be stored in a viable state indefinitely provided the temperature is maintained below the glass transition temperature, which for practical purposes may be considered to be $-130\,^{\circ}C$. For a review of this process see reference 2.

Most routine cryopreservation procedures employed in culture collections (e.g. the ATCC, the Deutsche Sammlung von Mikroorganismen und Zellkulturen (DSMZ) and the ECACC) have been based on the use of cooling rates in the region of $1\,^{\circ}C/min$. Such rates are relatively easy to achieve on a consistent basis, using either programmable freezing machines, or passive cooling in, for example, insulated boxes in a $-80\,^{\circ}C$ freezer, or by suspending vials in a cryogen such as the vapour phase above liquid nitrogen (LN).

An alternative approach to cryopreservation centres on the achievement of a 'glassy' state at low temperatures for the long-term maintenance of viability. It has been known for a long time from a theoretical basis that it should be possible to vitrify aqueous solutions [3], but only at extremely high cooling rates (tens of thousands of $^{\circ}C/min$), which would be impractical for routine cell banking and may also be unrealistic with LN due to the insulation provided by nitrogen gas 'boil off'. It has been shown that for many CPAs such as DMSO, glycerol and other polyols, concentrations in the region of 40–60% w/v will allow low-temperature vitrification (at cooling rates of several hundred $^{\circ}C/min$). These high solute mixes increase viscosity (favouring glass formation) and have a strong colligative action (reducing the statistical likelihood of ice nucleation during rapid cooling). This approach has been successfully applied to the cryopreservation of small numbers of valuable cells, such as animal or human embryos in reproductive medicine [4], where very small volumes of medium can be cooled, for example in thin plastic straws [5]. The technique has been adapted for the storage of human embryonic stem (cell) (hES) lines [6]. One major issue is the toxicity of the CPA to particular cell types at these high starting concentrations (around 40% w/v), which undoubtedly have osmotic and chemical effects on cells. Additionally, insufficiently rapid warming may permit growth of ice crystals, which in turn may reduce post-thaw viabilities. For the above reasons, vitrification has not proven readily applicable for the routine cryopreservation of cells in a cell-banking environment.

The following sections describe the typical methodology for establishing a bank of a particular cell line.

6.2 Methods and approaches

To establish a cell bank, a well-mixed pool of a large number of cells is aliquotted and frozen as a single stock. This approach can provide replicate cultures for a number of years. However, if reproducible stocks of cells are needed over several decades, for example to support production of a viral vaccine or for a public service culture collection, an archive or 'master' stock is essential to enable the working stock to be regenerated periodically. This tiered master cell bank (MCB) and working cell bank (WCB) system enables multiple working cell banks to be generated, making identical cells available over decades.

Typically, the MCB receives the highest level of quality control and characterization, with lesser testing applied to the WCB [7]. Regulatory bodies have made provision for manufacturers to perform full characterization and testing on WCBs, but this is costly and would probably need to be justified to the regulator involved [8].

Before engaging in establishing cell banks, it is wise to consider a range of issues that may impact on the requirements for the size and use of each particular cell bank. These include:

- suitability of the current method for preservation of the culture in question

- anticipated timescale for use of the cells

- number of vials of cells required for the purpose, for quality control and to cover any contingencies

- application of the cells, which, in the case of cells used for manufacture of products or generation of transplantable cells, will require compliance with stringent process control, safety testing and quality control protocols.

The selection of healthy cultures with high viability and in the exponential phase of growth is important [9]. Attempts to cryopreserve low viability or differentiated cultures may result in very poor viability on recovery, or even loss of the culture.

Cell banking should facilitate the long-term use of a cell line while minimizing the passage number of the cells used. Banking also promotes homogeneity of cell phenotype, limiting the performance drift of cells often seen when cells are passed from laboratory to laboratory. When a cell bank is made, it is often desirable to confirm that the cells have the desired characteristics and have not changed their properties during the banking process. Such quality control assessments may take the form of a variety of tests, not all of which would always be necessary. Table 6.1 lists the most common tests. The key tests cover identity, viability and sterility. Additional homogeneity tests provide information about uniformity of the cell bank. Tests might include cell counts, viability testing, growth

Table 6.1 Quality control testing of cell banks (for details of typical tests see reference 10 and Chapter 9).

Property	Test	Test sample
Viability	Trypan blue staining	Cell samples direct from cell bank
Identity	STR DNA profile (if human); *COI* DNA sequence (if not human) - see Chapter 9, Section 9.2.6.1	Washed cell pellet direct from cell bank vial
Sterility	Microbiological broth culture	Sample of cells grown from a bank vial without antibiotics (including use of specialist media and other tests for mycoplasma)
Karyology	Giemsa-banded metaphase	'Spreads' of a suspension of fixed cells from a colcemid-treated culture from a bank vial
Molecular karyology	Array-based comparative genomic hybridization	Genomic DNA from pre- versus post-cell bank cultures
Cell surface markers	Immunocytochemistry or flow cytometry	Cells from bank vial cultured on a glass slide, or cell suspension from bank vial QC culture
Gene expression	RT-PCR/'gene chip'	Cells from a bank vial cultured under standard conditions
Homogeneity	Cell proliferation or specific assay	Cells direct from cell bank vials

rates, or functional assays. Homogeneity testing should be performed on a significant proportion of vials in the cell bank. The size of this proportion should be decided upon by the user, but usually the larger the cell bank, the smaller the proportion required. The use of bank vials for testing purposes will obviously impact the size of the remaining cell bank; therefore, the proportion of vials required for testing should be taken into account when deciding the size of the cell bank.

The processes involved in the cryopreservation of cells, their interrelation, the protocols of this chapter in which they are described, and the associated issues and concerns are illustrated in Figure 6.1.

6.2.1 Preparation for cryopreservation

This section covers the various steps that must be performed to ready a cell preparation for freezing for cryopreservation. *Protocol 6.1* covers the selection and initial preparation of the cells, *Protocol 6.2* describes the necessary harvesting technique and *Protocol 6.3* goes through the addition of the cryoprotective agent and the subsequent aliquotting.

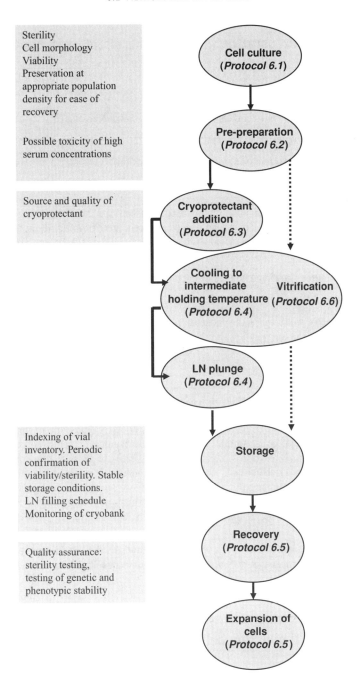

Figure 6.1 Cryopreservation flow diagram. The right-hand column gives a flow diagram of cryopreservation, highlighting key steps (*Protocols 6.1–6.6* are described in detail in the main text) for both a conventional and vitrification preservation regime. The left-hand column gives important supplementary considerations. (Adapted from reference 11.)

PROTOCOL 6.1   Preparation and selection of cell cultures for cryopreservation

Equipment and reagents

- Cell cultures in T-flasks, roller bottles or other suitable culture vessels

- Inverted microscope with phase-contrast optics and 10× and 20× objectives[a]

- Incubator (37 °C, with an atmosphere of CO_2 in air matched to the medium in use)

Method

1 Examine all cultures under the inverted microscope to evaluate the degree of confluency and general appearance[a]. In order to promote good recovery after freezing, do not use cultures that have been left to incubate under abnormal conditions or incubated after they have reached confluency.

2 Discard any cultures showing evidence of microbial contamination[b].

3 Discard all cultures showing significant morphological or behavioural abnormalities[b].

4 Return all selected cultures to the incubator while preparing to harvest the cells (*Protocol 6.2*).

Notes

[a] A range of cell morphologies may be observed during the normal healthy culture of a particular cell line and it is essential that the operator is familiar with these in order to be able to spot cells displaying abnormal morphologies. The appearance of cells with new, unusual morphologies may indicate problems with the culture system, transformation (possibly of viral origin) or the emergence of cellular cross-contamination as the contaminating cell population becomes established (see Chapter 4, Section 4.2.1.2, and Chapter 9). Use of phase contrast objectives will aid the operator in inspecting live cells. Use of an inverted light microscope, equipped with pre-centred phase plates to minimize the requirement for microscope adjustment, greatly simplifies the workflow. Additional objectives (4× and 40×) can be added to provide a convenient low-magnification view for screening of cultures and a high-magnification detailed view of individual groups of cells.

[b] It can be helpful to keep a record of cultures discarded and the reasons for this, as trends may appear with particular cultures if they are ageing or there is a contaminant in the original seed cultures.

PROTOCOL 6.2   Harvesting cell cultures for preservation

Equipment and reagents

- Selected cell cultures

- Sterile 0.25% trypsin/EDTA

(continued)

- Sterile Ca^{2+}- and Mg^{2+}-free PBS (CMF-PBS)
- Centrifuge (capable of achieving 80–100 g)
- Inverted microscope (with $20\times$ and $40\times$ objectives)
- Incubator (37 °C, with an atmosphere of CO_2 in air matched to the medium in use)
- Class II MSC
- Sterile pipettes, and hand or electric pipette pump
- Sterile 15- and 50-ml centrifuge tubes
- Culture medium (as recommended for the specific cell line but without antibiotics)[a]
- Isopropyl alcohol (IPA) spray and wipes
- Flask or pot for cell culture liquid waste.

Method

1 Switch on the Class II MSC some minutes before use and disinfect its work surfaces using the IPA spray and wipes[b].

2 Place the selected cultures FOR ONE CELL LINE ONLY in the Class II MSC along with the appropriate media and reagents required for harvesting[c].

3 For attached cells, drain the culture medium from each culture flask then add 5–10 ml of CMF-PBS to each flask, rinse the cell culture surface and discard the CMF-PBS to waste. For cells growing in suspension, transfer the cell suspension into sterile 50-ml centrifuge tubes and continue from step 7 below.

4 Add sufficient trypsin/EDTA to each flask to coat the cell monolayer (e.g. 1 ml for a 25 cm^2 T-flask). Ensure the trypsin/EDTA has washed over all the monolayer, then remove any excess and recap the culture vessel.

5 Incubate all vessels at 37 °C for up to 5 min, then check under the inverted microscope to ensure the cells have detached from the substrate[d,e].

6 Return flasks to the MSC and flush each with 5–10 ml of PBS and pool the cell suspensions in the minimum number of sterile centrifuge tubes.

7 Centrifuge the cell suspensions at approximately 100 g for 5 min.

8 Return the centrifuge tubes to the MSC, decant the supernatant and resuspend the cell pellets in the residual CMF-PBS by flicking the base of each tube.

9 Carefully resuspend and pool all pellets in one of the tubes using 10–20 ml of pre-warmed growth medium, and perform a cell count using a haemocytometer (Chapter 4, *Protocol 4.5*).

Notes

[a]It is good practice to avoid the use of antibiotics in cell culture unless cultures are at high risk of being contaminated, such as primary cell cultures. In addition, it is important to remember that, during the freezing process (*Protocol 6.4*), as external water freezes there is the potential for concentration of solutes in the unfrozen component of the ice matrix, and

(continued overleaf)

elevated concentrations of biologically active compounds such as antibiotics may have an adverse effect on the cells.

[b] For helpful guidance on the correct use of Class II MSCs see Section 4.2.1.1 of Chapter 4, and the Appendix 2 of reference 9.

[c] It is essential to only have one cell line in the MSC at any one time to avoid the possibility of cell line cross-contamination. All containers of media and reagents should be disinfected using a 70% IPA spray and wipes.

[d] For cultures grown without serum, a trypsin inhibitor (e.g. soybean trypsin inhibitor) must be employed to neutralize the trypsin once the cells have detached, or an alternative cell disaggregation method should be used. In such cases, if the supplier of the cell line can provide a suitable protocol, this should be adopted.

[e] If a few areas of cells remain attached the culture can be returned to the incubator and re-examined after another few minutes.

PROTOCOL 6.3   Addition of cryoprotectant and aliquotting of cell suspensions

Equipment and reagents

- Pre-warmed sterile growth medium

- Dimethylsulphoxide (USP or EP grade)[a]

- Cryovials (1.5-ml plastic with silicone washers and internal screw thread caps)

- 10- to 20-ml sterile pipettes

- Pipette pump

- IPA spray and wipes

- Flask or pot for cell culture liquid waste

Method

1 Having enumerated the viable cell concentration of the cell suspension in growth medium (see *Protocol* 6.2, above, and Chapter 4, *Protocol 4.5*), and working in the Class II MSC, dilute the suspension to a final cell concentration of 5×10^5 to 5×10^6/ml in ambient or pre-warmed medium containing sufficient dimethylsulphoxide to give a final concentration of 10% (v/v) in the cell suspension. A number of other common cryoprotectants are listed in Table 6.2[b].

2 If there is a large volume of suspension to cryopreserve, which could mean that the cells would be in DMSO for an extended period prior to freezing, it may be helpful to store the cells on ice both prior to aliquotting and after aliquotting but before freezing[c].

(continued)

3 Set out an appropriate number of pre-labelled cryovials microtubes in a rack (Figure 6.2)[d].

4 Aliquot the cells into the cryovials[e].

Notes

[a] DMSO should be obtained as a spectrally pure grade and stored at ambient temperature for no longer than 12 months.

[b] The optimum cell concentration for cryopreservation may vary with different cell lines, and advice should be sought from the supplier of the cells. However, typically cells may become damaged if frozen at concentrations higher than 5×10^6/ml. Equally, too low a cell concentration will also adversely affect recovery of viable cultures on thawing. Other cryoprotectants can be used in place of DMSO, with glycerol at 10% (v/v) being the most popular. However, cryoprotectants other than DMSO may not enter the cell as rapidly as DMSO, and incubation on ice in the presence of these crypotectants may be necessary before freezing.

[c] DMSO has a broad range of biological effects, which include induction of cell differentiation, and some cells are highly sensitive to its properties.

[d] Tube labelling should be done using an indelible marker or strongly adherent polyester labels that will survive handling, alcohol disinfection and storage at ultra-low temperature.

[e] Where there is a large number of vials to be aliquotted it may be necessary to cool the cell suspension and remix the cells with care periodically.

6.2.2 Freezing and thawing of cells

Once all the preparations have been completed, there follows the actual freezing process. The method employed for cells in vials is described in *Protocol 6.4*, and the corresponding thawing process is detailed in *Protocol 6.5*. An alternative freezing process, vitrification, which can be used in a number of situations where only small volumes of cells need to be frozen, is described (as used for hES cells) in *Protocol 6.6*.

Table 6.2 Common cryoprotectants.

Low molecular weight –OH solutes	Sugars	Polymers	Other compounds
Glycerol	Sucrose	Hydroxyethyl starch (HES)	Dimethyl sulphoxide (DMSO)
Ethylene glycol			
Propylene glycol	Raffinose	Polyvinyl pyrrolidone (PVP)	Acetamide
Methanol	Trehalose	Dextrans	
Ethanol	Sorbitol	Bovine serum albumin (BSA)	

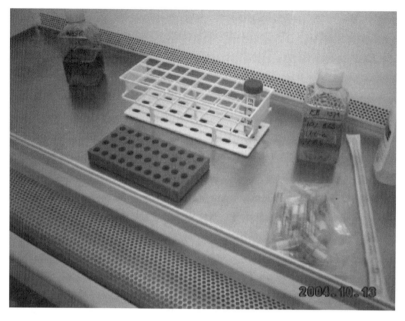

Figure 6.2 Preparing to freeze a small bank of cryopreserved cells in a Class II MSC (Photograph by G Stacey, courtesy of NIBSC-HPA).

PROTOCOL 6.4   Cryopreservation of aliquotted cells in vials

Equipment and reagents

- Device to cool cryovials (e.g. controlled rate freezer, (such as Cryo-7, Planer, UK); passive freezing device (e.g. passive freezing stage, Taylor Wharton; 'Mr Frosty'[TM], Invitrogen))[a]
- LN[b] and long-term storage refrigerator

Method

1 Prepare the freezing device(s), ensuring it/they are correctly set up at the starting temperature. Where appropriate, charge with sufficient LN to complete the cooling profile required.

2 Place cryovials in spaces in the freezing device[c].

3 Initiate cooling and, if possible with the equipment being used, monitor progress to ensure that the correct cooling profile is achieved. When cooling is complete, perform either step 4 or step 5.

4 Disconnect the device and transfer the cryovials to storage in the vapour phase of a LN refrigerator[d,e].

5 Disconnect the device and plunge the cryovials into fresh LN[f], then transfer the vials to storage in the vapour phase[g] of a LN refrigerator[e].

(continued)

Notes

[a] A variety of devices are available to provide carefully controlled cooling rates. These are all designed to try to achieve a moderately slow cooling rate of around 1 °C/min.

[b] Remember to always use the correct clothing, equipment and procedures to ensure the safe handling of LN (see Chapter 4, Section 4.2.8.1).

[c] Home-made freezing chambers can be made by placing cell vials in paper towelling and placing in one of the expanded-polystyrene boxes used for 500-ml containers, taping on the box lid and then leaving the box in a –80 °C freezer overnight.

[d] Holding cryopreserved vials at temperatures significantly above −130 °C for extended periods is not recommended. For example, many cell lines stored cryopreserved as above but at −80 °C will progressively lose viability, such that after a period of several weeks or months no viable cells may be recoverable on thawing.

[e] Many effective slow cooling regimes employ cooling at 1 °C/min to a 'safe' intermediate temperature (typically −70 °C), from which the ampoules can be directly transferred by plunging into LN, through the final glass transition temperature (at around –100 °C) to the storage temperature (in either the vapour[f] or liquid phase of nitrogen).

[f] Cells for clinical use should only be dunked into fresh LN which has not been in any contact with other biological materials.

[g] Although storage in the liquid phase is common practice in many (research) laboratories, storage in the vapour phase is essential if the cells or their products are intended for *in vivo* use, due to the potential for adventitious agents to move through the LN from one (poorly sealed) vial to another [12].

PROTOCOL 6.5 Recovery of cells from frozen storage, and quality control

Equipment and reagents

- Solid carbon dioxide pellets ('dry ice')

- Culture flasks

- Sterile growth medium (pre-warmed)

- Water bath or bead bath at 37 °C

- Sterile pipettes, and hand or electric pipette pump

- IPA spray and wipes

- Flask or pot for cell culture liquid waste

- Incubator (37 °C, with an atmosphere of CO_2 in air matched to the medium in use)

- Inverted microscope (with 20× and 40× objectives)

(continued overleaf)

Method

1 On the day of thawing, recover the required vial(s) from the cell bank storage location and store on dry ice (−78 °C) until needed.

2 Thaw one vial at a time by warming rapidly (in a 37 °C water bath or bead bath) until the ice has disappeared. Check that the vial's closure is still secure.

3 Dry the vial; then disinfect the outside of the vial by spraying with IPA before placing it in the Class II MSC. Wait for all the IPA to evaporate, then either transfer the vial contents directly to pre-warmed (pre-gassed where appropriate) culture medium in a T-flask or to pre-warmed culture medium in a centrifuge tube, centrifuge at 80–100 g for 5 min and then resuspend the cells in fresh pre-warmed culture medium and transfer to a T-flask pre-equilibrated with the relevant CO_2/air atmosphere[a].

4 Seal the flask and incubate in the cell culture incubator.

5 Examine the cells under the inverted microscope after a few hours to check the cells are refractile and that there is no evident contamination. For adherent cultures, check that the cells are beginning to attach to the culture surface and appear free of microorganisms[b,c].

6 When the cells have been in culture at least overnight, check under the microscope that the culture has typical morphology and remains free of microorganisms.

7 Expand the culture by passage without antibiotics, to provide appropriate cells for quality control. Examples of the tests that should be employed, and the corresponding test samples, are given in Table 6.1. The quality control of cell lines is discussed at length in Chapter 9.

8 All quality control and safety data should be collated and reviewed before a particular cell line is made available for general use either within or outside an organization.

Notes

[a] DMSO has a broad range of biological effects, which include induction of cell differentiation, and some cells are highly sensitive to its properties. Following thawing, the cryoprotectant should be removed (unloaded) and the cell culture expanded. The cryoprotectant can be immediately diluted out of the cell suspension by adding fresh tissue culture medium (e.g. 20 ml to 1 ml of cell suspension). Alternatively, the cryopreservative can be removed by centrifugation of the thawed cell suspension and resuspension of the cells in fresh medium. Cryopreserved cells following thawing are particularly susceptible to osmotic injury, which can be reduced through the adoption of stepwise unloading, whereby a small amount of fresh medium is added to the thawed cryovial to dilute the cryoprotectant by c. 50%. The partially unloaded cells can then be further diluted through resuspension in fresh medium as they are transferred to a culture flask. Whichever specific procedure is adopted the benefit of removal of the cryoprotectant must be weighed against the susceptibility of the cells in question to loss of viability if centrifuged while in a potentially fragile state.

[b] Motile microorganisms may be seen easily under the inverted cell culture microscope. Non-motile bacteria will be more difficult to identify due to cell debris and Brownian motion, which can make non-viable particles appear to be motile to the inexperienced observer.

[c] Some cell lines may take longer than others to attach, and in certain cases overnight incubation may be required before attachment is seen. The experienced operator should know what is normal for the cell line in question.

PROTOCOL 6.6   Vitrification of hES cells[a, b]

Equipment and reagents

• Routine culture equipment for hES line

• 5-ml cryovial to hold straws

• Sterile pipettes

• Nunc four-well plate (cat. No.176740)

• 1- to 2-l LN Dewar vessel

• Cryocanes

• Cutting tool [fine drawn glass pipette tip, needle or other manipulator allowing scoring, pick-up and transfer of colonies (specific tools are often operator specific, however dissecting needles (Agar Scientific UK Ltd) and IVF micromanipulator tools (Hunter Scientific UK Ltd) are commonly used)

• Vitrification straws

• Vitrification solutions (see below)

Solutions for vitrification

Solution 1: ES-HEPES medium

	Volume	Supplier	Cat. No.
DMEM	15.6 ml	Invitrogen	21969-035
Fetal bovine serum (FBS)	4 ml	—	—
1M HEPES	0.4 ml	Invitrogen	15630-049
Total volume = 20 ml			

Solution 2: 1 M sucrose solution (filtered)

	Volume	Supplier	Cat. No.
Sucrose powder	3.42 g	Sigma	S7903
ES-HEPES medium	8 ml	—	—
FBS	2 ml	—	—
Total volume = 10 ml			

(continued overleaf)

10% vitrification solution (VS1)

	Volume	Supplier	Cat. No.
ES-HEPES medium	2 ml	—	—
Ethylene glycol	0.25 ml	Sigma	E9129
DMSO	0.25 ml	BDH	103234L
Total volume = 2.5 ml			

20% vitrification solution (VS2)

	Volume	Supplier	Cat. No.
ES-HEPES medium	0.75 ml	—	—
1 M sucrose solution	0.75 ml	—	—
Ethylene glycol	0.5 ml	Sigma	E9129
DMSO	0.5 ml	BDH	103234L
Total volume = 2.5 ml			

Method

1 Prepare the vitrification solutions listed above.

2 To a Nunc four-well plate add 1 ml of ES-HEPES to well 1, 1 ml of VS1 to well 3 and 1 ml of VS2 to well 4. Incubate at 37 °C, 5% CO_2 for at least 1 h (Figure 6.3).

3 Transfer the prepared vitrification plate and the dishes of hES cells for cryopreservation to a dissecting microscope in a vertical unidirectional flow cabinet or (preferably) a Class II MSC.

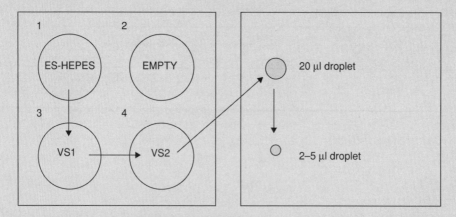

Figure 6.3   Preparing hES lines for vitrification.

(continued)

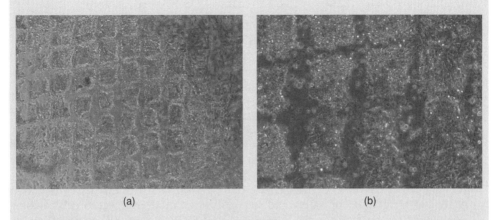

(a)   (b)

Figure 6.4 Colonies of hES cells on feeder layer, 'gridded' prior to subculture/preparation for vitrification. Showing colonies cut into a 2–3-mm^2 grid. (a) Low magnification, (b) high magnification (Photograph by C Machado, courtesy of NIBSC-HPA).

4 Use a cutting tool such as a fine drawn glass pipette tip to score across the undifferentiated colonies, dissecting them into four to six pieces (slightly larger than the 10–20 pieces used for subculture) (Figure 6.4).

5 Transfer colony pieces to well 1 of the vitrification plate. Leave for 1 min.

6 Meanwhile, transfer individual 20-μl drops of VS2 from well 4 of the plate to the inside of the four-well plate lid. The number of drops required is the same as the number of colony pieces to be frozen.

7 Transfer colony pieces from well 1 to well 3 of the plate. Leave for 1 min.

8 After 1 min, transfer the colony pieces in 10- to 20-μl of liquid, from well 3 to well 4. Leave for 25 s.

9 After 25 s, transfer each colony piece, in the smallest volume possible, to a separate 20-μl drop of VS2 on the plate lid.

10 Pick up the colonies in a 2- to 5-μl drop, and place this drop onto the plate lid.

11 Holding a vitrification straw at a 30° angle to the plane of the lid, immediately touch the narrow end of the straw to the side of the 2- to 5-μl drop. The droplet should be drawn up by capillary action.

12 Immediately plunge the end of the straw into LN.

13 Transfer the straw to pre-cooled 5-ml vials in the LN. Place 5-ml vials (labelled with cell name, date, passage level) onto cryocanes in LN, in a 1- to 2-l Dewar vessel.

Transfer vials in Dewar to LN cell storage refrigerator for long-term storage.

(*continued overleaf*)

> ***Notes***
>
> [a] Protocol courtesy of UK Stem Cell Bank (UKSCB)/National Institute for Biological Standards and Control (NIBSC)/Health Protection Agency (HPA).
>
> [b] The outcome of producing vitrified straws from hES cultures is highly variable and largely dependent on operator skill/experience. Straws vitrified on the same day, from the same cultures can show very different viabilities and levels of differentiation on thawing. The less experience someone has, the less likely they are to obtain viable cells on thawing of the straws. The use of standard freezing protocols (where possible) is likely to be preferable, especially for those new to stem cell culture; there is liable to be a greater chance of recovering viable cells, and there may be greater homogeneity between frozen vials.

For general considerations on safety and standardization issues relating to cell banking see references 13, 14 and 15. Further discussion of the principles of cell cryopreservation and thawing can be found in reference 11.

6.2.3 Considerations for thawing cryopreserved material

The most common method of warming is to plunge vials containing the cells into a water bath at 37 °C, with gentle shaking to avoid local thermal gradients being formed in the water bath. For samples of around 1 ml volume, this restricts the maximum warming rates to a few hundred degrees Celsius per minute. The importance of warming rates stems from the necessity to thaw the sample of cryopreserved cells through temperatures where biophysical events may occur: a glassy matrix 'devitrifies' at around -100 to -80 °C; water molecules begin to mobilize within the ice–solute matrix above -60 °C; and ice recrystallization becomes significant (particularly inside the cells) above -40 °C. These temperature ranges are only approximate, but physical events dictated by the phase diagram of the particular ice/solute/CPA mixture in a given cell sample are unavoidable. Warming rates that can be achieved in routine cell banking are on 'the edge' of potential failure, reinforcing the need for care and vigilance in all steps of the thawing protocol [11]. The need for rapid warming is particularly acute where samples have been stored using vitrification protocols. Thawing should be achieved rapidly by transferring the cryovial of vitrified straws to a Dewar vessel containing LN, and when a culture vessel of pre-warmed culture medium is ready, directly dip the end of one straw under the surface of the medium and as soon as the vitrified cells are thawed expel them (using a Gilson pipette fitted with a filter P20 tip, if they do not flow out unassisted) into the culture medium. Practical cell vitrification methods are limited to 'quasi-vitrification'; establishment of a low temperature glass within which are a potentially high number of ice nucleation centres, and in this situation thawing must be rapid.

6.2.4 Safety issues in cell banking

The use of cryogens (e.g. LN) has greatly increased the application of cryobanking for cell resources. However, considerations of safety, and the potential for cross-infection between individual cell containers (e.g. vials), need to be foremost in the minds of those involved in cryopreservation activities. Obvious dangers when handling cryopreserved specimens relate to skin 'burns' resulting from touching extremely cold materials, or potential explosions of vials or containers caused by rapid expansion of LN into the gas phase on warming. These problems can be readily overcome as long as correct use of safety gloves and eye protection is routinely employed and the risk of accident mitigated through the use of appropriate techniques and equipment. Where large volumes of LN are used in single or multiple containers, consideration must be given to potential oxygen deprivation in the local atmosphere if the liquid cryogen spills and vaporizes. Good ventilation and use of an oxygen depletion monitor in the storage room should minimize any risk to staff. However, provision of personal oxygen depletion monitors can provide additional warning of reduced oxygen levels. Another issue, perhaps not so obvious, is the need to ensure that samples stored at low temperatures cannot be infected with agents, especially viral agents, inadvertently released into the storage tanks from other (infected) samples [12]. For further information on infectious hazards from cell lines see reference 14.

6.3 Troubleshooting

- Trypsinization is often a cause of loss of cells prior to preservation. If the cell monolayer remains largely intact after extended incubation in trypsin/EDTA, then additional fresh trypsin/EDTA should be applied. If this is unsuccessful, and this problem has not been experienced with that cell line previously, then the source of the trypsin/EDTA should be checked and the solution replaced, or it may be that the culture has been allowed to become confluent or to remain in the confluent state for too long, in which case the culture should be abandoned. If this is the first occasion that the particular cell line has been passaged, it may be that an alternative disaggregation method may be needed and the supplier of the cells should be contacted for further information. It is good practice not to cryopreserve cell cultures that have become confluent as cells in the subconfluent state are more likely to be in the exponential phase of growth with a low cytoplasm–nucleus volume ratio and thus more amenable to preservation. Confluent cells may also have begun to undergo differentiation, and recovered cultures may not be representative of the original [16].

- Any occurrence which indicates that the cells are not performing as they should on a number of occasions should prompt an investigation to ensure that the cells are really what they are supposed to be. As a first line of investigation, any genetic profiling data should be compared with other sources of the same cell line (and the actual tissue of origin from the original donor, if available) to check whether cell line cross-contamination has occurred.

- Insufficiently rapid thawing can frequently lead to low viable cell yields. Do not forget to shake the vials while thawing in order to minimize the formation of thermal gradients and thus maximize the rate of heat transfer. On the other hand, shaking that is too vigorous can also cause problems.

- Although DMSO is a good CPA to use initially when trying to define suitable conditions for the cryopreservation of a particular cell line, it is not suitable for all cell lines. Some alternative CPAs are given in Table 6.2. In particular, cell lines that differentiate in the presence of DMSO must be frozen in a different CPA, such as glycerol (which is probably the second most commonly used CPA after DMSO). However, remember that glycerol and a number of other CPAs do not enter cells as quickly as DMSO, so the cells may need to be incubated on ice in the presence of the CPA prior to freezing.

- Other things that can be tried in order to optimize cell cryopreservation are:

  - adjusting the concentration of the CPA

  - using a controlled-rate freezing device to more closely control freezing conditions

  - changing the rate and/or temperature profile of the freezing step

  - changing the approach taken to diluting and removing the CPA following thawing, by (for example)

    - using slow (drop wise) dilution with medium

    - removing the CPA immediately after dilution by the use of centrifugation

    - avoiding centrifugation if this damages fragile cells

    - diluting the CPA to a greater extent (if not removed immediately).

Acknowledgement

We would like to thank Lesley Young of the UKSCB for her valuable input and advice on the cryopreservation of hES lines.

References

★ 1. Polge, C., Smith, A. and Parkes, A. (1948) Revival of spermatozoa after vitrification and dehydration at low temperatures. *Nature (London)*, **164**, 666–667. – *First published report of the successful use of a cryoprotectant to facilitate successful freezing and thawing of mammalian cells:*

★★ 2. Pegg, D.E. (2007) in *Cryopreservation and Freezedrying Protocols*, 2nd edn (J.G. Day and G.N. Stacey), Humana Press, Totowa, USA, pp. 39–57. – *Good review of fundamental principles.*

3. Luyet, B. and Gehenio, P.M. (1940) *Life and Death at Low Temperatures*, Biodynamica Press, Normandy, MO, USA.

4. Bernard, A. and Fuller, B. (1996) Cryopreservation of human oocytes: a review of current problems and perspectives. *Hum. Reprod. Update*, **2**, 193–207.

5. Chen, S-U., Lien, Y-R., Chen, H-F. *et al.* (2000) Open pulled straws for vitrification of mature mouse oocytes preserve patterns of meiotic spindles and chromosomes better than conventional straws. *Hum. Reprod.*, **15**, 2598–2603.

6. Reubinoff, B.E., Pera, M.F., Vajta, G. and Trounson, A.O. (2001) Effective cryopreservation of human embryonic stem cells by the open pulled straw vitrification method. *Hum. Reprod.*, **16**, 2187–2194.

★ 7. Hay, R.J. (1988) The seed stock concept and quality control for cell lines. *Analyt. Biochem.*, **171**, 225–237. – *Early reference on the quality control testing of tiered cell banks.*

8. Knezevic, I., Stacey, G., Petricciani, J. and Sheets, R. (2010) Evaluation of cell substrates for the production of biologicals: Revision of WHO recommendations: Report of the WHO Study Group on Cell Substrates for the Production of Biologicals, 22–23 April 2009, Bethesda, USA. *Biologicals*, **38**, 162–169.

★★★ 9. Coecke, S., Balls, M., Bowe, G. *et al.* (2005) Guidance on good cell culture practice: a report of the second ECVAM task force on good cell culture practice. *Altern. Lab. Anim.*, **33**, 1–27. [This can also be downloaded from http://www.esactuk.org.uk/ . Click on the Best Practice (GCCP) tab.] – *Essential reading; key guidance document on Good Cell Culture Practice.*

10. Stacey, G.N. (2007) Quality control of human stem cell lines, in *Culture of Embryonic Stem Cells* (eds J. Masters, B. O. Palsson and J. Thomson), Springer, Dordrecht, The Netherlands, pp. 255–275.

★★ 11. Fleck, R.A. and Fuller, B. (2007) Cell preservation, in *Medicines from Animal Cell Culture* (eds G.N. Stacey and J.M. Davis), John Wiley & Sons, Ltd, Chichester, UK, pp. 417–432. – *Review of the fundamental principles of cell freezing.*

12. Tedder, R.S., Zuckerman, M.A., Goldstone, A.H. *et al.* (1995) Hepatitis B transmission from contaminated cryopreservation tank. *Lancet*, **346**, 137–140.

13. ICH Q5A (R1) Viral Safety Evaluation of Biotechnology Products Derived from Cell Lines of Human or Animal Origin. http://www.ich.org/LOB/media/MEDIA425.pdf (Accessed September 2010).

★★ 14. Stacey, G. (2007) Risk assessment of cell culture procedures, in *Medicines from Animal Cell Culture* (eds G.N. Stacey and J.M. Davis), John Wiley & Sons, Ltd, Chichester, UK, pp. 569–588. – *Good general review of risk assessment in cell culture.*

★★ 15. Stacey, G. (2007) Standardization of cell culture procedures, in *Medicines from Animal Cell Culture* (eds G.N. Stacey and J.M. Davis), John Wiley & Sons Ltd, Chichester, UK, pp. 589–603. – *Good general review of cell standardization procedures.*

16. Terasima, T. and Yasukawa, M. (1977) Dependence of freeze–thaw damage on growth phase and cell cycle of cultured mammalian cells. *Cryobiology*, **14**, 379–381.

7

Primary Culture of Specific Cell Types and the Establishment of Cell Lines

Kee Woei Ng[1], Mohan Chothirakottu Abraham[2], David Tai Wei Leong[3], Chris Morris[4] and Jan-Thorsten Schantz[5]

[1] Division of Materials Technology, School of Materials Science & Engineering, Nanyang Technological University, Singapore
[2] John Stroger Hospital of Cook County, Chicago, USA
[3] Cancer Science Institute, National University of Singapore, Singapore
[4] Health Protection Agency Culture Collections, Centre for Emergency Preparedness and Response, Porton Down, Wiltshire, UK
[5] Department of Plastic and Hand Surgery, Klinikum rechts der Isar der Technischen Universität München, Munich, Germany

7.1 Introduction

Cells of all tissues derive from one of the three distinct primary embryonic germ layers of mesoderm, ectoderm and endoderm. It is essential when culturing primary mammalian cells to know where a particular cell type originates from, as it gives information about cell characteristics and physiological behaviour. Cells originating from the mesoderm include the mesenchymal lineages (osteogenic, chondrogenic, adipogenic and myogenic cells), the urogenital cells of kidney, the circulatory system and the connective tissue of the skin. Cells from the ectoderm include the epithelial lining of the skin, gut and neural tissue. Endodermal cells include those from the liver (hepatocytes) and the pancreas, and

Animal Cell Culture: Essential Methods, First Edition. Edited by John M. Davis.
© 2011 John Wiley & Sons, Ltd. Published 2011 by John Wiley & Sons, Ltd.

the respiratory tract. Protocols for culturing representative cell types from the three germ layers are described in this chapter.

When a primary culture from a tissue sample is prepared, it is often heterogeneous in nature. For example, in a keratinocyte culture it is common to find some fibroblasts, endothelial cells or melanocytes. Such heterogeneity is generally not a problem but should be taken into consideration when designing experiments to study specific characteristics of primary cells. The use of media formulations known to be selective for the specific cell type of interest may help to eliminate other cell types.

Primary cells generally have a limited lifespan in culture. It is therefore important that a proper experimental plan be drafted even before the primary tissue is obtained. This plan should be designed to prevent the possibility of problems occurring, such as running out of the cells necessary to continue or repeat an experiment. A common strategy to maintain a primary cell line for long-term studies is to immortalize it (see Section 4.1.1.2.*viii* of Chapter 4), such as by transfection with telomerase reverse transcriptase (TERT) using viral vectors [1].

The primary culture of a selection of different cell types is described below. With advances over the years it has become possible to culture an ever-increasing number of different cell types, and it is beyond the scope of this book to describe them all. However, the protocols given here illustrate some of the approaches that can be taken. These should serve as an introduction and background for the reader interested in a particular cell type, who may then need to obtain more details from the latest literature to meet specific needs. The reader is also encouraged to refer to a number of established cell culture textbooks for more comprehensive understanding of the topic [2–9].

7.2 Methods and approaches

All reagents and equipment described in this chapter and that come into contact with cells should be sterile, and all reagents (as far as possible) should be cell culture tested. Understanding of the principles of good practice, general laboratory safety and biosafety, none of which is explicitly covered in this chapter, are important prerequisites before primary cell culture work is carried out (Figure 7.1). Readers are expected to acquire these from Chapter 4 and other sources before attempting to carry out any of the procedures described here. All human tissues should be handled as hazardous material because of the potential risk of adventitious virus transmission (Chapter 4, Section 4.2.8.2).

It is important to note that, in most countries, ethical approval will be required from the local Ethics Committee or Institutional Review Board before work can be started to obtain tissues from either humans or animals for any research or diagnostic purposes. In addition, for human tissues, informed consent MUST – for most purposes – be obtained from the donor (or in the case of a cadaver, the appropriate relative(s)) before tissue samples can be taken or retained. In many countries this is covered by law, with severe penalties for transgression. For example, in England and Wales such activities fall under the Human Tissue Act 2004 (or for certain materials, the Human Fertilisation and Embryology Act 2008) and throughout the European Union are covered by the EU Tissues and Cells Directives. Make sure you are aware of current requirements in your locality BEFORE starting work.

Figure 7.1 Processing a fresh tissue biopsy within a Class II microbiological safety cabinet. The priority should always be to ensure sterility of the biopsy, especially while transporting it between the operating theatre and the cell culture laboratory.

7.2.1 Primary culture of mesenchymal cells (mesoderm)

The cells from the mesoderm form connective tissues such as bone, cartilage, fat and muscles as well as endothelium. These cells are usually spindle shaped and are anchorage dependent. They are usually quite robust and relatively easy to culture. In a co-culture system they tend to be the dominant cell type. Primary mesenchymal cells can be subcultured for multiple passages before senescence sets in. Mesenchymal cells can be isolated via enzymatic or explant methods [7]. Methods are described below for the isolation and culture of adipocytes (*Protocol 7.1*), chondrocytes (*Protocol 7.2*), dermal fibroblasts (*Protocol 7.3*), osteoblasts (*Protocol 7.4*), and human umbilical vein endothelial cells (HUVECs) (*Protocol 7.5*).

7.2.1.1 Adipocytes

PROTOCOL 7.1   Isolation and growth of adipocytes

Equipment and reagents

- Adipose tissue biopsy or liposuction aspirate (Figures 7.2 and 7.3)
- DMEM containing 4500 mg/l glucose, supplemented with 10% FBS, 100 U/ml penicillin and 100 µg/ml streptomycin

(continued overleaf)

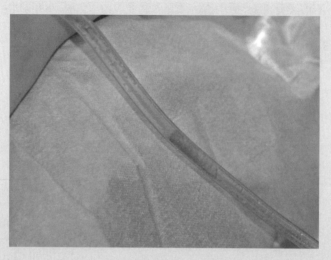

Figure 7.2 Collection of liposuction aspirates in a routine liposuction procedure. Liposuction aspirates are collected in sterile bottles and transferred to the cell culture laboratory to be processed as soon as possible to ensure maximum cell viability.

- Ca^{2+} and Mg^{2+}-free phosphate-buffered saline (CMF-PBS)
- 0.075% collagenase A (Roche) in CMF-PBS
- 160 mM NH_4Cl (Sigma) in CMF-PBS
- Sterile pipettes, micropipettes, scissors, beaker, magnetic stirrer bar, centrifuge tubes, cell culture flasks (75 cm^2 or bigger, depending on number of cells obtained)
- Magnetic stirrer base
- Vortex mixer

Method

1 Wash adipose tissue with CMF-PBS to remove blood.

2 Cut up adipose tissue with sterile scissors to obtain tissue pieces that are as small as possible.

3 Transfer tissue pieces into a sterile 50-ml tube (fill up to the 10-ml mark) and vortex in 20–30 ml of sterile CMF-PBS to further encourage tissue disaggregation.

4 Decant and discard the liquid and vortex again with 20–30 ml of fresh CMF-PBS.

5 Repeat step 4. Again decant and discard the liquid.

6 Add 0.075% collagenase A solution (enough to allow tissue pieces to float freely), transfer to a sterile beaker along with a sterile stirrer bar, cover and stir using a magnetic stirrer for 30 min at 37 °C.

7 Pour the digested mixture into a centrifuge tube leaving behind large tissue debris, and centrifuge at 200 g for 10 min. Decant and discard the supernatant.

(continued)

8 Add 10 ml of 160 mM NH_4Cl to the pellet, and incubate for 10 min at room temperature to lyse red blood cells.

9 Centrifuge at 200 g for 10 min, decant and discard the supernatant.

10 Resuspend pellet with a micropipette using 1 ml of complete DMEM.

11 Plate in complete DMEM at 5000–10 000 cells per cm^2 in 75 cm^2 tissue culture flasks, and place in a humidified incubator at 37 °C and 5% CO_2.

12 Change medium for the first time after 3–4 days and every 2 days subsequently.

13 Subculture at 80% confluence.

The typical morphology of an adipocyte culture is shown in Figure 7.3.

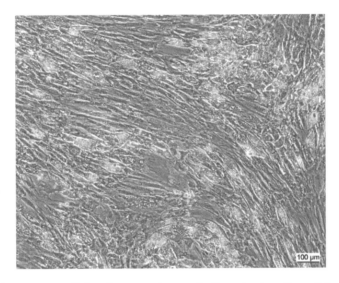

Figure 7.3 Phase contrast light microscopy image of adipocytes in culture. Fat vacuoles within mature adipocytes appear as clear round globules that can be clearly visualized by staining with Oil Red O.

7.2.1.2 Chondrocytes

In the body, there are three different types of cartilage: elastic (ear, nose), fibrous (rib, intervertebral disc) and hyaline (articular). The majority of the cartilage consists of extracellular matrix (ECM) built up from collagen and glycosaminoglycans (GAGs). The most reliable and efficient way to obtain a chondrocyte culture from a cartilage biopsy is via enzymatic digestion. Although an explant culture system also works, a much longer time is needed and usually fewer cells are obtained. The typical morphology of a chondrocyte culture obtained from enzymatic digestion is shown in Figure 7.4.

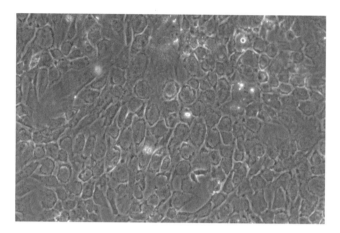

Figure 7.4   Phase contrast light microscopy image of a confluent monolayer of chondrocytes in culture, showing typical polygonal morphology. (Magnification x200)

PROTOCOL 7.2   Isolation and growth of chondrocytes

Equipment and reagents

- Cartilage biopsy

- Ham's F-12 Medium supplemented with 10% FBS, 100 U/ml penicillin, 100 µg/ml streptomycin, 25 µg/ml ascorbic acid and 2 mM L-glutamine

- Ca^{2+} and Mg^{2+}-free phosphate-buffered saline (CMF-PBS) containing 200 U/ml penicillin and 200 µg/ml streptomycin

- 2 mg/ml Collagenase Type II (Sigma) in Ham's F-12 medium (without supplements) or CMF-PBS

- 90-mm sterile Petri dishes

- Sterile pipettes, micropipettes, forceps and scalpel, cell culture flasks ($75\,cm^2$ or bigger, depending on number of cells obtained)

- 50-ml sterile centrifuge tubes

- 70-µm cell strainers (Falcon©, Becton Dickinson)

Method

1 Collect the freshly harvested cartilage pieces in CMF-PBS containing 200 U/ml penicillin and 200 µg/ml streptomycin. Rinse two times in CMF-PBS with penicillin and streptomycin, discarding the washes.

2 Place the cartilage pieces in a sterile 90 mm Petri dish, and using the sterile forceps and scalpel, mince into chips of 1–2 mm.

3 Transfer the chips into 50-ml tubes and add 10–15 ml of 2 mg/ml collagenase solution.

(continued)

4 Incubate the chips at 37 °C and 5% CO_2 for 16–24 h depending on the chip size.

5 Transfer the entire mixture into 50-ml centrifuge tubes.

6 Add 20–30 ml of fresh culture medium (containing serum).

7 Vortex each tube for 5 min.

8 Filter the suspension through 70-μm nylon mesh and discard the retained material.

9 Centrifuge the cells at 200 g for 10 min.

10 Wash. the pelleted cells by resuspending them in complete medium and centrifuging as in step 9. Perform this washing step three times in total.

11 Resuspend the pellet and plate in complete DMEM at 5000–10 000 cells per cm^2 in 75-cm^2 tissue culture flasks, and place in a humidified incubator at 37 °C and 5% CO_2.

12 Change medium for the first time after 3–4 days and every 2 days subsequently.

13 Subculture at 80% confluence.

7.2.1.3 Dermal fibroblasts

Fibroblasts (Figures 7.5 and 7.6) are generally regarded as the easiest mammalian cells to culture. They are relatively undifferentiated, and consequently the use of special reagents to induce differentiation is unnecessary. In addition, fibroblasts are robust and will attach and proliferate on a wide variety of culture vessels and biomaterials. In *Protocol 7.3*, the isolation and maintenance of fibroblasts derived from human dermis is described.

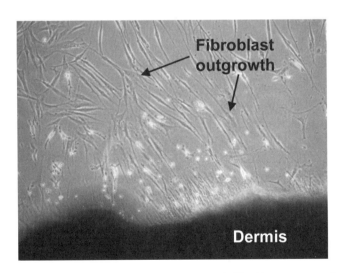

Figure 7.5 Phase contrast light microscopy image of fibroblast outgrowth from an explant culture of dermis. Using *Protocol 7.3*, fibroblasts should appear between 1 and 2 weeks after establishing the explant cultures. (Magnification: x100)

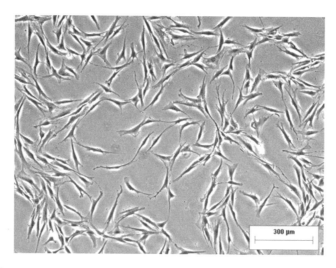

Figure 7.6 Phase contrast light microscopy image of isolated dermal fibroblasts in culture, showing typical spindle morphology.

PROTOCOL 7.3   Isolation and growth of dermal fibroblasts

Equipment and reagents

- Skin biopsy

- Medium: DMEM (4500 g/l glucose) supplemented with 10% FBS, 100 U/ml penicillin and 100 μg/ml streptomycin

- 2 mg/ml Collagenase type I (Gibco©, Invitrogen) in serum-free DMEM or PBS without Ca^{2+} and Mg^{2+} (CMF-PBS)

- 100-μm cell strainers (Falcon©, Becton Dickinson)

- Sterile pipettes, micropipettes, scissors, forceps and scalpel, cell culture flasks (75 cm² or bigger, depending on number of cells obtained)

- 50-ml sterile centrifuge tubes

Method

A. Explant method

1 Separate epidermis and dermis either by mechanical means using sterile forceps and scissors or by enzymatic digestion (see *Protocol 7.6*).

2 Cut dermis into pieces not bigger than strips of 2–3 mm × 5 mm.

3 Transfer samples into culture flasks or Petri dishes containing just enough medium to wet the entire vessel surface so that samples do not float.

4 Distribute samples over the culture surface to maximize the separation between pieces.

(continued)

5 Leave the culture undisturbed in the incubator (37 °C, 5% CO_2) for 3–5 days.

6 Change medium for the first time after 3–5 days and every 2 days subsequently.

7 Subculture at 80% confluence.

B. Enzymatic digestion method

1 Separate epidermis and dermis either by mechanical means using sterile forceps and scissors or by enzymatic digestion (*see Protocol 7.6*).

2 Mince dermal samples manually to obtain pieces that are as small as possible and transfer into a Petri dish or multiwell plate containing 2 mg/ml collagenase solution.

3 Incubate overnight at 37 °C, 5% CO_2.

4 Filter supernatant through a 100-μm cell strainer to remove tissue debris.

5 Centrifuge at 200 g for 10 min and plate in compete medium to give the desired cell population density (typically 1000–4000 viable cells/cm² for routine culture) in culture flasks.

6 Change medium for the first time after 3–5 days and every 2 days subsequently.

7 Subculture at 80% confluence.

7.2.1.4 Osteoblasts

PROTOCOL 7.4 Isolation and growth of osteoblasts

Equipment and reagents

• Biopsies of long bones of animal or human origin. Both trabecular (cancellous) or cortical bone will work (Figures 7.7 and 7.8).

• Media: Medium 199, BGJb, DMEM (4500 g/ml glucose), DMEM:Ham's F12, or McCoy's 5A, supplemented with 10% or 15% FBS, 100 U/ml penicillin, 100 μg/ml streptomycin and 2 mM L-glutamine

• PBS without Ca^{2+} and Mg^{2+} (CMF-PBS)

• Bone digest solution: 200 U/ml Collagenase (Worthington Biochemicals CLS-2) and 2 mg/ml trypsin (Sigma T4799) in CMF-PBS. Adjust to pH 7.2–7.5, filter sterilize, divide into 10-ml aliquots and store at −20 °C

• 70-μm or 100-μm cell strainers (Falcon©, Becton Dickinson)

• Sterile equipment: bone rongeur or strong scissors, Petri dishes, 25 and 75 cm² culture flasks, 50-ml centrifuge tubes, scalpel and No. 22–25 blades, forceps.

• 0.25% Trypsin/EDTA solution

(continued overleaf)

Method

1 Collect bone biopsies in CMF-PBS or another physiological saline solution.

2 Clean off any soft tissue from the outer surfaces and rinse in fresh CMF-PBS.

3 Select a whole bone and cut lengthways to open the cavity. Cut transversely into 5-mm pieces using a scalpel or strong scissors/bone rongeur for larger bones.

4 Reduce bone pieces in size to 2–3 mm and transfer to a 50-ml tube with 10 ml of growth medium without serum.

5 Vortex for 10 s and allow the pieces to settle.

6 Decant the medium and add another 10 ml of growth medium without serum.

7 Mix by swirling and let the pieces settle. Steps 5–7 should be repeated until the bone is free of marrow.

8 Establish a culture via explant, enzymatic digestion or wash out procedure (see below).

A. Explant method

1 Mince the samples manually into smaller pieces and transfer into a 75-cm^2 flask or 90-mm Petri dish. Use sufficient pieces to allow space for outgrowths of cells, i.e. cover no more than 20% of the surface evenly with bone pieces.

2 Add sufficient growth medium with serum so that all the pieces of bone are completely submerged. Incubate at 37 °C, 5% CO_2.

3 Leave the flask or dish undisturbed for 5–7 days before checking for outgrowth.

4 Replace the medium and double the volume when outgrowth is observed.

5 Check and replace the medium every 7 days until the culture is confluent. Cells exhibiting fibroblast or polygonal morphology may be seen in cultures.

6 Subculture at 80% confluence.

B. Enzymatic digestion

1 Mince the samples manually into smaller pieces and transfer to a 50 ml tube.

2 Wash in 20 ml of CMF-PBS and allow the pieces to settle.

3 Repeat the wash and decant the CMF-PBS.

4 Add 5 ml of bone digest solution and incubate the tube in a water bath at 30 °C for 15 min. Mix by swirling the tube every 5 min.

5 Discard the supernatant and add another 5 ml of bone digest solution.

6 Maintain at 30 °C for 15 min, mixing every 5 min.

7 Collect the supernatant in a fresh 50-ml tube and add 10 ml of growth medium.

8 Add another 5 ml of bone digest solution to the remaining bone pieces and incubate at 30 °C for 15 min.

(continued)

9 Centrifuge the first supernatant collected in step 7 at 80–100 g for 3 min, decant the supernatant and resuspend the cell pellet in 5 ml of growth medium.

10 Count and inoculate in 75-cm^2 flasks or larger, at between 5000 and 50 000 cells/cm^2.

11 Collect the second digest (from step 8) and repeat steps 9–10 to collect more cells.

12 Digestion of the remaining bone pieces (steps 8-10) can be repeated for up to 1–2 h from the first digest.

13 Incubate the flasks at 37 °C, 5% CO_2 until 90% confluence and subculture using trypsin/EDTA (see, for example, Chapter 4 *Protocol 4.3*).

C. Wash-out method

1 Mince the sample manually into smaller pieces and transfer into a 50-ml centrifuge tube.

2 Add 15 ml of culture medium and vortex the sample on a vortex mixer for 5 min.

3 Take the supernatant and centrifuge at 200 g for 10 min.

4 Discard the supernatant and resuspend the pellet in culture medium.

5 Count and inoculate into 75 cm^2 flasks or larger, at between 5000 and 50 000 cells/cm^2 and incubate at 37 °C, 5% CO_2.

6 The first medium change can be performed after 5 days when the original medium starts to turn orange. Subsequent medium changes should be carried out every 2–3 days.

7 Subculture at 90% confluence using 0.25% trypsin/EDTA.

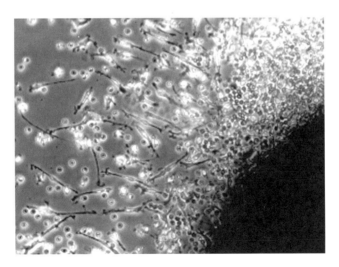

Figure 7.7 Phase contrast light microscopy image of an explant culture of bone, showing outgrowth of spindle-shaped osteoblasts and numerous red blood cells that remain round and non-adherent. (Magnification: x100)

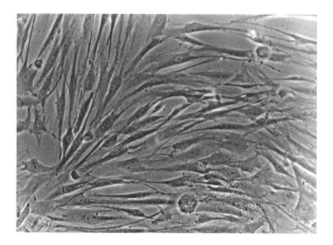

Figure 7.8 Phase contrast light microscopy image of mature osteoblasts in culture. Osteoblasts are typically spindle shaped and similar to fibroblasts in appearance. (Magnification: x200)

7.2.1.5 Umbilical vein endothelial cells

The method described in *Protocol 7.5* is a relatively simple procedure that has the advantage of providing large numbers of endothelial cells. However, these can usually only be passaged a maximum of three to five times before cell division stops.

Umbilical cords are normally obtained via a hospital maternity unit, which should be able to supply some information about the patient's medical history. In addition, the patient should preferably be pre-screened for infections. It is advisable for those working with umbilical cords to be immunized against Hepatitis B.

PROTOCOL 7.5   Isolation and growth of human umbilical vein endothelial cells (HUVECs)

Equipment and reagents

- Whole umbilical cords

- 1% Gelatin (Sigma G9382 or G1393) solution dissolved in sterile injectable-grade water. Sterilize by autoclaving at 121 °C for 20 min if made from powder. Store at 4 °C for up to 3 months. Prepare flasks by adding 3–5 ml of this solution to 25 cm² (or 10–12 ml to 75 cm²) flasks and incubate at 37 °C for 1–2 h or overnight at 4 °C. Prepared flasks can be kept at 4 °C for up to 1 week.

- Cord collection medium: HBSS with 20 mM HEPES, 100 U/ml penicillin, 100 μg/ml streptomycin and 50 μg/ml gentamycin. Prepare fresh.

- Basic culture medium: Medium 199 supplemented with 20% FBS and 2 mM L-glutamine.

(continued)

- Basic culture medium with growth factors: basic culture medium supplemented with 25 μg/ml EGF (Sigma E2759 or E9640) and 10 μg/ml heparin (Sigma H3393). Store in aliquots at –20 °C.

- 0.25% Trypsin/EDTA

- 0.5 mg/ml collagenase from *Clostridium histolyticum* (Roche 1074059 or Sigma C9722) in serum-free Medium 199. Filter sterilize if necessary. Prepare fresh or aliquot and store at –20 °C. As a guide, 10 ml per 20 cm of cord will be needed.

- Dulbecco's PBS A (DPBSA) 1× solution.

- Sterile equipment: 25- and 75-cm² culture flasks coated with gelatin, plastic 250-ml centrifuge tubes for collection of cord, dissecting board, scalpel and blades (no. 22–25), hypodermic needles, 10- and 20-ml syringes, Luer female to male adaptor, 50-ml centrifuge tubes, fine cord, 0.2-μm filters, small electrical crocodile clips or artery clamps, aluminium foil, cling film, sterile paper towel, forceps, scissors, plastic pipettes, sterile gloves, 70% ethanol.

Method

1 Collect whole umbilical cords as soon as possible after delivery, in a sterile vessel (250 ml centrifuge tube) containing enough cord collection medium at 4 °C to completely cover the cord.

2 Transport at 4 °C to the laboratory. Cords can be stored for up to 48 h. All subsequent steps should be performed in a Class II microbiological safety cabinet (MSC).

3 Prepare sufficient collagenase solution for the length of cord to be processed, and warm to 37 °C.

4 Cover the dissection board with sterile aluminium foil.

5 Soak paper towel in ethanol and wipe the entire length of cord to remove blood.

6 Cut off at least 3 cm from one end of the cord to expose the umbilical vein and two arteries. The vein can be distinguished from the arteries by the fact it has a thinner wall and larger lumen.

7 Insert a Luer adaptor into the vein. Cut away 2 cm of tissue and adjacent arteries to expose a length of vein with adaptor.

8 Tie the adaptor into the vein with cord, ensuring that the knots are tight. It may be necessary to tie two ligatures.

9 Close the other end of the cord using a crocodile clip or artery clamp. It is advisable not to exceed lengths of 30 cm as these can be difficult to work with.

10 Fill a 20-ml syringe with warm DPBSA and inject it into the cord via the Luer adaptor until the vein shows signs of distention. This is to check for any leaks and flush out blood from the vein. Remove the syringe and discard the wash to waste.

11 Insert another Luer adaptor into the other end of the vein, and repeat the procedure outlined in steps 8 and 9. Wash the vein with another 10–20 ml of warm DPBSA having placed an empty syringe in the opposite end of the vein. While the vein is full of DPBSA gently massage to loosen any blood clots. Remove the syringes and discard to waste.

(continued overleaf)

12 Fill a syringe with warm collagenase and connect to one of the adaptors. Place an empty syringe in the opposite end of the vein. Slowly inject the collagenase until the lumen is distended.

13 Wrap the cord in cling film which has been sprayed with 70% ethanol and place in an incubator at 37 °C for 10 min.

14 Remove and massage the vein to loosen the cells.

15 Withdraw the collagenase with one of the syringes and place in a 50-ml centrifuge tube.

16 Flush the vein with 10–20 ml of warm DPBSA and massage again to obtain as many cells as possible.

17 Add the solution to the 50-ml tube.

18 Centrifuge the collected solution at 80–100 g for 5 min.

19 Decant the fluid and then loosen the cell pellet by gently 'thumbing' the bottom of the tube.

20 Add 5 ml of basic medium with growth factors. A small sample (0.1–0.2 ml) can be used to count the cells to evaluate yields for future reference (Note: there will be erythrocytes in the suspension).

21 Transfer the cells to a gelatin-coated 25-cm^2 flask, having poured off the excess gelatin just before use. Incubate at 37 °C, 5% CO_2 overnight.

22 Check that the cells have attached and discard the medium. Wash. the cells with basic medium and add 5 ml of fresh medium with growth factors.

23 Every second day check the cells for growth and replace 50% of the medium with fresh medium with growth factors. Once the cells have reached 90–95% confluence they can be subcultured or used.

24 Subculture cells using 1–2 ml of trypsin/EDTA and collect them in 20 ml of basic medium.

25 Centrifuge at 80–100 g for 5 min and resuspend the pellet in fresh medium with growth factors. Divide the cells into two new gelatin-coated flasks[a].

Note

[a]Cell growth may be improved if 25–50% of the conditioned medium is added to new flasks. Cells can usually be passaged between three and five times before growth stops.

7.2.2 Primary culture of ectodermal cells

7.2.2.1 Epithelium

Epithelial cells, such as those lining the gut, and keratinocytes in the epidermis, are generally more difficult to culture than mesenchymal cells. These cells have cobblestone morphology and tend to grow in sheets [10]. They are more sensitive to plating densities and need cell-to-cell contact for growth. Based on current strategies, epithelial cells can be cultured in either serum-containing (*Protocol 7.6*) or serum-free (*Protocol 7.7*) conditions. Serum-free conditions are easier to manage but do not allow for repeated passaging of the cells. On the other hand, culturing epithelial cells in serum-containing formulations will permit longer-term passaging, but requires co-culturing in the presence

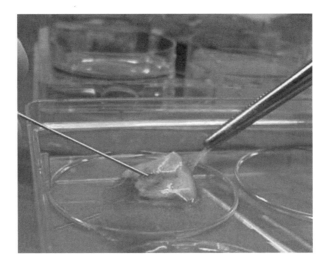

Figure 7.9 Separating the epidermis from the dermis of a skin biopsy after dispase treatment. A sterile needle is used to hold down the dermis while a pair of fine forceps is used to gently peel away the epidermis. A well-digested sample will present little resistance in removing the epidermis, allowing good separation of the two tissues so that homogeneous populations of both keratinocytes and fibroblasts can be obtained.

of feeder cells, typically fibroblasts. These feeders secrete growth factors that support the growth and proliferation of epithelial cells. A commonly observed problem in the culture of epithelial cells is overgrowth by stromal cells such as fibroblasts and vascular endothelial cells, even in the absence of feeder cells. Another possible undesirable outcome encountered when culturing epithelial cells is epithelial–mesenchymal transition (EMT), a condition where epithelial cells appear to trans-differentiate into mesenchymal cells as indicated by a change in cell shape from polygonal to spindle-like. This is accompanied by a transition in the expression profile of intermediate filaments such as from keratin to vimentin. Such issues can be overcome by using cell-specific growth formulations to discourage the growth of unwanted cell types, or by using sorting techniques such as fluorescence-activated cell sorting (FACS) to eliminate them [11]. General techniques are described here (*Protocol 7.6*) for the isolation and culture of keratinocytes from a skin biopsy (Figure 7.9).

PROTOCOL 7.6 Isolation and growth of keratinocytes on a feeder layer

Equipment and reagents

- Freshly excised human skin, for example from circumcisions

- Skin transport medium: DMEM supplemented with 10% FBS, 100 U/ml penicillin, 100 µg/ml streptomycin and 2.5 µg/ml amphotericin B.

(continued overleaf)

- 3T3 cells: from the ECACC. These are used to establish feeder layers onto which the primary keratinocytes are plated.

- 3T3 medium – DMEM supplemented with: 10% FBS, 4 mM L-glutamine, 100 U/ml penicillin, 100 µg/ml streptomycin. Store at 4 °C. After 4 weeks the medium should have fresh L-glutamine added from stock, as this component is unstable. FBS should be batch tested for optimum cell growth.

- Keratinocyte growth medium: a mixture of Ham's F12 and DMEM (2:1 v/v) supplemented with 10% FBS, 4 mM L-glutamine, 100 U/ml penicillin, 100 µg/ml streptomycin, 5 µg/ml insulin, 0.4 µg/ml hydrocortisone, 10 ng/ml FGF, 10^{-10} M cholera enterotoxin, 5 µg/ml transferrin, 2×10^{-11} M sodium tri-iodothyronine and 1.8×10^{-4} M adenine. Prepare fresh if possible, as the medium shelf life is 1 wk at 4 °C. Store the supplements concentrated at –20 °C. Prepare stock supplements in CMF-PBS, containing 0.1% bovine serum albumin. Some supplements require dissolving as follows:

  Insulin in 0.05 M HCl; hydrocortisone in ethanol; adenine in NaOH, pH 9.0; dissolve sodium tri-iodothyronine initially in 1 part of HCl, then add 2 parts ethanol.

- Mitomycin-C stock solution: dissolve in sterile water to a concentration of 400 µg/ml. Store at 4 °C in the dark. The solution is stable for 3–4 months.

- 0.25% trypsin/EDTA solution

- 70% ethanol

- Dispase medium: Add 2 mg/ml Dispase (Boehringer Mannheim) to 3T3 medium and filter sterilize. Prepare fresh for each use.

- Sterile equipment: forceps, scalpel, hypodermic needles, iris scissors, medical gauze, tissue culture flasks, pipettes, centrifuge tubes, Petri dishes and sterile 100 µm cell strainers (Falcon©, Becton Dickinson).

Method

Culture of 3T3 cells

1 Resuscitate the cells from frozen condition and culture as instructed by the supplier. Cells are grown in 3T3 medium at 37 °C in a 5% CO_2, 95% humidity incubator. Once they reach 80–90% confluence[a], passage them as below.

2 Remove medium from the flask and wash cells with an equivalent volume of CMF-PBS.

3 Add trypsin/EDTA at room temperature to the flask at approx 2 ml per 25 cm^2.

4 Swirl over the surface quickly and decant about half the volume. Incubate the cells at 37 °C until all cells have rounded up and detached, i.e. 3–5 minutes.

5 Collect the cells by pipetting medium over the surface, for example 5 ml for a 25-cm^2 flask and 10 ml for a 75-cm^2 flask. Mix gently, take a small sample (0.1–0.2 ml) for counting, and transfer the rest to a 15- or 50-ml centrifuge tube. Note the total volume of cells. NOTE. The presence of serum in the medium will neutralize the trypsin activity.

6 Centrifuge at 80–100 g for 5 min.

7 Decant the supernatant and resuspend the cell pellet in fresh medium. Calculate the number of cells required to seed into flasks or Petri dishes at approx 3000 cells per cm^2. Cells should

(continued)

reach confluence in approximately 3–5 days; if the time is outside these limits adjust the seeding density accordingly[a].

Production of feeder layers

1 Use flasks of 3T3 cells, prepared as described above, when they reach 50% confluence. Change the medium for fresh medium, and incubate for a further 24 h.

2 Add between 1 μg and 10 μg of mitomycin-C per ml of medium[b], and incubate for a further 12 h.

3 Wash. the cells three times with fresh medium and then incubate the cells with the final wash for approx 15–20 min at 37 °C.

4 Trypsinise the cells as detailed in steps 2–5 of the *Culture of 3T3 cells* section above, and seed into fresh flasks at approximately 25 000 cells per cm^2 in keratinocyte growth medium (1–1.5 ml medium/5 cm^2 of plastic surface area).

5 Incubate at 37 °C for approx 12 h to allow the cells to attach and spread.

Preparation of keratinocyte cultures

1 Place skin biopsy directly into a sterile universal container with enough skin transport medium at 4 °C to prevent the tissue from drying out[c].

2 Sterilize the skin by removing it from the transport medium and submerging briefly in 70% ethanol three times.

3 Shake dry and place skin samples into a sterile 90-mm Petri dish.

4 Trim away the hypodermis, that is the adipose and loose connective tissue, using fine forceps and iris scissors, until only the epidermis and the relatively dense dermis remain.

5 Cut the skin into long 2- to 3-mm thin strips using a sterile scalpel, with the skin epidermis face down.

6 Place these into a fresh universal container with at least enough Dispase medium to just cover them, and incubate either overnight at 4 °C or for 2–4 h at 37 °C.

7 Remove the strips from the dispase medium, dab off excess medium on the inside of the lid of a fresh 90-mm Petri dish, and then place them into the Petri dish.

8 Peel the epidermis away from the dermis with two sterile hypodermic needles, or one hypodermic needle and a pair of fine forceps (Figure 7.9). The epidermis is a semi-opaque thin layer, whereas the connective tissue of the dermis will have absorbed fluid and will appear as a thick, swollen, slightly gelatinous layer. This should come away easily. If sections remain attached, then either the strips were too thick or further incubation in Dispase is required. (NOTE: This should not be a problem after incubation overnight at 4 °C.)

9 Place the epidermis strips into 5 ml of trypsin/EDTA solution at 37 °C and shake rapidly for 1–2 min.

10 Add 15 ml of DMEM + l0% FCS to inactivate the trypsin. Pipette the medium and cells through a 100-μm cell strainer into a sterile 50-ml centrifuge tube to obtain a cell suspension. The strips should remain on the strainer.

(continued overleaf)

11 Pellet the single-cell suspension by centrifugation at approximately 50–100 g for 5 min.

12 Resuspend in keratinocyte growth medium and count.

13 Seed at approx 20 000–50 000 viable cells per cm^2 on to the pre-plated feeder layers which should be at 70–75% confluence.

14 Change the medium to remove any non-attached cells after 2–3 days, i.e. when the viable keratinocytes have attached.

15 Maintain the keratinocytes by changing the medium every 3–4 days thereafter. Cells will typically grow as circular colonies surrounded by feeder cells. Once they reach 70–80% confluence they must be subcultured or they will start to differentiate and die. Use the same procedure as for passaging 3T3 cells, as described in the *Culture of 3T3 cells* section, but allow 10–15 min trypsin incubation time for the cells to detach. Pipetting medium over the surface may be necessary to dislodge the cells.

16 Seed keratinocytes onto fresh feeder layers at between 10 000 and 50 000 cells per cm^2 (Figure 7.10).

17 Cryopreserve keratinocytes (see Chapter 6) in LN (or LN vapour) as described in steps 18–20, but only use cells from flasks that are approximately 50–60% confluent to ensure that they are actively growing.

18 Trypsinise and count the cells (see step 15).

19 After centrifugation, resuspend the cells in freezing medium made up of 90% FBS and 10% DMSO at a concentration of 2–4 $\times$ 10^6 cells per ml.

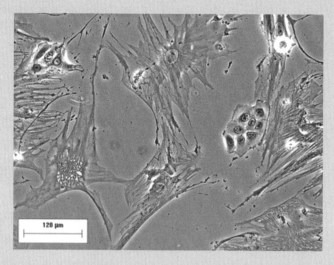

Figure 7.10 Phase contrast light microscopy image of keratinocyte colonies proliferating on a post-mitotic fibroblast feeder layer. These colonies appear as tight clusters of small polygonal cells and will expand as continuous sheets to overgrow the feeders.

(continued)

20 Aliquot 1 ml into each cryovial and preferably freeze them in a programmable freezer (Planer). Alternatively place the cyovials in a 'Mr Frosty' (Nalgene) container and keep this in a −80 °C freezer overnight. Subsequently transfer the cryovials to nitrogen storage, i.e. below −130 °C.

Notes

[a] Avoid 3T3 cell cultures becoming confluent as this can increase their resistance to mitomycin-C such that they can still divide after treatment and overgrow the keratinocytes.

[b] The concentration of mitomycin-C necessary to completely inhibit growth of 3T3 cells, while remaining viable, has to be tested with each batch. Therefore, the concentration range of 1–10 μg of mitomycin-C per ml of medium is a guide based on wide experience. *Note: Mitomycin-C may be carcinogenic. Refer to the manufacturer's safety data sheet before use and employ all necessary precautions when handling.*

[c] For keratinocyte cultures to have the maximum proliferation potential they need to be taken from young children or infants. One of the best sources is foreskin from circumcised babies.

PROTOCOL 7.7   Isolation and growth of keratinocytes in serum-free medium

Keratinocytes can also be grown under serum-free conditions without the use of a feeder layer (Figure 7.11). The reagents used and the procedures are as described in *Protocol 7.6* except where stated otherwise.

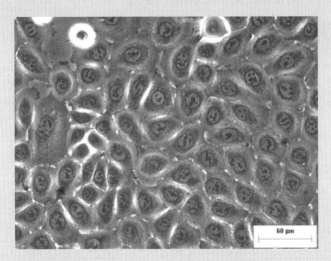

Figure 7.11   Phase contrast light microscopy image of a keratinocyte colony proliferating in serum-free medium, showing typical tight, cobblestone-like cell morphology.

(continued overleaf)

Equipment and reagents

The list of equipment and reagents required is similar to that in *Protocol 7.6* except that the items used for preparation of feeder cells are no longer needed. Additional reagents required are:

- Keratinocyte serum-free growth medium: for example Defined Keratinocyte-SFM (Gibco®, Invitrogen) with proprietary supplements

- Defined Trypsin Inhibitor (Invitrogen)

Method

1 Obtain keratinocyte single-cell suspensions as described in *Protocol 7.6*.

2 Plate isolated keratinocytes at a population density of at least 15 000 per cm^2 in serum-free defined keratinocyte medium with proprietary supplements, according to manufacturer's instructions.

3 Leave the culture flasks undisturbed for at least 3 days before performing the first medium change.

4 If cells do not attach, wait for a few more days before changing the medium.

5 Subculture at 70–80% confluence, using trypsin/EDTA as described in *Protocol 7.6*. However in place of serum-containing medium, use Defined Trypsin Inhibitor at 1:1 volume ratio to neutralize the trypsin.

7.2.3 Primary culture of endodermal cells

7.2.3.1 Hepatocytes

The liver is the second largest organ of the body after the skin, and performs a variety of important physiological functions. The histomorphological structure is complex, with cells organized in lobuli centred on the Glisson-Tris, consisting of the bile duct and vascular supply. Next to hepatocytes, endothelial cells, Ito cells (important in fat-metabolism) and Kupffer cells (which are phagocytic) are present. *Protocol 7.8* focuses on the isolation and culture of hepatocytes from a whole liver obtained from a small laboratory animal.

PROTOCOL 7.8   Isolation and growth of hepatocytes

Equipment and reagents

- Fresh whole liver

- Hepatocyte growth medium: Ham's F-12 supplemented with 0.2% bovine serum albumin and 10 µg/ml bovine insulin

- CMF-PBS

(continued)

- 0.05% w/v Collagenase in CMF-PBS

- Portal cannula

- 60 µm cell strainers

- Perfusion pump (e.g. mini peristaltic pump), needles and tubing for delivery of perfusates

- Cell culture vessels

Method

1 Obtain access to liver in freshly killed animal and commence isolation procedure as soon as possible.

2 Place and fix a portal cannula in position along the portal vein.

3 Make a cut in the lower vena cava.

4 Perform pre-perfusion with CMF-PBS while the liver remains *in situ* for the first 2–3 min.

5 Start the perfusate flow at a rate of 50 ml per minute.

6 Transfer the liver to a Petri dish and place in a position similar to its *in situ* orientation while pre-perfusion is carried out.

7 Perform perfusion by recirculating 0.05% collagenase for a further 10 min following the first 10 min of pre-perfusion with CMF-PBS. Terminate when the vena cava ruptures.

8 Liberate cells from the connective vascular tissue and resuspend in fresh hepatocyte growth medium.

9 Incubate cell suspension at 37 °C, 5% CO_2 for 30 min in a Petri dish.

10 Filter cell suspension through 60 µm cell strainers to further remove the connective tissue debris.

11 Centrifuge filtrate at 50 g for 1 min to obtain a pellet.

12 Resuspend and wash cell suspension twice with hepatocyte growth medium.

13 Plate onto culture vessels at 10–15 × 10^4 cells per cm^2. This cell concentration is necessary in order to maintain the hepatocyte phenotype.

14 Change medium every 1–2 days, taking special care to be gentle so as not to dislodge cells.

7.3 Troubleshooting

This section addresses some of the common problems encountered when culturing primary cells, their causes and possible solutions. However, many of the issues apply equally to the culture of established cell lines. Many of these problems can be avoided by working according to the *Good Cell Culture Practice* guidelines [12].

It is important that the individual researcher keeps good records of the cells in culture, the materials used, when and how cells were passaged and which medium was used with what supplements (see reference 12, and Chapter 11). Every cell culture scientist should adopt a routine of checking the cultures on a daily basis in order to immediately recognize problems in the culture and react to them (see Chapter 4, Section 4.2.2.1*i*).

A quick reference guide to culture problems is given in Table 7.1. Usually the problems can be traced to one or several of the four aspects discussed below:

7.3.1 Cells

Cells can behave differently due to nutritional deficiencies, the presence of toxic products or microbial contamination, or senescence. These changes in behaviour can be recognized as changes in cell shape and/or growth kinetics. When such changes are observed, it is advisable to check the glutamine or serum level in the culture medium to make sure that nutritional deficiency is not the cause. The accumulation of toxic products, and senescence, can also lead to changes in cellular phenotype and increasing doubling times. Multinucleated cells and intracellular debris are also commonly observed in such cases. Microbial contamination can include bacteria, fungi, yeast and mycoplasma, of which bacteria and fungi contaminations are the most common. When Phenol Red-containing culture medium suddenly turns yellow (indicating a rapid drop in pH), bacterial or fungal contamination is almost invariably the cause. The best solution is usually to discard the culture, disinfect the environment thoroughly and start again with a new sample, but for precious or irreplaceable cultures strategies are available by which an attempt can be made to eradicate the contamination (see Chapter 9, Section 9.2.5).

7.3.2 Materials

These can be divided into cell culture hardware or cell culture media and supplements. Not all cells will adhere well on the same type of culture vessel surface. Some cell types will develop irregular growth patterns on different surfaces. It is therefore advisable to identify the best cell culture surface for different cell types. The design of the culture vessel (T-flask, cell factory, Petri dish, roller flask, bioreactor) may also affect cellular behaviour. Cross-comparisons using different cell culture vessels from various suppliers will help.

Cell culture media and supplements are the most immediate factors in the cells' environment that will affect cellular behaviour. Serum, for example, can cause significant differences in cellular behaviour due to possible batch variations. For cell types that are sensitive to serum differences, batch testing of serum may be necessary to ensure consistent results.

7.3.3 Culture environment

The cell culture environment is largely determined by the incubation conditions and their interaction with the medium and culture vessel system being used, as discussed in Chapters 4. Most mammalian cell culture media require 5% CO_2 at 37 °C, although DMEM was originally designed for use with 10% CO_2. As discussed in Chapters 4 and 9, the incubator is a frequent source of culture contaminants. Thus it is advisable to regularly

Table 7.1 Quick reference guide for troubleshooting

Problem	Possible cause	Suggested solution
Unexpected cell death	Low CO_2 level in incubator	Change CO_2 tank
		Check for leaks
		Check/calibrate CO_2 sensor
		Avoid opening incubator doors unnecessarily or for extended periods of time
	Temperature fluctuations in incubator	Monitor incubator temperature
		Check temperature sensor
	Wrong medium formulation	Check suitability of medium formulation (concentration and quality of supplements, required CO_2 concentration)
	Cell damage during processing (cryopreservation, thawing, subculturing etc.)	Check suitability of protocols (formulations of reagents, centrifugation force used, length of time spent outside incubator or in the presence of potentially damaging agents (e.g. trypsin) etc.)
	Build-up of toxic metabolites in medium	Change medium more frequently
Rapid shift in pH of culture medium	Wrong CO_2 level in incubator	Change CO_2 tank
		Check for leaks
		Check/calibrate CO_2 sensor
	Overly tight caps on cell culture flasks	Loosen caps to allow gas exchange
		Use flasks with caps that have built-in filters
	Insufficient buffering	Add 10–25 mM HEPES buffer
	Contamination with bacteria, yeast or fungus	Attempt to salvage culture by using antibiotics and/or antimycotics (see Chapter 9, Section 9.2.5)
		Discard and start with fresh sample
Increase in cell population doubling time	Depletion or absence of essential supplements such as serum, glutamine or growth factors	Replace medium with freshly prepared medium
		Ensure that media bottles are stored in proper conditions (e.g. at 4 °C in the dark) to prevent degradation of supplements
	Switch in serum source	Carry out batch testing of serum

(*continued overleaf*)

Table 7.1   (continued)

Problem	Possible cause	Suggested solution
	Low level contamination with bacteria, fungus or mycoplasma	Isolate culture, grow a sample in antibiotic-free medium and monitor for up to 2 weeks for bacterial and fungal contamination Test a sample for the presence of bacteria, fungi or yeasts using microbiological culture techniques (see Chapter 9, Section 9.2.3.2) Test for mycoplasma (see Chapter 9, Section 9.2.3.3). If test turns out positive, discard cultures and conduct a major clean-up of cell culture lab, change all filters in the MSC and start over again. Commercial mycoplasma removal kits are available (see Chapter 9, Section 9.2.5) but will have to be evaluated for effectiveness.
	Senescence	Discard culture and isolate fresh cells from new tissue biopsy, or get from frozen stocks a sample having a lower passage number. Establish an immortalized cell line from fresh material or cells of a lower passage number. if this does not change the cellular characteristics needed for subsequent experiments
Contamination with bacteria, yeast, fungus or mycoplasma	Contaminated tissue biopsy	Check that tissue harvesting process is done under aseptic conditions Ensure that transportation of tissue biopsy from the operating theatre to the cell culture laboratory is done in an air-tight sterile container with a suitable medium containing antibiotics and/or antimycotics Sterilize the surface of the tissue biopsy upon arrival in the cell culture laboratory using sodium hypochlorite solution or 70% ethanol. Process biopsy in medium/saline containing antibiotics and/or antimycotics
	Contaminated equipment	Ensure proper maintenance and servicing of autoclave (see Chapter 2, Section 2.2.1.1), and proper maintenance and regular cleaning of incubators, water baths, micropipettes, centrifuges, pipette aids and so on.

(*continued*)

Table 7.1 (continued)

Problem	Possible cause	Suggested solution
	Contaminated consumables	Use only properly sterilized plasticware. Do not reuse cell culture vessels designed to be discarded after a single use. Check reagents/supplements (particularly serum).
	Poor laboratory techniques	Ensure that new members of the laboratory are given proper training before embarking on independent cell culture work
		Seek advice and learn from more experienced cell culture scientists

clean (and if possible sterilize) the incubator and ensure that all cultures are checked for the presence of microbial contamination on a daily basis (see Chapter 4, *Protocol 4.1*). It is also important to calibrate your incubator and to regularly check the level of CO_2 in its atmosphere. There should always be at least one spare CO_2 tank connected to the incubator (see Chapter 1, Section 1.2.2.2).

7.3.4 Protocols and techniques

There are various methods of isolating primary cells from a tissue specimen (enzymatic digestion, mechanical disruption of tissues, explant cultures). Different cells have different isolation and culture preferences and therefore care should be taken to satisfy any special needs of particular cell types. Examples of parameters that may affect cell culture results include formulations of freezing medium, cell plating densities and even maximum allowable centrifugation force.

References

1. Freshney, R.I. and Freshney, M.G. (1996) *Culture of Immortalized Cells*, Wiley-Liss, New York, NY.

2. Boulton, A.A., Baker G.B. and Walz, W. (1992) *Practical Cell Culture Techniques*, Humana Press Inc, Totowa, NJ.

3. Cohen, J. and Wilkin, G. (1996) *Neural Cell Culture: A Practical Approach*, IRL Press at Oxford University Press, Oxford, UK.

4. Davis, J.M. (2002) *Basic Cell Culture: A Practical Approach*, 2nd edn, IRL Press at Oxford University Press, Oxford, UK.

★★ 5. Freshney, R.I. (2010) *Culture of Animal Cells: A Manual of Basic Technique and Specialized Applications*, 6th edn, Wiley-Liss Inc, Hoboken, New Jersey, USA. – *Comprehensive reference for cell culture techniques.*

6. Jones G.E. (1996) *Human Cell Culture Protocols*, Humana Press Inc, Totowa, NJ.

7. Koller, M.R., Palsson, B.O. and Masters, J. R.W. (2000) *Human Cell Culture: Volume* **V**: *Primary Mesenchymal Cells*, Kluwer Academic Publishers, Japan.

8. Pollard, J.W. and Walker, J.M. (1997) *Basic Cell Culture Protocols (Second Edition) Volume 75*, Humana Press Inc, Totowa, NJ.

★★ 9. Ng, K.W. and Schantz, J.T. (2010) *A Manual for Primary Human Cell Culture*, 2nd edn, World Scientific Publishing Co, Singapore. – *For protocols of more primary cell types.*

10. Wise, C. (2002) *Epithelial Cell Culture Protocols*, Humana Press, Totowa, NJ.

11. Basu, S., Campbell, H.M., Dittel, B.N. *et al.* (2010) Purification of specific cell population by fluorescence activated cell sorting (FACS). *J. Vis. Exp.*, **41**, pii: 1546.

★★★ 12. Coecke, S., Balls, M., Bowe, G. *et al.* (2005) Guidance on good cell culture practice: a report of the second ECVAM task force on good cell culture practice. *Altern. Lab. Anim.*, **33**, 261–287. [This can also be downloaded from http://www.esactuk.org.uk/ . Click on the Best Practice (GCCP) tab.] – *Essential reading for anyone involved with cell culture.*

8

Cloning

John Clarke[1], Alison Porter[2] and John M. Davis[3]

[1]Haemophilia Centre, St Thomas' Hospital, London, UK
[2]Lonza Biologics plc, Slough, Berkshire, UK
[3]School of Life Sciences, University of Hertfordshire, Hatfield, Hertfordshire, UK

8.1 Introduction

A clone is a population of cells that are descended from a single parental cell. Clones may be derived from continuous cell lines or from primary cultures, but in either case the purpose of cloning is the same: to minimize as far as possible the degree of genetic and phenotypic variation within a cell population. This is done by isolating a single cell under suitable conditions and then allowing it to multiply to produce a sufficient number of cells for the required purpose.

8.1.1 Uses of cloning

Cloning cells from continuous lines has a number of applications:

(a) Many continuous lines are genetically unstable and their properties may alter during passage. Cloning can be used to isolate cultures with properties more closely resembling those of the original population (but these may have to be recloned at intervals if instability is pronounced; see Section 8.1.2). Conversely, cloning procedures may be used to isolate variants. Examples of the latter may include karyological and biochemical variants and cells that exhibit different levels of product secretion or different susceptibilities to viruses. Treatment with mutagens can be used in order to increase the proportion of variant cells.

(b) Hybridoma cells are cloned to generate cultures that secrete monoclonal antibodies.

(c) Variation within a continuous cell line may be studied by examining the properties of panels of clones established at different passage levels.

(d) Cells transfected with DNA do not necessarily form populations with homogenous genetic constitutions, and cloning can enable cells to be selected and cultures developed with the required characteristics.

(e) In pharmaceutical production, it may be desirable from a regulatory standpoint that a cell line used to derive a product can be defined as originating from a single cell.

In general, the cloning of primary cells is less successful than the cloning of established cell lines, as they tend to have a low colony-forming efficiency (CFE, see Section 8.2.2) and 'normal' cells can only undergo a limited number of population doublings (see Chapter 4, Section 4.1.1.2), which may prevent the generation of a sufficient number of cells for future use. Nevertheless, it is sometimes possible to isolate specific cell types from a mixed primary population, and to develop clones large enough for subsequent studies and free of unwanted cell types (often fibroblasts) that might otherwise overgrow the culture.

8.1.2 Limitations of cloning

Although a cloning step(s) during cell line development is included to ensure that a cell line is derived from a single cell, it does *not* guarantee subsequent homogeneity of the derived cell population. An inherent characteristic of many cell lines is that they are phenotypically heterogeneous in culture; a trait which often cannot be eliminated even by multiple rounds of cloning. Some cloned cell lines are capable of differentiation *in vitro* and environmental conditions (e.g. presence or absence of growth factors, inducers, etc.) may change the phenotype or balance of phenotypes within the population. Furthermore, the genetic instability that may have called for the cloning will in many instances not be eliminated by cloning, with the result that although initially genetically homogeneous, the 'clonal' population again becomes heterogeneous as the cells multiply. This may necessitate recloning at regular intervals to minimize the degree of heterogeneity. Also, in any cell population, there will be a tendency for mutations and epigenetic changes to accumulate with time. In addition, in at least some cell types stochastic processes can lead to very rapid generation of phenotypic heterogeneity [1]. Thus the homogeneity of a cloned cell population will depend on many factors, including the intrinsic properties of the cell line, the elapsed time in culture since cloning and also on the culture environment.

Many different approaches have been taken to the problem of isolating single cells under conditions in which they will continue to proliferate, and protocols for many of these approaches are given later in this chapter. Each method has its advantages and disadvantages, but there is one problem common to them all: there is no way to be *certain* that the derived population originates from a single cell.[1] As there is always a slight but

---

[1] Although Wewetzer and Seilheimer [2] claim that by using their micromanipulation method one *can* be sure that single cells have been isolated, any such claim must be dependent on the fine detail of the technique used; and as none of the current authors have hands-on experience of this technique they will reserve judgement on the accuracy of this claim. However, their experience of other micromanipulation techniques (see, for example, *Protocol 8.4,* and in particular Note b) would cause them to be doubtful of such a claim.

finite chance that one or more cells are present along with the target cell, cloning always deals with the *probability* of a colony being clonal in origin. A reliable estimation of this probability is important in certain situations, for example when a cell line is to be used in the production of a product for therapeutic use.

A commonly used method for cloning is limiting dilution (see *Protocol 8.2*). For calculation of a probability of monoclonality, it relies on the assumption that cells are distributed according to the Poisson distribution. At least two rounds of cloning by this method are normally carried out to increase the probability of obtaining a monoclonal colony [3]. However, a number of factors, related to the assumption of how the cells are distributed, can lead to incorrect estimates of the probability of monoclonality. First, the assumption that cells are distributed according to the Poisson distribution was based on models using artefacts such as ion-exchange beads in place of cells [4]. Such artefacts may not be a good representation of cells. For example, unlike the beads, some cell species tend to adhere to each other and this could affect the distribution; an increased proportion of colonies could arise from two or more cells. Secondly, although the cells may be distributed according to the Poisson distribution, it does not necessarily follow that colonies that develop from them also follow this mathematical model [5].

Examples of possible problems include:

(i) Potential toxic effects of one cell on a neighbouring cell(s), leading to a higher proportion of the growing colonies actually being monoclonal than would be predicted by the Poisson distribution.

(ii) The possibility of an increased chance of survival of colonies arising from more than one cell compared to a colony arising from a single cell, therefore increasing the proportion of colonies arising from two or more cells. Some evidence has been reported that this may occur with certain cells [6].

(iii) Potential errors when microscopic observation is being used to eliminate those colonies arising from more than one cell. Colonies may mistakenly be classified as arising from a single cell if the cells are in close proximity, or if some cells are temporarily dormant (depending on when the observation is performed). This would result in fewer colonies than estimated being monoclonal.

(iv) Inaccurate estimates of clonality for colonies having a desired characteristic. When colonies are screened for a particular phenotype, then the probability that a colony displaying that characteristic has actually arisen from a single cell decreases with the rarity of the desired phenotype [7].

Cells with a poor CFE are likely to have been distributed at a high average number of cells per well. i and ii may be particularly problematic if occurring in conjunction with this scenario.

The limiting dilution technique of cloning is attractive because it is inexpensive and simple. However, in some systems the numbers obtained, from the mathematical analysis to which it may be subjected, may bear little relationship to what is actually occurring in the cloning plate due to the invalidity of the assumptions on which they are based. It is therefore suggested that in many cases the careful use of a suitable technique in which the seeding of a single cell is observed (see, for example, *Protocol 8.4*) will give greater assurance that true clones are isolated. On the downside, such techniques are very labour intensive, can only be used with limited numbers of cells, and may not be applicable in all situations.

8.1.3 Special requirements of cells growing at low densities

During cloning, cells must be capable of growth and multiplication at extremely low population densities (at least during the first few divisions after plating) but this is a very non-physiological situation. *In vivo*, animal cells normally exist at population densities of up to 10^9 cells/cm^3 in the presence of autocrine and paracrine factors and hormones, as well as considerable cell–cell and (in most cases) cell–matrix interaction. Many of these environmental factors are important for cell growth and proliferation. Thus the culture conditions employed during cloning must attempt to mimic the essentials of the *in vivo* environment, but in a situation where individual proliferating cells can be spatially isolated from one another. This places much greater demands on the culture medium and its additives than is the case during 'normal' (i.e. higher population density) culture.

All culture media, sera and additives to be used during cloning should be tested for their ability to support growth of the relevant cells at low population density. The use of a test for CFE (see *Protocol 8.1*) is a good, quantitative way of doing this, and facilitates optimization of the medium/serum/additive mixture. In general, enriched media containing cell growth factors and other growth-enhancing substances [8] may improve clonal growth, as will high concentrations (e.g. 20% or more) of FBS. However, it is important to note that, during the development of cells that could potentially be used in the manufacture of biopharmaceuticals, the use of animal or human sera (or indeed any other animal-derived materials, such as trypsin) is undesirable from a regulatory standpoint. In such cases it may prove possible to use other, non-animal-sourced substitutes, and/or it may be that lower CFEs may have to be tolerated. In other situations – where the use of serum is acceptable – it is essential to test individual serum batches with the particular cell line(s) of interest, as there is considerable batch-to-batch variation and a batch of serum that is highly effective at supporting the clonal growth of one cell line may be inhibitory to another.

Low CFE and slow initial cell proliferation may also be improved by adding conditioned medium to the cultures. This is medium harvested from normal cultures (generally of the cells being cloned), after growth to approximately 50% confluency or 50% normal maximum concentration, and then (to remove any viable cells) it is passed through a filter of 0.45-μm or 0.2-μm porosity prior to use. Improvements in CFE may also be obtained by the use of layers of feeder cells treated to prevent replication (see for example Chapter 7, *Protocol 7.6*).

It must be stressed that actively growing cultures are essential in order to obtain as high a CFE as possible, and consequently cloning must only be performed using cells from such cultures (i.e. for established cell lines, cells in log phase – see Chapter 4, Section 4.2.5.2).

8.2 Methods and approaches

8.2.1 Development of techniques

Cell cloning techniques were devised very early during the development of modern cell culture. Sanford *et al.* [9] isolated single cells within capillary tubes and Wildy and Stoker [10] isolated single cells in droplets of medium under liquid paraffin. In both cases, conditioning factors could accumulate in proximity to the isolated cells and promote clonal growth. The serial dilution procedure developed by Puck and Marcus [11] was less technically demanding. These authors also used feeder layers of X-irradiated, non-replicating cells to supply growth factors. Cooper [12] devised a procedure involving the distribution of small volumes of a dilute suspension of cells among the wells of 96-well tissue culture trays, and MacPherson [13] devised a technique in which individual cells were taken up in extra-fine Pasteur pipettes and inoculated into the wells of 96-well trays. The simpler and more rapid 'microspot' technique utilizing these trays was introduced by Clarke and Spier [14]. Other techniques include colony formation in semi-solid media [15], in hanging drops [16] and on glass coverslip fragments [17], and separation of cells by fluorescence-activated cell sorting [18].

8.2.2 Choice of technique

The procedure to be used should be selected with particular reference to the CFE of the cell population. CFE is defined as

$$\text{Number of colonies obtained/Number of individual cells plated}$$

and is usually expressed as a percentage. The CFE may vary from less than 1% for some primary cells, to practically 100% for some established cell lines. CFE may be assessed using *Protocol 8.1*.

PROTOCOL 8.1 Determination of colony-forming efficiency

As described, this procedure is only suitable for cells that will attach (at least to some extent) to the surface of the tissue culture vessel. Cells growing free in suspension will require the use of a semi-solid medium (see *Protocols 8.7 and 8.8*) in order to localize the growing colonies.

Equipment and reagents

• Complete medium suitable for the cells under test (including serum, other supplements and conditioned medium, as appropriate)

• Tissue culture flasks, Petri dishes or multiwell dishes (4- or 6-well)

• Inverted microscope

• Equipment for performing a viable cell count (see Chapter 4, *Protocol 4.5*).

(continued overleaf)

Method

1 Resuspend adherent cells by standard subculture method; use suspension cells directly from culture.

2 Count viable cells. Prepare suspensions of 100, 1000 and 10 000 cells/ml in all combinations of medium and serum under investigation.

3 Inoculate suitable volumes of the cell suspensions into multiwell trays, Petri dishes, or tissue culture flasks.

4 Incubate at appropriate temperature (dependent on the species of origin of the cells – see Chapter 4, Section 4.1.2.2). If using Petri dishes or multiwell trays, use a humidified atmosphere containing the appropriate CO_2 concentration for the medium being used (see Chapter 4, Table 4.3). Flasks should be pre-gassed with the relevant CO_2/air mixture.

5 Using an inverted microscope, inspect the cells after 4 days and then after every 2 days; with a marker pen, mark on the base of the container the positions of colonies that appear.

6 When distinct colonies have formed, count the total number and calculate the CFE [(colonies formed/total cells seeded) × 100%].

7 If the CFE appears very low or cell growth appears slow, consider repeating the test using feeder cells and/or modified medium constituents.

8 If only a limited number of clones are required and well-separated colonies form, these may be isolated for subculture at this stage by the use of cloning rings (see *Protocol 8.5*).

For cells with a high CFE, the 'microspot' procedure (*Protocol 8.3*) is very effective and results in the isolation of single cells in wells of 96-well tissue culture trays. The procedure is easy to perform and allows direct identification of wells containing single cells. The 'micro-manipulation' procedure (*Protocol 8.4*) also allows this, and facilitates the direct selection of individual cells with particular observable characteristics. However, it is technically more difficult and time-consuming.

However, if CFE is low (less than 5%), the above techniques would require a large number of 96-well trays to generate a significant number of clones, and regular inspection of all the wells would be very protracted and tedious. In this situation, the various other techniques described would be more convenient, and these can in fact be used with cells of high or low CFE simply by adjusting the number of cells placed in each plate or well.

Some of the techniques described below are applicable to any cell type (Section 8.2.3), whereas some are only suitable for use with attached cells (Section 8.2.4), and others are generally only suitable for cells capable of growing in suspension (Section 8.2.5).

8.2.3 Methods applicable to both attached and suspension cells

8.2.3.1 Limiting dilution

Probably the most widely used method of cloning, this technique depends on distributing cells into multiwell plates at a sufficient dilution such that there is a high probability that any colony which grows subsequently is derived from a single cell.

i. Theoretical considerations   If a number of single cells are distributed randomly and independently into a large number of wells, then the fraction of the total number of wells that will contain a particular number of cells is described by the Poisson distribution:

$$F_r = \frac{(c/w)^r}{r!} e^{-(c/w)}$$

where r is the number of cells in a well, F_r the fraction of the total number of wells containing r cells, c the total number of cells distributed and w the total number of wells into which they were distributed. The ratio c/w can be replaced by the term u, which thus represents the average number of cells per well. The above equation then becomes:

$$F_r = \frac{u^r}{r!} e^{-u}$$

It is u, the average number of cells per well, that is the parameter usually used to describe the way a limiting dilution cloning is performed (e.g. 'The cells were cloned at 0.1 cells per well'). By knowing u, it is possible to calculate the fraction of wells containing a given number of cells.

Thus the fraction of wells containing no cells is

$$F_o = \frac{u^0}{0!} e^{-u} = e^{-u}$$

Similarly, the fraction containing one cell is

$$F_1 = \frac{u^1}{1!} e^{-u} = u e^{-u}$$

The fraction containing two cells is

$$F_2 = \frac{u^2}{2!} e^{-u} = \frac{u^2}{2} e^{-u}$$

The fraction containing three cells is

$$F_3 = \frac{u^3}{3!} e^{-u} = \frac{u^3}{6} e^{-u}$$

and so on.

Using these equations, it is possible to estimate the fraction of wells containing a given number of cells in any cloning procedure performed at a known average number of cells per well, as illustrated in Table 8.1.

If one assumes that growth will occur in any well receiving at least one cell, then the probability that any colony (chosen at random) is actually derived from only a single cell

Table 8.1 Limiting dilution cloning – fraction of wells (F_r) containing a given number of cells (r): variation with average number of cells per well (u).

No. of cells per well (r)	Average number of cells per well (u)		
	1.0	0.3	0.1
0	0.368	0.741	0.905
1	0.368	0.222	0.090
2	0.184	0.033	0.005
3	0.061	0.003	0.000

is equal to the ratio of the number of colonies derived from one cell to the number of colonies in total. This is numerically equivalent to:

$$\frac{F_1}{1 - F_o} = \frac{ue^{-u}}{1 - e^{-u}}$$

and representative figures can be calculated from the data in Table 8.1. This is illustrated in Table 8.2. It should be noted that even at an average of 0.1 cells per well there is still approximately a 5% chance that any particular colony will *not* be a clone. Thus where clonality is important, it is recommended that cells cloned by limiting dilution are actually cloned more than once at a low average number of cells per well. The statistics of multiple rounds of cloning are dealt with in reference 3. It should be stressed that all the foregoing is dependent on the initial assumptions being true, namely that only single cells are being distributed (i.e. that there are no clusters of cells), and that colonies are selected at random. Clearly, if wells are monitored during cell growth and those showing more than one focus of growth are excluded, then the chances of picking a clone are increased. However, the problems and limitations already discussed in Section 8.1.2 should always be borne in mind.

For a fuller account of limiting dilution techniques, including derivation of the relevant statistics from first principles, see reference 19.

Table 8.2 Limiting dilution cloning – for cloning at a given average number of cells per well (u), the probability that any colony picked at random will be derived from a single cell

	Average number of cells per well (u)		
	1.0	0.3	0.1
Probability	0.582	0.857	0.951

ii. Practical aspects   A general procedure for cloning cells by limiting dilution is given in *Protocol 8.2*.

PROTOCOL 8.2   Cloning by limiting dilution

Equipment and reagents

- 96- and 24-well tissue culture trays

- Other culture vessels as required (e.g. 25 and 75 cm^2 tissue culture flasks)

- Complete medium suitable for the cells (including serum, other supplements and conditioned medium, as appropriate)

- Inverted microscope

- Actively growing cells

- Equipment for performing a viable cell count (see Chapter 4, *Protocol 4.5*).

Method

1 Resuspend adherent cells by standard subculture method; use suspension cells directly from culture. Make every effort to obtain a monodisperse (single-cell) suspension.

2 Perform viable cell count.

3 For an expected CFE of less than 5%, prepare 10 ml of each of three cell suspensions with concentrations of 5000, 1000 and 500 cells/ml. For a CFE of 5–10%, prepare suspensions with concentrations of 500, 100 and 50 cells/ml. For a CFE of over 10%, prepare suspensions with concentrations of 50, 10 and 5 cells/ml.

4 Add 100 µl of one cell suspension to each well of a 96-well tray. Repeat for each dilution, using separate trays. If feeder layers are required, these should be prepared in the trays in advance.

5 Add 100 µl of complete medium to each well. Alternatively, if considered necessary, add 100 µl of conditioned medium, or 100 µl of 1 : 1 conditioned medium–fresh medium, to each well.

6 Incubate, at the appropriate temperature for the cells, in a humidified atmosphere containing the correct CO_2 concentration for the medium in use.

7 Using the inverted microscope, inspect the wells after 4 days and then at 2-day intervals[a].

8 Mark the wells in which a single colony appears. Ensure that there is only one centre of growth as the colony develops.

9 In some cases, depending on the growth characteristics of the cells, as colonies grow it may be useful to feed the cells at intervals with medium. If growth is vigorous, 50% of the medium in a well might be changed every 4–7 days. In media containing Phenol Red, the medium should be changed sufficiently frequently that it never turns bright yellow, as excess acidity will kill the cells. However with other cells, feeding may retard or completely inhibit growth. The precise regime will depend on the observed behaviour of the colonies, the known requirements of the cells, and the characteristics of the medium. Feeding may disrupt cell colonies that are not strongly adherent, giving rise to new foci of proliferation. With such

(continued overleaf)

cells, feeding should be avoided, or delayed (if possible) until the operator is confident that only one colony is growing in the well.

10 Select those plates in which only a limited number of wells show cell growth. When colonies of a suitable size have formed, resuspend the cells by trypsinization if adherent, or careful pipetting if non-adherent.

11 Subculture each colony into an individual well in a 24-well tray, which may contain feeder cells and/or conditioned medium if considered necessary. The total volume of medium per well should not exceed 1 ml.

12 Incubate (and feed if required).

13 When sufficient cells are present (i.e. when more than 50% of the area of the well is covered), subculture into larger culture vessels.

14 Freeze stocks of cloned cells in LN as soon as a sufficient number of cells are available (see Chapter 6).

15 In many cases, it will be necessary to repeat the foregoing procedure once or twice, for the reasons discussed previously.

Note

[a]There is always a danger that, if one disturbs the plates too frequently, colony formation may be inhibited by, for example, the inevitable rise in pH that occurs when a bicarbonate-buffered culture is removed from the appropriate CO_2-containing atmosphere present in the incubator. Conversely, the vibrations caused by such handling could cause cells to be dislodged from a single colony and form their own separate colonies within the culture vessel. This would mean that, on subsequent observation, two or more colonies would be seen and thus the well's contents be deemed (incorrectly) to be non-clonal. Consequently, the frequency of observation has to balance these factors with the need to detect independent colonies growing in the same well, and the need to detect the presence of a colony and subculture it into a larger culture vessel before it exhausts the medium in its well.

8.2.3.2 Microspot technique

This technique (*Protocol 8.3*) is simple and convenient, lending itself well to aseptic technique and giving low risk of adventitious contamination. An enhanced antibiotic regime is not usually required.

PROTOCOL 8.3   Cloning of cells by separation within microspots in 96-well plates

Equipment and reagents

- Sterile Pasteur pipettes
- Inverted microscope
- 96- and 24-well flat-bottomed tissue culture trays

(continued)

- 25- and 75-cm² tissue culture flasks
- Complete medium (including conditioned medium if required).

Method

1 Resuspend adherent cells by standard subculture method; use suspension cells directly from culture. Make every effort to obtain a monodisperse (single-cell) suspension.

2 Perform viable cell counts. Adjust the cell concentration to between 500 and 1000 cells/ml using complete medium.

3 Insert the tip of a Pasteur pipette into the cell suspension. Allow a small volume to rise into the tip by capillary action.

4 Tap the tip of the loaded pipette in the centre of the base of each well of a 96-well tissue culture tray. The action should deposit a droplet ('microspot') of about 1 μl of suspension. The droplet should spread a little, but not touch the sides of the well[a].

5 Examine the wells microscopically for the presence of a single cell. Use 50× magnification[b] initially, then 100× to confirm[c]. Mark the wells that contain a single cell, using a waterproof marker.

6 Occasionally, surface characteristics of the plastic bases of individual wells prevent droplets from spreading, causing dark areas to appear around the perimeter. This will also occur where a droplet touches the side. Cells within such areas may sometimes be seen with a higher magnification and/or by adjusting the illumination. If the whole droplet still cannot be rigorously inspected, that well must not be used.

7 When each tray has been inspected, add 200 μl of medium to each marked well. Alternatively, if deemed necessary, add 100 μl of complete medium plus 100 μl of conditioned medium, or 200 μl conditioned medium containing resuspended feeder cells.

8 Follow *Protocol 8.2*, steps 6–14.

Notes

[a] Because of the small size of the droplets, drying can occur during the microscopic examination of the wells in step 5 if not performed rapidly enough. Thus when the technique is first used it is recommended that droplets are only placed in half of the wells prior to microscopic examination. Once medium has been added to the appropriate wells (step 7) return to step 4 and place droplets in the remaining unused wells.

[b] Stated magnifications are overall magnifications, so for example 50× might be achieved using a 5× objective and a 10× eyepiece.

[c] It is beneficial to have a second observer to verify that only a single cell is present at step 5. A video monitor connected to the microscope will facilitate this.

8.2.3.3 *Cloning by micro-manipulation*

The observation of the seeding of a single cell can be a reliable means of ensuring colonies are clonal. One technique for achieving this is the use of micro-manipulation (*Protocol 8.4*). As cells are viewed during this cloning process, selection of those with observable characteristics such as a particular morphology is also possible. Practice may initially be required by an operator to develop the dexterity to achieve consistent results. An adequate period of time, with no interruptions, will then be needed to carry out the procedure.

PROTOCOL 8.4   Cloning by micro-manipulation

Equipment and reagents

- Inverted microscope, with digital camera or equivalent, standing in a Class II MSC

- Monitor: connected to the digital camera or equivalent

- Micro-manipulator with micro-syringe and accessories, including pre-formed micro-pipettes with 30° tip angle (Research Instruments Ltd., Falmouth, UK)

- Sterile, bacteriology-grade Petri dishes (60 and 35 mm diameter)

- 96-well half-area plates (Corning, code 3696)

- 96- and 24-well tissue culture plates

- 25 and 75 cm^2 tissue culture flasks

- Medium (including serum and/or conditioned medium as required).

Method

A micro-manipulator can be fitted to the side of the stage of an inverted microscope, with pre-formed micro-pipette(s) and an oil- or air-filled micro-syringe with flexible tubing attached to the micro-manipulator. The micro-manipulator is then used to move a micro-pipette horizontally and vertically in order to place it where desired. The micro-syringe and flexible tubing permit the precise control of liquid aspirated into and expelled from the micro-pipette.

1 Assemble and attach the digital camera, micro-manipulator and micro-syringe, and so on, to the microscope, according to the manufacturer's instructions.

2 To six well-separated wells in a sterile 96-well half area plate add 100 μl of medium. Store this plate in a CO$_2$ incubator until required. (The use of well-separated wells is recommended in order to avoid confusion and inadvertent cross-contamination.)

3 Use standard subculture method to resuspend adherent cells; use suspension cells directly from culture. Make every effort to obtain a monodisperse (single-cell) suspension.

4 Prepare a cell suspension for cloning: the concentration of cells should be low enough such that only one cell is normally visible in the microscope's field of view.

5 Add 5 ml of this suspension to a 60-mm Petri dish. To two further 35-mm Petri dishes, add medium alone.

6 Aseptically attach a new sterile micro-pipette to the micro-syringe on the micro-manipulator.

7 Place the first 35-mm Petri dish (containing only medium) on the microscope stage.

(continued)

8 Lower the micro-pipette into the medium and, using the micro-syringe, draw a small volume into it.

9 Raise the micro-pipette and replace the 35-mm dish with the 60-mm dish containing the cell suspension.

10 Using the microscope to examine the cell suspension, locate a suitable cell which is well separated from others[a].

11 Lower the micro-pipette into the cell suspension. Use the controls of the micro-manipulator to manoeuvre the tip towards the chosen cell. Then, using the micro-syringe, gently draw the cell into the tip of the micro-pipette.

12 Raise the micro-pipette, and replace the dish with the second 35-mm Petri dish containing only medium.

13 Lower the micro-pipette into the medium, and gently expel the cell. Examine the contents of the dish in the vicinity of the micro-pipette. After confirming that only a single cell has been aspirated, gently draw it back into the micro-pipette[a,b].

14 Raise the micro-pipette and replace the Petri dish with the half-area plate stored in the CO_2 incubator.

15 Lower the micro-pipette into one of the wells containing medium, and expel the single cell into it.

16 Raise the micro-pipette and return the half-area plate to the CO_2 incubator.

17 Repeat steps 6–16 for the other five wells that contain medium in the half-area plate, using a new micro-pipette and 35-mm Petri dishes each time.

18 Allow the cells to settle to the bottom of the wells in the half-area plate (this should take approximately half an hour) before carefully removing 50 μl of medium from the top of each well.

19 Using an inverted microscope, confirm the presence of a single cell in each well[a,b].

20 Repeat steps 2–19 for any further plates required.

21 Continue by following steps 9–14 of *Protocol 8.2* (although plate selection will not be required at step 10).

Notes

[a]Verification, by a second person, of the presence of only a single cell at steps 10, 13 and 19 is beneficial. The monitor assists in this process.

[b]Although the observations performed at steps 13 and 19 give very great confidence that only a single cell has been isolated, *certainty* is impossible. In particular, the nature of microscopy – which gives limited depth of field over a limited area – means that it is always conceivable that a (second) cell has been transferred at the same time but not been observed, for example by adherence to the outside of the micro-pipette. Although this is highly unlikely, this consideration means that even using this technique the best we can say is that there is a *very high probability* that colonies derived from cells isolated in this way are indeed clonal.

8.2.3.4 *Fluorescence-activated cell sorting*

Cells can be separated on the basis of their inherent properties and the properties of particular surface molecules (either intrinsic or bound to the surface by the scientist) using fluorescence-activated cell sorting [18]. This technology can therefore be used to select cells with desired properties including, when used in conjunction with other techniques, their rate of secretion of molecules of interest [20–22]. The cell sorting procedure can be further adapted to deflect single cells into the wells of multiwell plates and therefore used as a method for cloning cells. The combination of these two attributes results in the ability to screen and clone much larger numbers of cells than classical cloning methods, in one automated process [22]. This method of cloning can also remove the need to perform the multiple rounds of cloning often required by the limiting dilution method.

On the downside, this method of cloning requires extremely expensive and sophisticated equipment and a highly skilled operator. It may also be limited by a specific ligand being unavailable or the requirement for optimization of the method for a particular cell line. Also, like all the other (very much cheaper) cloning methods, there is still a non-zero probability that a colony thus isolated will be derived from more than one cell, due to the presence of more than one cell in the droplet(s) deflected.

8.2.3.5 *Automated cell cloning*

Automated systems that select and expand 'clonal' cells are available on the market. For example, the Quixell™ from Stoelting (http://www.stoeltingco.com) is an automated micro-manipulation system [2, 23]. Although the equipment itself may be expensive, it may have the advantage of removing the need for multiple rounds of cloning that can be required when using limiting dilution cloning.

Other examples include the ClonePix system from Genetix (http://www.genetix.com) and the CellCelector™ from Aviso (http://www.aviso-gmbh.de). In these systems, cells are trapped in a semi-solid medium and colonies allowed to form. Predetermined limits, such as size, shape and distance from neighbours, are used to identify clonal colonies. Because of the viscosity of the semi-solid medium, secreted protein is retained in the vicinity of the colony, thus enabling measurement of the amount of protein secreted by that colony. As with fluorescence-activated cell sorting, this enables the selection and cloning of much larger numbers of cells, in one automated process. It can also remove the need for multiple rounds of cloning. However, the equipment is again expensive and, as with fluorescence-activated cell sorting, can still carry a non-zero probability that an isolated colony will be derived from more than one cell because of the presence of more than one cell in the initial formation of the colony.

8.2.4 Methods for attached cells

8.2.4.1 *Cloning rings*

In this technique (*Protocol 8.5*), cells are grown at low population density in conventional plastic or glass tissue culture vessels. Once discrete colonies form, cloning rings are used to isolate individual colonies and permit their trypsinization and removal for subculture. Cloning rings are small, hollow cylinders, generally made from glass, stainless steel or PTFE, and can be of any size that is convenient. They may be cut from suitable tubing and the ends smoothed with a file or stone. Alternatively, they may be purchased from Bellco (cloning cylinders, code 2090) who supply them in a range of sizes in either stainless steel or borosilicate glass.

Stainless steel rings of 8 mm inside diameter, 2 mm wall thickness, and *c.* 12 mm height have been found to be particularly satisfactory as their weight seats them firmly in the silicone grease, permitting a good seal to be maintained during the cell detachment procedure (*Protocol 8.5,* steps 7–11). The height of cloning rings must be chosen with reference to the dimensions of the vessel within which they will be used, as they should not be so tall that they prevent closure of the vessel (*Protocol 8.5,* step 10).

PROTOCOL 8.5   Cloning of attached cells using cloning rings

Equipment and reagents

- Sterile forceps

- Silicone grease

- Glass Petri dishes or small glass beakers

- 24-well tissue culture trays

- Complete medium including serum (and conditioned medium if required)

- Inverted microscope

- Four-well or six-well culture plates, tissue culture grade Petri dishes (90 mm) or 'peel-apart' T-flasks (75 or 150 cm^2)

- Cloning rings

- Trypsin/EDTA solution.

Method

1 Sterilize the cloning rings and silicone grease in separate glass Petri dishes or foil-covered glass beakers.

2 For established cell lines or primary cultures: seed cells into four- or six-well trays (or selected alternative container) at 10 000, 1000, 100 and 10 cells/container. These quantities may be varied according to the expected CFE of the cells. They should be increased in proportion to the size of the container if Petri dishes or peel-apart flasks are used. The intention is to obtain discrete, well-separated colonies.

3 For hybrid or recombinant cells, culture under appropriate selective conditions. Cloning will generally be performed at the end of a hybridization or transfection procedure. The number of cells seeded per dish will be governed by the proportion of hybrids or recombinants expected, and their CFE in the selective medium.

(continued overleaf)

4 For all cells: incubate, at the temperature appropriate for the cells, in a humidified atmosphere containing the correct CO_2 concentration for the medium in use.

5 Inspect after 4 days and then at 2-day intervals. Mark the position of colonies that appear to have arisen from one growth centre and are well separated from other cells.

6 Remove the medium from the cultures, rinse with PBS and drain thoroughly.

7 Using sterile forceps, pick up a cloning ring and dip its base in sterile silicone grease. Ensure that the grease is evenly distributed.

8 Place the ring around a marked colony and press down, moving slightly to obtain a good seal.

9 Repeat steps 7 and 8 for the other selected colonies.

10 Fill the rings with trypsin/EDTA solution. Leave for 20 s, then remove most of the solution, leaving just a thin film[a]. Use a new pipette for each ring. Close the culture vessel and incubate at 37 °C, inspecting periodically to monitor the detachment process.

11 When the cells have detached, fill each ring with complete medium. Carefully pipette the medium up and down to suspend the cells, and transfer the suspension to a well of a 24-well tray. These trays may contain feeder cells or conditioned medium if required. The total volume of medium per well should not exceed 1 ml. Repeat the process with the contents of each ring, using a new pipette each time. Place the tray in the incubator and incubate as in step 4.

12 When sufficient cells are present (more than 50% of the area of the well is covered), subculture into 25-cm^2 flasks, and subsequently into 75-cm^2 flasks.

13 Stocks of cloned cells should be frozen in LN as soon as a sufficient number of cells are available (see Chapter 6).

Note

[a] Cells that are difficult to detach from the substrate my require more trypsin be left in the cloning ring, up to a maximum of c. 100 μl (in an 8-mm i.d. ring). The activity of this trypsin should be quenched with serum prior to cell harvest.

8.2.4.2 Cloning on a hydrophilic FEP surface

Another cloning technique applicable only to attached cells involves growing cells at low density on hydrophilic FEP (PTFE) membranes. This technique originally employed Petriperm dishes (from Heraeus or Sartorius) and as the authors' hands-on experience is with these dishes, *Protocol 8.6* describes their use. However, such dishes no longer appear to be available from the suppliers named above. Nevertheless, it may be possible to adapt the technique to employ other cell culture vessels containing hydrophilic FEP (PTFE) membranes, such as Millipore's Biopore™ Millicell inserts for Petri dishes or multiwell plates.

The principle of the technique is very simple; if cells are plated at a sufficiently low population density such that the resulting colonies are well separated, individual colonies can be removed while still attached to the substrate simply by cutting the FEP foil with a scalpel.

PROTOCOL 8.6   Cloning on hydrophilic FEP

Equipment and reagents

- 60-mm Petriperm 'Hydrophil' dishes (or similar small cell culture vessels with a hydrophilic FEP growth surface)
- Two pairs of sterile forceps
- Sterile scalpel (with thin, pointed blade)
- Sterile Petri dishes (90 mm or 100 mm diameter), or other appropriate sterile vessel
- Sterile 50-ml centrifuge tubes
- Basal medium
- FBS
- Conditioned medium (if required)
- Sterile 5-ml or 10-ml pipettes
- PBS
- Trypsin/EDTA solution (Sigma).

Method

A. Plating the cells

1 Using sterile forceps, aseptically remove the Petriperm dishes from their package and place each one in a 90-mm or 100-mm sterile Petri dish. This is necessary to ensure that the bottom of the Petriperm dish remains sterile during incubation.

2 Trypsinize the inoculum as appropriate for the cells in use. Make every effort to obtain a monodisperse (single-cell) suspension.

3 Perform a viable count on the cells.

4 Using basal medium, dilute the cells to 1000 cell/ml[a].

5 To a 50-ml centrifuge tube, add:

   (a) 10 ml of FBS

   (b) 10 ml of conditioned medium (replace with basal medium if not required).

   (c) 29 ml of basal medium.

   (d) 1 ml of diluted cell suspension.

6 Close the tube and mix gently by inversion.

(continued overleaf)

7 Pipette 5 ml[b] of this cell suspension into each of the Petriperm dishes. This must be done carefully, avoiding, for example, whirlpools that would tend to concentrate cells in the centre of the dish.

8 Incubate the dishes and examine regularly for the growth of colonies.

B. Picking the colonies

1 Add an appropriate amount of medium to the receiving vessels, and gas if necessary[c].

2 Place a pair of sterile forceps, and a sterile scalpel bearing a thin, pointed blade, under the lid of a large Petri dish. These items should be returned to the dish when not in use in order to preserve their sterility.

3 Remove the medium from the Petriperm dish. Rinse once with PBS, discard the rinse and replace the lid.

4 Hold the plate vertically so that the colonies are visible, with the bottom of the culture surface towards you.

5 Using the sterile scalpel, cut around each colony leaving a small intact bridge of FEP foil.

6 Using the sterile forceps, tear away the pieces of foil bearing the colonies and place each of them in the well of a micro-titre plate. Work rapidly and keep the micro-titre plate covered to prevent drying of the colonies.

7 Trypsinize the cells in the micro-titre wells, monitoring the trypsinization under the microscope.

8 After trypsinization, inactivate the enzyme by the addition of a small volume of serum, serum-containing medium or other trypsin inhibitor.

9 Transfer the cells from each well into a separate receiving vessel or culture well (prepared in step 1 of this section), and incubate as appropriate.

Notes

[a] This cell concentration is sufficient for cells with a high CFE (50–100%), yielding 50–100 colonies per plate. Proportionately higher concentrations should be used with cells displaying a lower CFE.

[b] The volume given (5 ml) is suitable for a 60-mm diameter ($= 28\,cm^2$) dish. If using vessels with a different growth surface area, the volume should be adjusted proportionally.

[c] The size of tissue culture vessel used to receive each colony after trypsinization will depend on the degree to which the cells can be safely diluted. Some colonies can be transferred straight into 25-cm^2 flasks, but this is unusual and it would be more common to use 35-mm Petri dishes or wells of a multiwell plate. If in doubt, it is safer to err on the side of smaller volume, and examine the cells more frequently.

8.2.5 Methods for suspension cells

8.2.5.1 Cloning in soft agar

The method described here (*Protocol 8.7*) uses a base layer of 0.5% agar, overlaid with cells suspended in 0.28% agar. These concentrations can be changed somewhat if required without adversely affecting the performance of the system, and it may be possible to omit the base layer for some cells.

PROTOCOL 8.7 Cloning in soft agar

Equipment and reagents

- Agar (tested for non-toxicity on the cells in use)
- Medium containing 15% FBS
- 90-mm Petri dishes
- 24-well tissue culture trays
- Sterile pipettes
- Inverted microscope
- Sterile Pasteur pipettes.

Method

1 Prepare a 0.5% (w/v) stock agar solution by adding 2 g of agar powder to 30–40 ml of tissue culture grade water. Autoclave this, then allow it to cool to 55 °C. Add medium (containing 15% FBS, and pre-warmed to 55 °C) to make a total volume of 400 ml. This should be sufficient to clone 10 cell lines.

2 To each of three 90-mm Petri dishes add 12–14 ml of molten 0.5% agar. Leave for 20 min at room temperature to allow the agar to gel.

3 Add 0.8 ml of medium to each of four wells in a 24-well culture tray.

4 To the first of these wells, add 2–4×10^3 cells in 0.2 ml of medium[a].

5 Pipette this suspension up and down a few times to mix the cells, then transfer 0.2 ml to the next well.

6 Repeat step 5 to make two more serial 1:5 dilutions. Discard 0.2 ml from the final well.

7 Allow the stock 0.5% agar solution to cool to 47–45 °C, then add 1 ml to each well, mix and rapidly transfer the contents of each well to separate agar-containing Petri dishes (prepared in step 2). Incubate the dishes as appropriate for the cell line and medium being used.

8 Using an inverted microscope, check daily for cell growth, and discard plates containing too many colonies, or no cells. When well-separated colonies containing 20–100 cells are seen, pick them individually from the plate using a Pasteur pipette. Transfer them to 1 ml of medium (containing 15% FBS) in separate wells of a 24-well tray for further growth[b].

Notes

[a] This cell suspension MUST consist of single cells. If a significant proportion of cell clumps are present, the resultant colonies will not be clones.

[b] The best size of colony to transfer, and the volume into which it should be transferred, will vary from cell line to cell line. The figures given here generally work well with mouse hybridomas.

8.2.5.2 Cloning in methylcellulose

In principle, this technique (*Protocol 8.8*) is similar to cloning in agar, but has the advantage that neither the cells nor the medium need be subjected to the elevated temperatures required to keep agar liquid. Although not specifically described here, methylcellulose can also be used as an overlayer on an agar base.

PROTOCOL 8.8 Cloning in methylcellulose

Equipment and reagents

- Powdered basal medium

- Wide-mouthed conical glass flask

- Methylcellulose powder (e.g. Methocel® MC, 3000–5500 mPa/s, Sigma, code 64630)

- 100-ml Duran bottles (Schott)

- Conditioned medium

- 35-mm bacteriological-grade Petri dishes

- 100-mm Petri dishes

- Sterile Pasteur or capillary pipettes.

Method

A. Preparation of 2% (w/v) methylcellulose stock solution

1 Prepare a 2× solution of the basal medium to be used (i.e. prepare from powder as for single-strength medium, but in half the volume).

2 Sterilize (either by hot air or autoclaving) a 500-ml wide-mouthed conical flask capped with aluminium foil and containing a large magnetic stirrer bar. If autoclaved, the flask should only contain a minimal amount of water after sterilization.

3 Weigh the flask, complete with stirrer bar and cap. Note this weight, then add to the flask 100 ml of tissue culture grade water.

4 Boil gently over a Bunsen burner for 5 min. Remove the foil cap and sprinkle 4 g of methylcellulose powder onto the surface of the water. It is important that none of it touches the walls of the flask. Recap the flask.

5 Heat the contents until they just start to boil. Remove the Bunsen burner and swirl the flask gently to help mixing until the suspension (which is white and opaque at this stage) has ceased to boil.

6 Repeat step 5 until 5 min have elapsed since the suspension first started to boil.

7 Remove the flask from the heat and plunge it into an ice/water slurry. Swirl continuously until all the contents of the flask have become viscous. This should be accompanied by a partial clearing of the flask's contents, which should now be translucent.

(continued)

8 Add 100 ml of the $2\times$ basal medium prepared earlier. Swirl the flask until the entire contents are mobile then place on a magnetic stirrer and stir at $4\,^{\circ}\mathrm{C}$ for 1 h.

9 Weigh the flask and add sterile water until the total weight of the flask, stirrer bar, cap and contents is equal to 204.5 g plus the starting weight noted in step 3.

10 Stir overnight at $4\,^{\circ}\mathrm{C}$, then dispense (by careful pouring) into two 100 ml Duran bottles. Store at $-20\,^{\circ}\mathrm{C}$.

B. Plating the cells

1 Thaw the 2% stock methylcellulose solution and allow to equilibrate in a $37\,^{\circ}\mathrm{C}$ water bath.

2 Using low speed centrifugation, pellet the cells to be cloned. Resuspend in a mixture containing (by volume) 53.3% FBS, 26.6% conditioned medium, and 20% basal medium, to a cell concentration of about 110 cells/ml[a].

3 Place 7.5 ml of this suspension in a 50-ml centrifuge tube and, using a sterile syringe without a needle, add 12.5 ml of the warmed methylcellulose solution. Mix initially by inversion, then by gentle vortexing. (The concentration of the FBS will now be 20%, and that of the conditioned medium 10%.)

4 Using a syringe fitted with a 19-gauge or wider needle, place 1-ml aliquots of this cell suspension in 35-mm bacteriological-grade Petri dishes. Tip the dishes to distribute the suspension over the whole surface. The volumes given in step 3 should be sufficient for 18 or 19 dishes.

5 Place two of these dishes, along with a third, open 35-mm Petri dish containing distilled water, in a 100-mm Petri dish and incubate under appropriate conditions in a humidified incubator.

C. Picking the colonies

1 Examine the plates at regular intervals (e.g. every 2–3 days) for the presence of colonies visible to the naked eye.

2 Examine colonies under the microscope. Check the colony is of a suitable size for picking (usually 0.5–1.0 mm diameter) and comprises healthy cells with a good morphology. Check that there are no overlapping colonies, or other colonies within about 1.5 mm.

3 Mark the position of colonies suitable for picking using a felt-tipped pen on the underside of the Petri dish.

4 Remove individual colonies from the plate by aspirating into a Pasteur pipette or capillary pipette. (It may be helpful to view the process with the aid of a free-standing magnifying glass or dissecting microscope.)

5 Place the cells in a suitable tissue culture vessel (35-mm dish, or well of a 24- or 48-well plate) with a maximum of 1 ml of tissue culture medium. It may be helpful to continue to use conditioned medium and high FBS concentrations at this stage.

6 Expand the cells as required. Check for the presence of the required properties and freeze a number of ampoules of any suitable cells as soon as possible.

(continued overleaf)

> **Note**
>
> [a]This cell concentration should be sufficient for a cell line with a high CFE (50–100%), yielding 20–40 colonies per plate. Proportionately higher concentrations should be used with cells displaying a lower CFE.

This whole procedure, as applied to the cloning of hybridomas, has been described in greater detail elsewhere [24].

8.3 Troubleshooting

- As discussed in Section 8.1.2, it is impossible to be *absolutely certain in all cases* that a population has arisen from a single cell, however carefully the cloning procedure is designed and performed. The realistic goal must be to *maximize the chances* that authentic clonal populations are produced. The micro-spot and micro-manipulation procedures are designed to isolate single cells in ways which allow the isolated cell to actually be observed. Here, a second operator may be employed to check the observation and confirm, as far as possible, that only a single cell is present in each selected well. In the limiting dilution procedure, it is rarely possible to identify wells that contain only one cell at the outset. Instead, there should be frequent observation during early stages of growth to identify the wells that contain a single colony which appears to be developing from a single growth centre. In the procedures where whole colonies are picked (cloning rings, petriperm dishes, soft agar, and methylcellulose), the colonies selected should preferably be discrete, compact and well-separated from others, to minimize the chance of including cells from other colonies.

- Some cells may be difficult to grow from single-cell isolates and/or low densities. Factors that may influence cell multiplication under such conditions are discussed in Section 8.1.3. Clonal cell growth must be encouraged by starting from actively growing cultures (preferably in log-phase), and by use of the optimal combination of medium, serum (where appropriate – see Section 8.1.3) and/or other supplements, with inclusion of conditioned medium or feeder cells if appropriate. If the optimal regime is not already established, it should be determined before cloning is initiated. Optimization may be usefully combined with determination of the CFE of the cells (*Protocol 8.1*).

- Some cell types may be inherently genetically unstable and/or phenotypically variable in culture, as discussed in Section 8.1.2. This variability may recur as the cell lines develop after cloning. Therefore, the relevant biological features of such cell lines should be monitored regularly. This will allow evaluation of their stability and their continued suitability for use, and indicate whether further rounds of cloning are necessary. Frozen cell banks should be established early after cloning, as soon as sufficient cells are available (see Chapter 6).

- Microbial contamination of cell cultures must obviously be prevented. Several techniques described in the present chapter (including single-cell isolation by micro-manipulation, and colony picking) necessitate working with cultures in vessels that may have to be kept open for some time, and particular care must be taken while performing such manipulations. These should always be carried out in a microbiological safety cabinet (Chapter 1, Section 1.2.3.1, and Chapter 2, Section 2.2.5.5), and the other general precautions and good practice described in Chapters 1, 2, 4 and 9 must always be followed. It may be appropriate to employ (or enhance) antibiotic cover during the initial stages, as long as the cells in use are not adversely affected. This may then be discontinued (or reduced) during subsequent culture development.

- Finally, remember that for any technique to yield clones, the starting cell suspension MUST comprise single cells.

References

1. Smith, J.R. and Whitney, R.G. (1980) Intraclonal variation of proliferative potential of human diploid fibroblasts: stochastic mechanism for cellular aging. *Science*, **207**, 82–84.

2. Wewetzer, K. and Seilheimer, B. (1995) Establishment of a single-step hybridoma cloning protocol using an automated cell transfer system: comparison with limiting dilution. *J. Immunol. Methods*, **179**, 71–76.

★★ 3. Coller, H.A. and Coller, B.S. (1986) Poisson statistical analysis of repetitive subcloning by the limiting dilution technique as a way of assessing hybridoma monoclonality, in *Methods in Enzymology*, vol. **121** (ed. J.J. Langone and H. van Vunakis), Academic Press, London, pp. 412–417. – *Review of the statistics of multiple rounds of limiting dilution cloning.*

4. Coller, H.A. and Coller, B.S. (1983) Statistical analysis of repetitive subcloning by the limiting dilution technique with a view toward ensuring hybridoma monoclonality. *Hybridoma*, **2**, 91–96

★★★ 5. Underwood, P.A. and Bean, P.A. (1988) Hazards of the limiting-dilution method of cloning hybridomas. *J. Immunol. Methods*, **107**, 119–128. – *Good demonstration of the potential pitfalls in applying the statistics of the Poisson distribution to real cells in limiting dilution cloning.*

6. McCullough, K.C., Butcher, R.N. and Parkinson, D. (1983) Hybridoma cell lines secreting monoclonal antibodies against foot-and-mouth disease virus (FMDV). II. Cloning conditions. *J. Biol. Stand.*, **11**, 183–104.

7. Staszewski, R. (1984) Cloning by limiting dilution: an improved estimate that an interesting culture is monoclonal. *Yale J. Biol. Med.*, **57**, 865–868.

8. Ham, R.G. and McKeehan, W.L. (1979) Media and growth requirements, in *Methods in Enzymology*, vol. **58** (eds W. B. Jakoby and I. H. Pastan), Academic Press, London, pp. 44–93.

9. Sanford, K.K., Earle, W.R. and Likely, G.D. (1948) The growth in vitro of single isolated tissue cells. *J. Natl. Cancer Inst.*, **9**, 229–246.

10. Wildy, P. and Stoker, M. (1958) Multiplication of solitary HeLa cells. *Nature*, **181**, 1407–1408

11. Puck, T.T. and Marcus, P.I. (1955) A rapid method for viable cell titration and clone production with HeLa cells in tissue culture. The use of X-irradiated cells to supply conditioning factors. *Proc. Natl. Acad. Sci. USA*, **41**, 432–437.

12. Cooper, J.E.K. (1973) Microtest plating, in *Tissue Culture – Methods and Applications* (eds P. F. Kruse and M. K. Patterson), Academic Press, London, pp. 266–269.

13. MacPherson, I.A. (1973) Soft agar technique, in *Tissue culture – Methods and Applications* (eds P. F. Kruse and M. K. Patterson), Academic Press, London, pp. 241–244.

14. Clarke, J.B. and Spier, R.E. (1980) Variation in the susceptibility of BHK populations and cloned cell lines to three strains of foot-and-mouth disease virus. *Arch. Virol.*, **63**, 1–9.

15. MacPherson, I. and Montagnier, L. (1964) Agar suspension culture for the selective assay of cells transformed by polyoma virus. *Virology*, **23**, 291–294.

16. Rittenberg, M.B., Buenafe, A. and Brown, M. (1986) A simple method of cloning hybridomas in 20μl hanging drops, in *Methods in Enzymology*, vol. **121** (eds J. J. Langone and H. Van Vunakis), Academic Press, New York, NY, pp. 327–331.

17. Paul, J. (1975) *Cell and Tissue Culture*, 5th edn, Churchill-Livingstone, Edinburgh, p. 255.

18. Shapiro, H.M. (2003) *Practical Flow Cytometry*, 4th edn, Alan R. Liss, New York, NY.

★★ 19. Lefkovits, I. and Waldmann, H. (1999) *Limiting Dilution Analysis of Cells of the Immune System*, 2nd edn, Oxford University Press, Oxford. – *Good description, derived from first principles, of the statistics of the Poisson distribution as applied to limiting dilution cloning.*

20. Weaver, J.C., McGrath, P. and Adams, S. (1997) Gel microdrop technology for rapid isolation of rare and high producer cells. *Nat. Med.*, **3**, 583–585.

21. Holmes, P. and Al-Rubeai, M. (1999) Improved cell line development by a high throughput affinity capture surface display technique to select for high secretors. *J. Immunol. Methods*, **230**, 141–147.

22. Browne, S.M. and Al-Rubeai, M. (2007) Selection methods for high-producing mammalian cell lines. *Trends Biotechnol.*, **25**, 425–432.

23. Caron, A.W., Massie, B. and Mosser, D.D. (2000) Use of a micromanipulator for high-efficiency cloning of cells co-expressing fluorescent proteins. *Methods Cell Sci.*, **22**, 137–145.

★★ 24. Davis, J.M. (1986) A single-step technique for selecting and cloning hybridomas for monoclonal antibody production. In *Methods in Enzymology*, vol. **121** (eds J.J. Langone and H. Van Vunakis), Academic Press, London, pp. 307–322. – *Thorough description of the technique of cloning in methylcellulose, as applied to the isolation of hybridomas.*

9

The Quality Control of Animal Cell Lines and the Prevention, Detection and Cure of Contamination

Peter Thraves and Cathy Rowe

European Collection of Cell Cultures, Porton Down, Wiltshire, UK

9.1 Introduction

Animal cell cultures are widely used across a range of scientific disciplines. Critical to their use as a start material or reagent is the requirement to maintain cell cultures in a defined and contaminant-free state. Quality control procedures have been employed by academic, commercial and government-sponsored organizations to validate the contamination-free status of cell cultures. Established guidelines describe formalized procedures and validated tests for procured tissues from which cell cultures are derived, and for the maintenance of such cultures [1–4]. Quality control procedures are not seen as an 'optional extra' in the characterization of a new cell culture but as a qualification stage prior to any subsequent work with that cell line. Technical staff in an organization using animal cells must be trained and vigilant in the prevention, control and elimination of contamination in cell cultures throughout the recovery from storage, growth and cryopreservation stages. Although routine observation is the first indicator of the overall health of a cell culture, not all contaminations are overt. An imperceptible contamination

---

Animal Cell Culture: Essential Methods, First Edition. Edited by John M. Davis.
© 2011 John Wiley & Sons, Ltd. Published 2011 by John Wiley & Sons, Ltd.

can become established before any gross indications become evident and remedial action can be taken.

When a cell line is contaminated, procedures must be in place to dispose of the culture or eradicate the problem in a manner that is safe to other cell cultures in the laboratory and to the technical staff.

Cell culture collections and other bio-resource centres exist to provide quality-controlled cell lines, and recipients of their cell cultures must be assured of the contamination-free status of the cell line. In contrast, where cell lines are being created *de novo* or being introduced into the laboratory from a source without quality control procedures in place, then quarantine culture facilities must be used [5].

9.2 Methods and approaches

9.2.1 Obtaining animal cell cultures

The increased use of animal cells in academic research, pharmaceutical production, and clinical and other studies has seen a parallel increase in the demand for cell lines. In addition to the development of cell culture collections there has also been an increase in the number of commercial organizations developing specialized cell types, including primary cultures and stem cell/progenitor cell lines. Both cell culture collections and commercial companies have responded to the recent surge in demand for cell lines for use in functional genomics studies using whole cell systems.

9.2.1.1 Cell culture collections

The resource centres listed at the end of this chapter (along with their web addresses) and some of the more specialized collections produce catalogues of material held. Their web-based catalogues allow their customers to view up-to-date listings of available stocks of cell cultures.

9.2.1.2 Quarantine and receipt of animal cell lines

All new cell lines brought into a cell culture area should be introduced into a quarantine laboratory specifically set aside and fully equipped for the purpose of handling unqualified cell lines, with appropriately trained personnel [5,6].

i. Accessioning scheme   The introduction of cell lines into the cell culture area(s) of any organization should be a tightly controlled process. All the major cell culture collections and research organizations, non-profit and commercial, now have officers who are tasked with the sourcing, acquisition and addition of new cell lines to their collection, in a process known as accessioning or acquisition.

The process of accessioning not only covers the shipment of a cell culture from one organization to another but also includes obtaining, checking and archiving all attendant documentation describing provenance, culture conditions, biohazard risk, ownership (of intellectual property) and any restricted use.

The properties of the cell culture and its maintenance should be recorded on a formal document for evaluation and future reference. In addition, and prior to dispatch of the cell line from the source laboratory or supplier, all cell lines should be subject to a bio-hazard risk assessment by the accessioning representative. The bio-hazard risk assessment will determine the appropriate level of containment to be used when handling the cell line [5] in terns of bio-containment and genetic manipulation to ensure that the facility is accordingly accredited or licensed to handle the cell lines. Genetically modified cell lines should be carefully reviewed to assess the extent and the details of the genetic manipulation that has been performed [6]. Prior to receipt, it is essential that every laboratory has defined procedures for the handling of new or incoming cell lines, that all staff are aware that such procedures exist, and that any staff likely to be handling such cultures are fully trained in these procedures.

ii. Management of cell lines in quarantine   A flow diagram of the scheme used at ECACC for the introduction of new cell lines into the collection is given in Figure 9.1.

The main points are listed below.

- All cultures should be handled in a quarantine laboratory separate from the main tissue culture area.

- Cultures should be in handled in a Class II MSC unless a higher level of containment is required. This will offer the operator protection, as the exact source of the cell line and its contamination status may not be known.

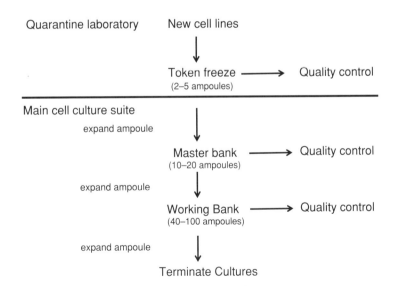

Figure 9.1 Hierarchical banking scheme for incoming cell lines as used at ECACC (adapted from reference 55).

- An initial assessment of potential microbial contamination should be made immediately, (broth for bacteria; PCR-based or other rapid methodology analysis for mycoplasma).

- An assessment should be made of the cell type in the culture by examination and photographic documentation.

- A token freeze of between three and five ampoules should be made as soon as possible after receipt of the cell line.

- A complete set of quality control tests should be applied to the token freeze. These include a cell count and viability assessment, sterility tests (bacteria, yeast, fungi, mycoplasma), species identification and DNA characterization by restriction enzyme analysis or short tandem repeat (STR) PCR analysis.

- After satisfying the above conditions, an ampoule from the token freeze can be transferred to the main cell culture laboratory for the production of MCBs and WCBs.

It is recommended that this approach be adopted for all cell lines received into the laboratory, irrespective of the source of the cell line. In the long term it can save both time and expense.

Many laboratories routinely use antibiotics in their cell cultures. This practice is not recommended for a number of reasons. From a contamination detection or surveillance viewpoint, antibiotics can suppress bacterial contamination to a level that is undetectable by microscopic examination and can lead to the development and spread of antibiotic-resistant microbial strains. In addition, although not eliminating any mycoplasma present, antibiotics, particularly the amino-glycosides, can reduce the level of infection below the level of detection of several currently used tests. It is essential that all quality control tests for microbial contamination be performed on cell cultures subcultured for at least two passages in antibiotic-free medium.

iii. Quarantine laboratory design
Optimal design criteria for the cell culture laboratory are described in Chapter 1. It is worth emphasizing that when establishing a cell culture laboratory, either within an existing facility or from new, an area (preferably an entire laboratory) should be set aside for quarantine purposes.

The quarantine laboratory should have the following features:

(i) It should be as far away from the main cell culture clean area as possible.

(ii) The air-handling system of the laboratory should operate under negative pressure with respect to the rest of cell culture area.

(iii) It should be self-contained with its own incubator, water bath, microscope, MSC, and so on.

(iv) Staff should have separate personal protective equipment (lab coats, along with disposable gloves, hairnets, shoe covers and face masks) for working in the quarantine laboratory.

(v) There should be an approved regimen in place for the routine and incident-related cleaning of the equipment and fabric (walls, floors, doors, etc.), and the fumigation of the MSC. All staff working in the quarantine laboratory must be fully trained in the required procedures.

(vi) Protocols for the disposal of contaminated cultures and associated reagents should also be in place.

9.2.2 Production of cell banks

As part of the accessioning process, when a culture is received into the cell culture laboratory it is essential that a clearly defined hierarchical banking scheme is in place. Starter (growing cultures) or frozen ampoules should be propagated following the instructions provided by the originator/depositor and a token freeze (three to five ampoules) prepared. As stated above, cultures from this token stock should undergo detailed microbial quality control and characterization. When the tests provide satisfactory results then a MCB (or seed stock) can be manufactured. All steps prior to the resuscitation of ampoules from the fully tested token freeze should be performed in the quarantine cell culture laboratory. It is on the MCB that the major authentication efforts, such as iso-enzyme analysis, karyology and DNA fingerprinting, should be applied as it is ampoules from this bank that are used to create the working banks [7, 8].

As outlined in Figure 9.1, one ampoule from the MCB is resuscitated and the culture propagated to produce a WCB. The number of ampoules required will depend on demand. Significant time and expense will be invested in the production and quality control of these banks, particularly MCBs, so the number of ampoules required should not be underestimated. For some industrial uses it is not uncommon to have MCBs containing as many as 200 ampoules and up to 1000 ampoules in the WCBs. When the working bank is almost depleted then a new one is created from a single ampoule of the MCB. In the event that the MCB is becoming depleted, a new MCB is generated using a single ampoule from the token bank, where available, or MCB if not.

If this hierarchical system of cell banking is used there should be no stock shortages of a cell line. In addition, this will minimize any changes in genotype and phenotype. The level of characterization and quality control required at each stage of hierarchical banking for a cell line will depend on the intended use of the cells and any necessary regulatory requirements (Section 9.2.7). Irrespective of the proposed use of the cells, it is recommended that, as an absolute minimum, each banking stage be tested for microbial contamination (bacteria, yeasts and other fungi, and mycoplasma) and the cell line's species be verified; numerous other tests will also normally be required (see Chapter 6, Table 6.1). The whole issue of cryopreservation and cell banking is covered in greater detail in Chapter 6.

9.2.3 Microbial quality control

Microbial quality control is an essential part of all routine cell culture practice and should not be neglected. It comprises the testing of cell cultures and associated reagents for a variety of microorganisms.

Table 9.1 Common sources of microbial contamination.

Organism	Source
Bacteria	Clothing, skin, hair, aerosols (e.g. due to sneezing or pipetting), insecure caps on media bottles and culture flasks
	Air currents
	Humidified incubators
	Purified water
	Insects
	Plants
	Contaminated cell lines
Fungi (excluding yeasts)	Damp wood or other cellulose products, for example cardboard
	Humidified incubators
	Plants
Yeasts	Humidified incubators
	Operators
Mycoplasma	Contaminated cell lines
	Serum
	Media
	Operators

9.2.3.1 Sources of contamination

These can be divided into three main categories:

- poor laboratory conditions
- personnel (particularly inadequately trained personnel)
- non-quality-controlled cell lines.

Poor laboratory conditions are combated by the implementation of good laboratory conditions as described in Chapter 1, and Section 9.2.1.2 above.

It is important that all staff, in addition to having good aseptic technique while working in the MSC, are aware of the various possible routes and likely sources of contamination (Table 9.1) and are familiar with best practice techniques. The most common sources of contamination are from laboratory personnel (particularly any with poor personal hygiene), and from cell lines received from external sources (although serum can also be a major source of mycoplasma). Wherever possible, obtaining cell cultures from a recognized cell culture collection is advisable.

9.2.3.2 Tests for bacteria, yeasts and fungi

When cell lines are cultured in antibiotic-free media, contamination by bacteria, yeasts or other fungi can frequently be detected by an increase in the turbidity of the medium and/or by a change in pH. Reagents added during the preparation of cell culture media, for example serum, glutamine or growth factors, or used in culture procedures, for example trypsin, can contribute to the risk of contamination. It is therefore good practice to set up quality control checks on culture media and reagents prior to use. The traditional method for the detection of bacteria and fungi involves microbiological culture using enrichment broths.

i. Detection of microbial contamination   Two types of microbiological culture medium have been suggested by the US Code of Federal Regulations [9] and the European Pharmacopoeia [10] for use in the detection of microbial contamination by cultivation. These are (a) fluid thioglycollate medium for the detection of aerobic and anaerobic bacteria and (b) soya bean–casein digest (tryptone soya broth (TSB)) for the detection of aerobes, facultative anaerobes and fungi. They are used as described in *Protocol 9.1*.

PROTOCOL 9.1   Detection of bacteria and fungi by cultivation

Equipment and reagents

- Personal protective equipment (latex medical gloves, laboratory coat, safety glasses)
- Water bath set to 37 °C
- Microbiological safety cabinet at appropriate containment level
- Centrifuge
- Incubator set at 32 °C
- Incubator set at 22 °C
- Soya bean–casein digest (TSB)
- Fluid thioglycollate medium (TGM)
- *Bacillus subtilis* (National Collection of Type Cultures (NCTC), UK)
- *Candida albicans* (NCTC)
- *Clostridium sporogenes* (NCTC)

Method

1 Culture cell line in the absence of antibiotics for at least two passages prior to testing.

2 Bring attached cells into suspension with the use of a cell scraper. Suspension cell lines may be tested directly[a].

3 Inoculate 2× thioglycollate medium (TGM) and 2× tryptone soya broth (TSB) with 1.5 ml of test sample.

4 Inoculate 2 TGM and 2 TSB with 0.1 ml of *C. albicans* (containing 100 colony forming units (cfu)).

(continued overleaf)

CH 9 THE QUALITY CONTROL OF ANIMAL CELL LINES

5 Inoculate 2 TGM and 2 TSB with 0.1 ml of *B. subtilis* (containing 100 cfu).

6 Inoculate 1 TGM with 0.1 ml of *C. sporogenes* (containing 100 cfu).

7 Leave 2 TGM and 2 TSB un-inoculated as negative controls.

8 Incubate broths as follows:

- For TSB, incubate one broth of each pair at 32 °C and the other at 22 °C for 14 days

- For TGM, incubate one broth of each pair at 32 °C and the other at 22 °C for 14 days

- For the TGM inoculated with *C. sporogenes* incubate at 32 °C for 14 days

9 Examine test and control broths for turbidity after 14 days.

Criteria for a valid result

All positive control broths show evidence of bacteria/fungi within 14 days of incubation and the negative control broths show no evidence of bacteria or fungi.

Criteria for a positive result

Test broths show turbidity.

Criteria for a negative result

Test broths should be clear and show no evidence of turbidity.

Notes

[a]The remainder of the Protocol (steps 3–9) must be carried out in a microbiology laboratory away from the cell culture laboratory.

9.2.3.3 Tests for mycoplasma

Mycoplasma infection of cell cultures was first observed by Robinson *et al.* in 1956 [11, 12]. The incidence of mycoplasma-infected cultures has been found to vary between laboratories and arises from poor aseptic technique and insufficient training in good cell culture practices. It is crucial that mycoplasma-free cultures and cell lines are used because the presence of mycoplasma can

- affect the rate of cell proliferation

- induce morphological changes

- cause chromosome aberrations

- influence amino acid and nucleic acid metabolism

- induce cell transformation.

Naturally, regulatory organizations insist that all cell cultures and additives used for the production of reagents for diagnostic kits as well as therapeutic agents are free of mycoplasma infection [13, 14].

i. Hoechst 33258 DNA staining method The fluorochrome dye Hoechst 33258 binds specifically to DNA causing fluorescence when viewed under UV light. A fluorescence microscope equipped for epifluorescence with a 340- to 380-nm excitation filter and 430-nm suppression filter is required for viewing of the stained cells. A detailed description of this technique is provided in *Protocol 9.2*. If test cells are stained directly with Hoechst 33258 it may prove difficult to differentiate between contaminated and non-contaminated cultures if the cells have a low cytoplasm–nucleus ratio. To resolve this problem, and for reasons of standardization, the assay is performed using a suitable monolayer culture as an indicator onto which the test sample is inoculated. This approach has the advantage of also allowing the screening of serum, cell culture supernatants and other reagents not containing cells.

Cell lines used as indicators should have a high cytoplasm/nucleus ratio, for example Vero cells (ECACC Product No. 84113001). Monolayers of indicator cell lines are cultured on coverslips for 2–24 hours prior to inoculation with the test sample. The indicator cells are inoculated at a concentration such that a semi-confluent monolayer is present at the time of staining. Some preparations may exhibit extracellular fluorescence due to disintegrating nuclei. Such fluorescent debris is usually not of a uniform size and is usually too large to be mycoplasma. Contaminating bacteria or fungi will also fluoresce if present but will appear much larger than mycoplasma. A comparative analysis with positive and negative slides is recommended. Positive control slides/coverslips can be prepared from control strains (e.g. *Mycoplasma hyorhinis* and *M. orale*) available from the NCTC, stock numbers NCTC10130 and NCTC10112, respectively. The recommended inoculum for these positive strains is about 100 cfu/dish.

Because of their infectious status these positive controls should be handled by fully trained personnel in specialist facilities separate from the rest of the cell culture laboratory. If such separate facilities are not available positive slides can be supplied commercially by the ECACC.

The main advantages of this method in comparison with the culture isolation method (*Protocol 9.3*) are the speed with which the results to be obtained, and the fact that the non-cultivatable strains such as those of *M. hyorhinis* can be detected. Although this staining method is not as sensitive as the culture method, it is generally considered that approximately 10^4 mycoplasmas per ml are required to produce a clear positive result by the Hoechst staining technique.

PROTOCOL 9.2 Testing for mycoplasma by indirect DNA staining (Hoechst 33258 stain)

DNA staining methods such as Hoechst staining techniques are relatively quick, with results available within 24 h, which compares favourably with 4 weeks for detection by culture. However, the staining of cultures directly with a DNA stain results in a much-reduced sensitivity (10^6 cfu/ml). This may be improved by co-culturing the test cell line in the presence of an indicator cell line such as Vero. This co-cultivation step results in a sensitivity of 10^4 cfu/ml

(continued overleaf)

of culture. This step also improves sensitivity by increasing the surface area upon which my-coplasma can adhere. Like detection by culture (*Protocol 9.3*), DNA staining methods are suitable for the detection of mycoplasma both from cell cultures and from cell culture reagents.

Equipment and reagents

- Media – pre-warmed to 37 °C (refer to the Cell Line Data Sheet or relevant publications for the correct medium)

- 70% ethanol in water

- Methanol

- Acetic acid, glacial

- Fresh Carnoy's fixative (1 : 3 glacial acetic acid: absolute methanol)

- Hoechst 33258 stain solution:[a]

  Stock solution = 0.1 mg/ml in water (filter sterilize using a 0.2-μm filter)

  Wrap container in aluminium foil, store at 4 °C

  Working solution-Prepare fresh each time as necessary

  Make 1 : 1000 dilution of stock in water to yield a final concentration of 0.1 μg/ml.

- Vero cells (ECACC Prod. No. 84113001)

- Mountant (autoclave 22.2 ml of 0.2 M citric acid with 27.8 ml of 0.2 M disodium phosphate. Add 50 ml of glycerol. Filter sterilize and store at 4 °C)

- *M. hyorhinis*, NCTC 10130

- *M. orale*, NCTC 10112

- Personal protective equipment (sterile gloves, laboratory coat, safety visor)

- Sterilizing oven, or Bunsen burner

- MSC of appropriate containment level

- CO_2 incubator set at 37 °C

- Microscope (UV epifluorescent)

- 35-mm plastic tissue culture dishes

- Cell scraper

- Microscope slides and 22-mm coverslips

- Aluminium foil

- Bottle for toxic waste

Method

1 For each sample and control, sterilize two coverslips in a hot oven at 180 °C for 2 h; or immerse in 70% ethanol and flame by passing through a blue Bunsen flame and

(*continued*)

allowing the ethanol to burn off completely. Also sterilize two coverslips to use as a negative control.

2 Place the coverslips in 35-mm culture dishes (one per dish).

3 Store until needed.

4 To prepare the Vero indicator cells, add 2×10^4 cells in 2 ml of antibiotic-free growth medium to each tissue culture dish.

5 Incubate at 37 °C in 5% CO_2 for 2–24 h to allow the cells to adhere to the coverslips[b].

6 Bring attached test cell lines into suspension using a cell scraper. Suspension cell lines may be tested directly.

7 Remove 1 ml of culture supernatant from duplicate 35-mm dishes and add 1 ml of test sample to each. Inoculate two dishes with 100 cfu of *M. hyorhinis* and two with 100 cfu of *M. orale*[c,d].

8 Leave duplicate tissue culture dishes un-inoculated as negative controls.

9 Incubate dishes at 37 °C in 5% CO_2 for 1–3 days[b].

10 After 1 day observe one dish from each pair for bacterial or fungal infection. If contaminated discard immediately. Leave the remaining dish of each pair for a further 2 days.

11 Remove the medium from the dish. Fix cells to coverslip by adding a minimum of 2 ml of freshly prepared Carnoy's fixative (1:3 glacial acetic acid: absolute methanol) to the tissue culture dish and leave for 3–5 min.

12 Decant used fixative to toxic waste bottle. Add another 2-ml aliquot of fixative to coverslip and leave for a further 3–5 min. Decant used fixative to toxic waste bottle.

13 Air dry coverslip by resting it against the tissue culture dish for 30--120 min.

14 Replace coverslip in dish and add a minimum of 2 ml of Hoechst stain[a]. Leave for 5 min shielded from direct light by aluminium foil.

15 Decant used stain, and any unused working stain solution, to toxic waste.

16 Add one drop of mountant to a pre-labelled microscope slide and place coverslip (cell side down) onto slide.

17 Keep slide covered with aluminium foil, allowing it to set for at least 15 min at 37 °C or for 30 min at room temperature.

18 Observe slide under UV epifluorescence at ×1000 magnification.

Criteria for a valid result[e]

Negative controls show no evidence of mycoplasma infection
    Positive controls show evidence of mycoplasma infection
    Vero cells clearly seen with fluorescing nuclei.

(*continued overleaf*)

Criteria for a positive result[e]

Cells treated with test samples are seen as fluorescing nuclei plus extra-nuclear fluorescence of mycoplasma DNA (small cocci or filaments).

Criteria for a negative result[e]

Cells treated with test samples are seen as fluorescing nuclei against a dark background. There should be no evidence of mycoplasma (Figure 9.2b,c).

Notes

[a] Hoechst stain is a toxic mutagen and should be handled and discarded with care and in accordance with local safety procedures.

[b] Culture dishes should be placed in a sealed box or in large Petri dishes to reduce evaporation.

[c] These positive controls must be handled in a laboratory remote from the main tissue culture laboratory. Although this procedure recommends the setting up of positive controls, this may be neither feasible nor desirable in a cell culture facility with limited resources. If this is the case, then positive slides can be purchased from ECACC. If positive controls are not being used then it is strongly recommended that you get an independent testing laboratory to periodically test your cell lines.

[d] Control organisms are available from the NCTC, Porton Down, UK.

[e] In some instances results may be difficult to interpret for the following reasons:

- bacterial/yeast/fungal contamination

- too much debris in the background

- broken nuclei as cells are all dead

- too few or no live cells.

ii. Culture isolation method Using specialized agar and broth media, most mycoplasma strains can be cultured *in vitro* with the exception of certain strains of *M. hyorhinis*. The methods are discussed in detail in reference 16. As is the case with microbial contamination detection methods, the culture method described in *Protocol 9.3* should be performed in a dedicated laboratory space separate from the main cell culture area.

Before use in this assay, both agar and broth media should be assessed for their ability to support the growth of species of mycoplasma known to contaminate cell cultures, for example *Acholeplasma laidlawii, M. arginini, M. fermentans, M. hominis, M. hyorhinis and M. orale*. These type strains are available from the NCTC. The main advantage of this method is that, theoretically, one viable mycoplasma per inoculum can be detected, compared with 10^4 per ml for the Hoechst staining method.

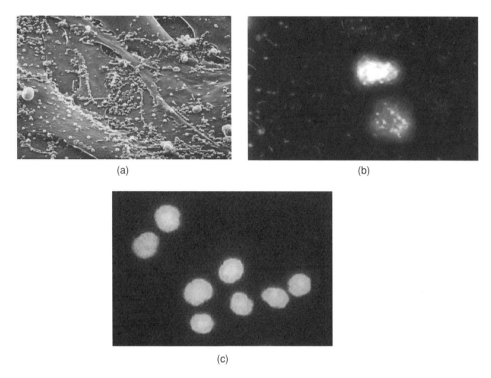

(a)

(b)

(c)

Figure 9.2 (a) Scanning electronmicrograph of a culture of human diploid fibroblasts infected with mycoplasma. Reproduced by courtesy of NIBSC - a Centre of the Health Protection Agency. (b) Mycoplasma infected Vero cells stained with Hoechst 33258 viewed under UV epifluorescence at ×1000 magnification through a blue filter. (c) Uninfected (negative control) Vero cells stained with Hoechst 33258 viewed under UV epifluorescence at ×1000 through a blue filter.

PROTOCOL 9.3 Detection of mycoplasma by culture[a]

Detection of mycoplasma by culture is the reference method of detection and has a theoretical level of detection of 1 cfu. However, there are some strains of mycoplasma that are non-cultivable (certain strains of *M. hyorhinis*). The method is suitable for the detection of mycoplasma in both cell cultures and cell culture reagents and results are obtained within 4 weeks. Mycoplasma colonies observed on agar plates have a 'fried egg' appearance (Figure 9.3).

Equipment and reagents

- 70% ethanol in water

- Mycoplasma Pig Agar plates (in 5-cm Petri dishes)

- Mycoplasma Pig Agar broths (in 1.8-ml aliquots)

(continued overleaf)

- *M. orale* NCTC 10112[b]
- *M. pneumoniae* NCTC 10119[b]
- Test samples (see *Protocol 9.2,* step 6)
- Personal protective equipment (sterile gloves, laboratory coat, safety visor)
- Water bath set to 37 °C
- MSC of appropriate containment level
- CO_2 incubator set at 37 °C
- Gas Jar (Gallenkamp)
- Gas Pak Anaerobic System (Gallenkamp)
- Gas Pak Catalyst (Gallenkamp)
- Gas Pak Anaerobic Indicator (Gallenkamp)

Method[a]

1 Inoculate two agar plates with 0.1 ml of test sample.

2 Inoculate one agar plate with 100 cfu *M. pneumoniae*.

3 Inoculate one agar plate with 100 cfu *M. orale*.

4 Leave one agar un-inoculated as a negative control.

5 Inoculate one broth with 0.2 ml of test sample.

6 Inoculate one broth with 100 cfu *M. pneumoniae*.

7 Inoculate one broth with 100 cfu *M. orale*.

8 Leave one broth un-inoculated as a negative control.

9 Incubate agar plates anaerobically for 14 days at 37 °C using a gas jar with anaerobic Gas Pak and catalyst.

10 Incubate broths aerobically for 14 days at 37 °C.

11 During incubation, between days 3 and 7 and again between days 10 and 14, subculture 0.1 ml of test broth onto an agar plate and incubate all plates anaerobically as above.

12 Observe agar plates after 14 days' incubation, at ×300 magnification using an inverted microscope, for the presence of mycoplasma colonies[c].

Criteria for a valid result

All positive control agar plates and broths show evidence of mycoplasma by the formation of characteristic colonies on agar plates and usually a colour change in broths.
    All negative control agar plates and broths show no evidence of mycoplasma.

Criteria for a positive result

Test agar plates/broths show characteristic colony formation.

(continued)

Criteria for a negative result

All test agar plates and broths show no evidence of mycoplasma.

Notes

[a]The whole protocol must be carried out in a laboratory remote from the main tissue culture laboratory.

[b]Control organisms (*M. pneumoniae* and *M. orale*) are available from NCTC, Porton Down, UK. *M. pneumoniae* is a potential pathogen and must be handled in a Class II MSC operating to ACDP Category 2 conditions (or equivalent outside the UK).

[c]Mycoplasma colonies have a characteristic colony morphology commonly described as 'fried egg" (Figure 9.3) due to the opaque granular central zone of growth penetrating the agar surrounded by a flat translucent peripheral zone on the surface. However, in many cases only the central zone will be visible.

iii. Other methods of mycoplasma testing PCR-based detection assays for mycoplasma are useful for screening a large number of samples quickly with results being obtained within a few hours [15]. They are also useful for detecting contamination with the non-cultivable strains of *M. hyorhinis*. The presence of contaminant mycoplasma is easily detected using agarose gel electrophoresis by verifying the bands of amplified DNA fragments from test samples against both positive and negative controls. It is recommended that PCR should be carried out in conjunction with culture isolation and DNA staining methods. One alternative to the Hoechst and cultivation methods that has become widely used as a screening method (and as an alternative to the PCR method) in cell culture laboratories is the rapid and easy luminometric detection assay MycoAlert.

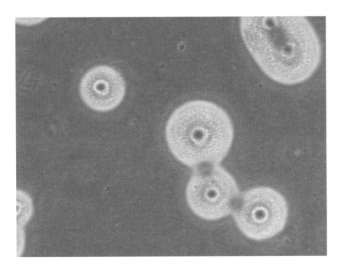

Figure 9.3 *Mycoplasma gallisepticum* colony on a pig agar plate. Note the typical "fried egg" morphology. Light micrograph, viewed under bright field illumination ×400 magnification.

The MycoAlert Assay™ kit contains all the reagents for a selective biochemical test that exploits the activity of certain mycoplasma enzymes. Viable mycoplasmas are lysed, and the enzymes released react with the MycoAlert substrate catalysing the conversion of ADP to ATP. By measuring the level of ATP in a sample before and after addition of the MycoAlert substrate, a ratio can be obtained indicating the presence or absence of mycoplasma. If mycoplasma enzymes are absent, the second reading shows no increase over the first. The concentration of ATP can be detected using a bioluminescent reaction based on luciferase activity. The emitted light intensity is linearly related to the ATP concentration and is measured using a luminometer.

9.2.3.4 Virus testing

There are three major concerns pertinent to the virus contamination of cell lines. These are:

(i) The safety of laboratory staff handling the cell lines and performing the procedures required for the reduction or elimination of the associated risk in experimental and other procedures.

(ii) The medicinal safety of any products derived from cells and their clinical application. This problem also impinges on regulatory matters which are discussed in more detail in Section 9.2.7.

(iii) The validity of any experimental data produced using contaminated cell lines.

i. Sources of viral contamination   Potential sources of viral contamination are:

- the original tissue used to prepare the cell line
- other infected cultures
- growth medium
- serum
- other reagents (e.g. trypsin)
- laboratory personnel.

Each of these is explored in more detail below.

- *Tissue-derived contamination*

The factors influencing the choice of starting material for cell line derivation should take into account the possibility of existing virus infection. The viruses potentially involved will depend on the species of origin, the tissue taken and the clinical history of the animal or patient. An evaluation of the viruses endemic to the population from which the tissue is isolated should be performed. A list of the commonly occurring viruses in humans is provided in Table 9.2.

Table 9.2 Some of the more common pathogenic viruses that can occur in humans.

Virus	Tissue involved	Persistence *in vitro*
Herpes simplex virus-1	General	+
Herpes simplex virus-2	General	+
Human cytomegalovirus	General	+
Epstein–Barr virus	General	+
Hepatitis B	General	+
Hepatitis C	General	+
Human herpes virus 6	General	+
Human immunodeficiency virus-1	General	+
Human immunodeficiency virus-2	General	+
Human T-cell lymphotropic virus-I	General	+
Human T-cell lymphotropic virus-II	General	+
Adenovirus	General	±
Reovirus	General	−
Rubella	General	−
Measles	General	+
Mumps	General	−
Human parvovirus	General	+
Varicella zoster	General	+
Respiratory syncytial virus	Respiratory	−
Influenza A	Respiratory	−
Influenza B	Respiratory	−
Parainfluenza	Respiratory	−
Rhinovirus	Respiratory	−
Coronavirus	Respiratory	−
Poliovirus	Enteric	−
Coxsackie A	Enteric	−
Coxsackie B	Enteric	−
Echovirus	Enteric	−
Rotavirus	Enteric	−
Norwalk virus	Enteric	−
Calicivirus	Enteric	−
Astrovirus	Enteric	−
Papillomavirus	Skin/epithelium	+
Poxvirus	Skin/epithelium	−

+, easily persists *in vitro*; −, poorly persistent *in vitro*.

A risk assessment must be made of the incidence of viral infection in the population from which clinical specimens are derived and in the laboratory staff engaged in tissue processing. For example, 80–90% of the human population has experienced Epstein–Barr virus (EBV) infection and may still carry the virus in cells from their peripheral blood, lymph nodes or spleen [17]. There is, therefore, a high risk of EBV contamination associated with human material. This is counterbalanced by the fact that most laboratory staff will also be EBV sero-positive (i.e. immune to reinfection). However, adequate precautions must be taken to ensure that sero-negative staff are not exposed to a risk of infection.

A survey of blood donors revealed a 0.2–0.5% incidence of hepatitis B virus (HBV) infection in healthy donors [18]. Thus staff of a laboratory receiving samples of human material are at a high risk of exposure to HBV-contaminated material. Staff may be protected by vaccination and the correct handling of clinical materials. On the other hand, the incidence of human immunodeficiency virus (HIV)/human T-lymphotrophic virus (HTLV) infection in non-high-risk groups is low, and thus the risk of handling contaminated clinical material remains correspondingly low. The risk of sample contamination increases with clinical samples from higher risk groups, including homosexuals, bisexuals, intravenous drug users and inhabitants of geographical areas where such viral infections are endemic [19].

The viruses listed in Table 9.3 have been shown to be present in rodent populations. Those of greatest concern are the first five, namely Hantavirus, lymphocytic choriomeningitis virus (LCM), rat rotavirus, reovirus type 3 and Sendai virus, all of which are known to be pathogenic in humans. Laboratory-derived infection of humans with Hantavirus and LCM has been reported [20].

• *Contamination by other cultures in the laboratory*

If there is inadequate quarantining and testing of incoming cultures (see Section 9.2.1.2), inadequate segregation of different cultures, and poor practices and handling of cells in the laboratory (e.g. handling more than one cell line in the MSC at any one time – see Chapter 4, Section 4.2.1.2), then the potential for a virally contaminated culture to cross-contaminate other cultures is clear.

• *Medium-derived contamination*

The potential for basal media to pass on viral contamination to cells has been greatly reduced in recent years by the move amongst commercial media suppliers to source all components, as far as possible, from non-animal sources. However, cases of basal medium contamination by minute virus of mice (MVM) have been reported, probably due to poor rodent pest control during the shipping of powdered medium [21].

• *Serum-derived contamination*

The risk of viral contamination of serum can be reduced (but not entirely eliminated) by careful sourcing – for example, from countries where certain viruses are not endemic – and testing of the serum pool and/or bottled product. For most applications, BVDV, infectious rhinotracheitis virus and para-influenza-3 virus are probably the major concerns relating to bovine serum. It is known, for example, that 50–90% of cattle in the USA are infected with BVDV. Extensive testing of serum for the presence of viruses is performed by reputable suppliers, but levels of detection are limited by the sampling

Table 9.3 Rodent viruses potentially pathogenic in man.

Virus	Source species	Detected by MAP/RAP test?[a]
Hantavirus	Mouse, rat	Yes
Lymphocytic choriomeningitis virus	Mouse	Yes
Rat rotavirus	Rat	No
Reovirus type 3	Mouse, rat	Yes
Sendai virus	Mouse, rat	Yes
Ectromelia virus	Mouse	Yes
K virus	Mouse	Yes
Kilham rat virus	Rat	Yes
Lactate dehydrogenase virus	Mouse	Yes
Minute virus of mice	Mouse, rat	Yes
Mouse adenovirus	Mouse	Yes
Mouse cytomegalovirus	Mouse	Yes
Mouse hepatitis virus	Mouse	Yes
Mouse polio virus	Mouse	Yes
Mouse rotavirus, for example epizootic diarrhoea of infant mice	Mouse	Yes
Pneumonia virus of mice	Mouse, rat	Yes
Polyoma virus	Mouse	Yes
Rat coronavirus	Rat	No
Retroviruses	Mouse, rat	No
Sialodacryoadenitis virus	Rat	Yes
Thymic virus	Mouse	Yes
Toolan virus	Rat	Yes

[a] MAP, mouse antibody production; RAP, rat antibody production.

required and the sensitivity of the techniques employed. The level of contamination with viable virus particles can be considerably reduced by treating serum with one of a variety of methods, of which gamma-irradiation is probably the best in terms of both virus kill and maintenance of the serum's growth-promoting activity. For a detailed discussion of this whole topic (and other aspects of serum and its use in cell culture) see reference 22.

- *Other reagents*

Trypsin is widely used as a detachment agent in many cell culture protocols. It is usually of porcine origin and has been shown to be a potential source of porcine parvovirus [23]. Other agents used in cell culture, and that are derived from a human or animal source (e.g. albumin, transferrin), must also be considered a potential source of viral contaminants from that species.

• *Personnel*

Virus contamination originating from individuals performing cell culture has, over the years, proven difficult to document. However, the potential clearly exists, as a number of virus types, including rhinoviruses, rotaviruses and respiratory syncytial virus, have been shown to be extremely stable in aerosols, on workers' hands and on the surfaces of MSCs [24].

ii. Methods of virus detection   A range of methods exists for the detection of viruses, from classical haemadsorption (where the presence of certain viruses can be detected by the biding of erythrocytes to infected cells; Figure 9.4) to recent PCR-based assays. A brief discussion of the techniques employed is provided below. No single method will detect all viruses; for example, not all viruses cause haemadsorption or exhibit a cytopathic effect. PCR-based techniques are fast replacing the classical cell-based assays. Exhaustive screening for viral contaminants is highly specialized and can be an expensive proposition. It is advisable to review the certificate of analysis for all culture reagents prior to purchase to determine the extent of viral detection assays performed before release.

• *Co-cultivation*

In this method a sample of a test cell line or clinical sample is incubated with semi-confluent monolayers of a range of cell lines susceptible to a wide variety of viruses. These co-cultivations are maintained by passaging for 3 weeks and the host (indicator) cell lines are checked for the presence of cytopathic effects and haemadsorption (Figure 9.4). The following cell lines are susceptible to a wide range of viruses and often used, depending on the species of the test cell line/sample:

Figure 9.4 Phase-contrast micrograph (×400) of an MDCK cell monolayer infected with Influenza A virus. Haemadsorption of the added (phase-bright) adult chicken erythrocytes can be clearly seen.

- BHK21 (hamster)
- WI 38 (human)
- HeLa (human)
- Vero (monkey)
- MDCK (canine)
- JM (sensitive to HIV-1)
- H9 (sensitive to HIV-1)
- T cells in primary culture (sensitive to HTLV-I).

The co-cultivation method can be misleading as not all endogenous viruses will demonstrate cytopathic effects, but it can be a useful tool for detecting a spectrum of viruses where the primers necessary for RT-PCR assays may be too specific and thus restrict the range of viruses that can be detected.

- *Electron microscopy*

Transmission electron microscopy has been used not only to detect viral contaminants but also to identify them and, in certain cases, indicate the level of contamination. This method allows a wide range of viruses and virus-like particles to be identified by their characteristic morphology (Figure 9.5). It has been applied to a variety of cell lines including mouse–mouse hybridomas, and recombinant CHO lines producing monoclonal antibodies and recombinant therapeutic proteins. A detailed discussion of the protocols involved is given in reference 25.

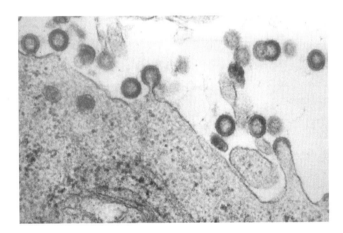

Figure 9.5 Transmission electron microscopy of human T lymphoblastoid cell line C1866 infected with simian immunodeficiency virus (SIV strain 32H). Budding of virus particles can be seen along the surface of the cell membrane 2 days post infection. (Magnification: ×84000) Image courtesy of © B.Dowsett, Health Protection Agency.

- In vivo *methods*

In vivo methods for the detection of viral contaminants may involve the inoculation of materials by a variety of routes into a range of laboratory animals of different ages. The animals are subsequently examined for evidence of adverse effects [26]. The murine antibody and rat antibody production tests are designed to detect a range of rodent viruses, as shown in Table 9.3. Briefly, the animals are inoculated with the test material and examined for the production of antibodies to the listed viruses. In the case of lactate dehydrogenase virus a raised level of enzymic activity is tested for. Many of these animal systems are being replaced by PCR-based assays and are used only as a final qualitative test.

- *Cell culture assays for murine retroviruses*

Murine leukaemia viruses (MuLV) may be ecotropic (infecting only rodent cells), xenotropic (infecting only non-rodent) or amphotropic (infecting several cell types) depending on the surface molecules of the virus and the cognate receptors on the animal cell. In the case of testing for replication-competent wild-type virus, xenotropic MuLV may be detected by causing the formation of foci in S^+L^- mink cells [27] and wild-type replication-competent ecotropic MuLV may be detected by the formation of syncitia and vacuoles in XC cells [28].

- *Reverse transcriptase assay for retrovirus detection*

This method (*Protocol 9.4*) uses polyethylene glycol to precipitate the protein and any virus particles present in cell-free lysates. The extracts are then assayed for reverse transcriptase (RT) activity by the incorporation of thymidine triphosphate (TTP), labelled with ^{3}H or a chemical/fluorescent tag, on to a poly-(rA):poly-(dT) primer template. The presence of host DNA polymerase, which may give high background incorporation on an RNA template, is detected by incorporation of the labelled TTP on to a poly-(dA):poly-(dT) primer template. These reverse transcription assays are usually performed using both RNA and DNA templates.

PROTOCOL 9.4   Detection of retroviruses in cell supernatants by reverse transcriptase assay

Equipment and reagents

- [^{3}H]-TTP (1.1 TBq/mmol)
- Positive control reverse transcriptase preparations (Roche Applied Science) (see step 7)

(continued)

- Centrifuge

- Supernatant sample

- PBS (pH 7.4)

- polyethylene glycol (PEG) 8000 solution (30% w/v) in PBS pH 7.4

- Virus solubilizing buffer: 0.8 M NaCl, 0.5% (v/v) Triton X-100, 0.3 mg/ml phenylmethylsulpho-nylfluoride, 20% (v/v) glycerol, 50 mM tris(hydroxymethyl)aminomethane (Tris)-HCl, pH 8.0, 4 mM dithiothreitol (DTT)

- Solution A (DNA template): 60 mM Tris-HCl, pH 7.5. 1.3 mM DTT, 0.1 mM ATP, 0.6 A_{260} units per ml poly(dA):poly(dT), 12 mM $MgCl_2$ (store at −20 °C).

- Solution B (RNA template): as solution A except that poly (dA):poly(dT) template primer is replaced by poly(rA):poly(dT) at the same concentration, (store at −20 °C).

- Solution C (DNA template): 60 mM Tris-HCl, pH 8.5, 1.3 mM DTT, 0.1 mM ATP, 0.6 A_{260} units per ml poly(dA):poly(dT), 0.3 mM $MnCl_2$ (store at −20 °C).

- Solution D (RNA template); as solution C except that poly(dA):poly(dT) template primer is replaced by poly(rA):poly(dT) at the same concentration (store at −20 °C).

Method

1 Take supernatant from cells in logarithmic phase of growth, as dying cells release host DNA polymerase, which gives a high background in the assay. Prepare cell-free supernatant (4 ml) by centrifugation at 400 g at 4 °C for 10 min.

2 Precipitate virus and protein from the supernatant by addition of 2.5 ml of 30% (w/v) PEG 8000 in PBS pH 7.4, and incubate at 4 °C for 18 h.

3 Collect precipitate by centrifugation at 800 g at 4 °C for 30 min, discard supernatant and resuspend pellet in 0.3 ml of virus solubilizing buffer.

4 Store samples in LN (vapour phase) until required.

5 Pipette 20-µl aliquots of [^{3}H]-TTP (1.1 TBq/mmol, 30 Ci/mmol) into 1.5-ml microfuge tubes (four for each test sample and two for each positive control). Place cotton wool plugs in the neck of tubes and dry in 37 °C incubator.

6 Add 150 µl of solution A to the first tube, 150 µl of solution B to the second, 150 µl of solution C to the third and 150 µl of solution D to the fourth.

7 For the positive controls, for example avian myeloblastosis virus (AMV) RT (Promega), add 150 µl of solution A to one tube and 150 µl of solution B to the second; for Moloney murine leukaemia virus (MoMLV) RT add 150 µl of solution C to one tube and 150 µl of solution D, to the second.

8 Seal all the tubes, place on a vortex mixer and mix gently. Allow any aerosol to settle for 5 min before opening tubes. Keep on ice.

9 Divide the contents of each tube into two aliquots of 75 µl each.

(continued overleaf)

10 Add 20 μl of test sample to appropriate tubes.

11 For positive controls add 30 units of AMV RT or 30 units of MoMLV RT in 20 μl of virus solubilizing buffer.

12 Seal the tubes and vortex. Incubate for 90 min in a water bath at 37 °C.

13 Terminate the reaction by the addition of 0.5 ml of cold 10% trichloroacetic acid (TCA) to each tube to precipitate any incorporated radioactivity.

14 Set up a Millipore filter box (or equivalent) with 2.4-cm GF/C filters (Whatman) and pre-wash filters with 20% TCA.

15 Pour contents of each tube onto a separate filter and rinse out each tube twice with 1 ml of cold 20%TCA and pour rinse on to relevant filter.

16 Wash each filter four times with 20 ml of cold 5% TCA.

17 Wash each filter once with 20 ml of cold 70% ethanol.

18 Remove filters and dry at room temperature.

19 Add each filter to a separate scintillation vial containing 5 ml of scintillation fluid.

20 Count on a beta-counter for 1 min.

21 Compare the activity bound to control (host polymerase) with assay filters (RNA templates). If their values are similar, or sample activity is below control activity, then the result is negative.

22 Positive controls should give an activity greater than 10 000 cpm and RNA/DNA template incorporation ratio of at least 5.

• *PCR method for detection of BVDV and other viruses*

BVDV crosses the placenta of pregnant cows and therefore is a common contaminant of FBS. PCR-based techniques provide more rapid results than the previously used co-cultivation and immuno-assay methods. PCR following an initial RT step can be performed in a single day and gives a clear qualitative (positive/negative) result. This method can readily determine the BVDV status of a cell line. For a detailed description of this topic and relevant protocols see reference 29. Most commercially available sources of serum are tested by the manufacturer prior to release and test results are provided as part of a Certificate of Analysis (C of A). PCR-based techniques can be used to detect any virus whose nucleic acid sequence is known. However, uncharacterized virus strains with divergent target sequences (non-conserved) may be missed by PCR due to poor/weak hybridization with the primers employed. In addition, the PCR-based assay does not give information on the potential infectivity of the viral contaminant detected. Accordingly, a function-based assay is advisable as a confirmatory test [30].

9.2.4 Transmissible spongiform encephalopathy

The risk of contamination of tissue culture cells with transmissible spongiform encephalopathy (TSE) agents as a result of the use of animal products as medium components has been considered to be low, in part because only a few (brain-derived) cell lines have been reported to be susceptible to TSE infection [31].

Nevertheless, in response to regulators' concerns, consideration must be given to the origin of FBS and other bovine sera. Most companies supplying bovine serum can supply from source countries with no reported cases. In addition, the infectious agent (a prion) is likely to be present predominantly in neural tissue, which is beyond the blood–brain barrier, and there is also a species barrier to consider. With these facts in mind it is considered that serum poses little risk of transmitting BSE [32]. However, those using bovine sera at any stage in the development or production of a therapeutic product should refer to the European guidance document on minimizing the risk of TSE transmission [33].

9.2.5 Eradication of contamination

If a cell culture is found to be contaminated with bacteria, fungi, mycoplasma or virus then the best method of elimination is to discard the culture. Fresh cultures can be obtained from seed stock, from a bio-repository or from the originator of that cell line. It is important to locate the source of the contamination (media, culture reagents, faulty safety cabinet, poor aseptic technique, etc.) to prevent a recurrence. As well as eliminating the contamination in any particular culture, it is important to eliminate the source of contamination by using completely new and tested media and reagents, cleaned and tested equipment and, where applicable, retrained staff. After any major contamination, the laboratory and MSC should, if possible, be fumigated according to the manufacturer's instructions (where applicable) (see also Chapter 2, *Protocols 2.5 and 2.6*) and all surfaces swabbed with a suitable disinfectant.

In the case of irreplaceable stocks it will be necessary to attempt to eliminate the contamination using antibiotics, using the method given in *Protocol 9.5*. This approach is possible for bacteria, mycoplasma, yeast and fungi. However, there are no reliable methods for the eradication of viruses from infected cultures. All elimination treatments must be performed in a facility separate from the main cell culture laboratory. It must be remembered that the majority of contaminating agents replicate faster than most cell cultures and so the chances of success are not high. Once the contaminant is identified and an antibiotic/antimycotic agent(s) has been chosen, the eradication process should be rigorously challenged to ensure that it is likely to be successful. This may require detailed validation using 'spiking' experiments to ensure that the methodology used is capable of eliminating the contaminant. Finally, all treated cultures must undergo a full regimen of quality control tests. At ECACC, after an eradication treatment we typically put treated cultures through a series of extended passages prior to retesting to confirm the absence of the contaminant.

PROTOCOL 9.5   Eradication of microbial contamination by antibiotic treatment

Equipment and reagents

- Personal protective equipment (sterile gloves, laboratory coat, safety visor)

- Water bath set to 37 °C

- MSC of appropriate containment level

- Centrifuge

- CO_2 incubator set at 37 °C

- Microscope

- Pipette

- Cell scraper

- 25-cm^2 culture flasks

- Appropriate growth media

- FCS

- Antibiotic solution (Table 9.4)

- Cell line to be treated.

Method

1 One ampoule is thawed and the cells resuscitated according to standard laboratory procedure[a].

2 Cells are maintained for at least two passages in 25-cm^2 flasks before treatment with an appropriate antibiotic. If cells are heavily contaminated with bacteria or fungi, give the culture several washes in serum-free medium and replace with the medium in which the cells are normally grown prior to antibiotic treatment. Each subculture should be performed at the lowest cell density at which growth occurs for the cell line.

3 Cells are maintained in the presence of antibiotics with relevant anti-microbial activity (Table 9.4) for a period of 14–21 days with the appropriate number of passages.

4 At 14 to 21 days the cells are passaged without antibiotics to three 25-cm^2 flasks.

5 (a) At the second passage without antibiotics one of the three flasks is tested for the presence of the relevant type of microbe (see Sections 9.2.3.2 and 9.2.3.3).

   (b) If this result is negative, the second flask is expanded for a token freeze of at least five ampoules. If the contaminant is still detectable then go back to step 1 and use a different antibiotic on an untreated sample of culture (to prevent the possibility of building up antibiotic resistance).

   (c) If the result at step 5(b) is negative, the remaining flask is subcultured continuously for a total of 10 passages, maintaining three flasks at each passage. One flask is tested at each passage and the results recorded. If the culture still tests negative after 10 passages

(continued)

the contamination is considered to have been eradicated. If contamination is detected again, go back to step 1 again and use a different antibiotic.

6 After 10 contamination free passages an ampoule from the token freeze is resuscitated, grown for two passages in antibiotic-free medium and tested for the presence of microbial contamination. If at any stage microbial contamination reappears then the eradication procedure will start again with an original culture, either with a different antibiotic or combination of antibiotics.

Note

[a]The whole protocol must be carried out in a laboratory remote from the main tissue culture laboratory.

Table 9.4 Antibiotics that can be used in the elimination of microbial contamination.

Antibiotic	Working concentration	Active against
Amphotericin B	2.5 mg/l	Yeasts and other fungi
Ampicillin	2.5 mg/l	Bacteria (Gp, Gn)
Cephalothin	100 mg/l	Bacteria (Gp, Gn)
Ciprofloxacin	10–40 mg/l	Mycoplasma, bacteria (Gp, Gn)
Gentamycin[a]	50 mg/l	Bacteria (Gp, Gn), mycoplasma
Kanamycin	100 mg/l	Bacteria (Gp, Gn), yeasts
Mycoplasma removal agent (MRA)[b]	0.5 mg/l	Mycoplasma
Neomycin	50 mg/l	Bacteria (Gp, Gn)
Nystatin	50 mg/l	Yeasts and fungi
Penicillin-G[d]	100,000 U/l	Bacteria (Gp)
Plasmocin[c]	25 mg/l	Mycoplasma
Polymyxin B	50 mg/l	Bacteria (Gn)
Streptomycin sulphate[d]	100 mg/l	Bacteria (Gp, Gn)
Tetracyclin	10 mg/l	Bacteria (Gp, Gn)

Gp, Gram positive; Gn, Gram negative.
[a]Only active against some mycoplasmas.
[b]Supplied by MP Biomedicals.
[c]Supplied by Invivogen.
[d]Usually supplied as a mixture of penicillin-G and streptomycin sulphate.

9.2.6 Authentication

Authentication of cell lines to confirm their identity and source of origin is an essential requirement in the management of cell stocks. Cross-contamination between cell cultures was first recognized in the late 1950s but has become more prevalent in recent times [8]. Surveys of the scientific literature have suggested that up to 20% of published work based on the use of human cell lines specifies lines that are known to be misidentified [34, 35].

This problem very often means that experimental data generated by the use of cross-contaminated or misidentified cell lines is invalid, and undermines the reliability of research based on its inclusion and use in future scientific endeavours [36]. The ECACC and other culture collections have an obligation to raise awareness of this problem in the cell culture community and many bio-resource centres now actively post details of misidentified cell lines on their websites. In a 2007 call for action to combat the use of misidentified and cross-contaminated cell lines, Dr Roland Nardone supported by a panel of cell culture experts including those at ECACC, submitted an open letter [35] to the Secretary of the US Department of Health and Human Services. The letter highlights the grave consequences of the continued failure of many users of cell cultures to take appropriate precautions to verify cell line identity and recommends approaches to end this failure. The recommendations include establishing cell culture practice compliance policies and education initiatives across the research community enforced by grant funding agencies and journals.

As more laboratories undertake authentication of their cell stocks, access to databases or libraries of authentication data is in demand. Bio-resource centres endeavour to make this information available, and the concept of having one super-database to house all available profiles has often been discussed. This will likely be explored in future years.

Cross-contamination is most likely to occur via poor cell culture practices, such as handling more than one cell culture at a time. It is vitally important that cell lines, their growth media and additives, paperwork and identification (labelling) be segregated, either in time or in space. In addition, it is vital that all tissue culture plastic and other materials are removed from the MSC, which is then disinfected before work with the next cell line. It is poor cell culture practice to work with more than one culture at one time, use the same bottle of medium and components for more than one cell line, and not to clearly separate the paperwork and labels for each cell line (see also Chapter 4, Section 4.2.1.2). Failure to do so can result in a culture being cross-contaminated with another cell line.

In the (somewhat unlikely) case where both cell lines in a contaminated culture replicate at the same rate, then their presence can be detected by differences in their morphology or determined by other criteria. If the contaminating cell line replicates faster than the host line, then within several population doublings the contaminating cell line will completely replace the original cell culture. Finally, where poor segregation of paperwork and labelling occurs a culture can be mislabelled. All of these problems are more likely to occur in cultures taken to high passage numbers or where slowly growing cultures are maintained for long periods of time.

Cross-contamination may well go unnoticed in laboratories using many cultures with similar morphologies, for example fibroblast cultures. In order to avoid the problem at the outset, cell lines must be obtained from accredited and quality-controlled sources such as the original depositor or an established cell culture collection rather than being passed from one laboratory to another.

It is important to emphasize that cell authentication is an essential part of quality assurance for both research and commercial use of cell cultures and should be of primary concern for everyone [34].

Three main methods are used at ECACC for cell line authentication: iso-enzyme analysis, cytogenetic analysis and DNA fingerprinting/STR analysis; outlines of each are given in the following sections.

9.2.6.1 Iso-enzyme analysis

Iso-enzymes are variants of polymorphic enzymes that catalyse the same biochemical reaction but have different electrophoretic mobilities. A standardized method, the Authentikit system (Innovative Chemistry), is available in kit form. In its basic format this provides for the detection of seven different enzyme reactions. In many cases a good indication of the species of origin can be obtained with the results of two enzymes, glucose-6-phosphate dehydrogenase and lactate dehydrogenase (Figure 9.6).

As the number of enzymes tested increases, a composite profile is developed and confidence in species identification is increased. However, it is difficult to achieve a unique identification for a cell line. Mixtures of cell lines from two species can also be detected provided the level of contamination is greater than 20%. Thus iso-enzyme analysis enables the rapid speciation of the cell cultures and is valuable in routine testing.

More recently two molecular methods have been introduced for species determination of cell cultures.

The first is a PCR restriction fragment length polymorphism (RFLP) assay based on the use of a pair of primers that anneal to a portion of the cytochrome b gene. The amplification product is digested with a panel of six restriction enzymes, and the pattern derived is resolved on a 3% high-resolution agarose gel. This protocol produces a unique restriction pattern, and the origin of animal cells as determined by this analysis is in agreement with iso-enzyme analysis results. Cytochrome b PCR-RFLP amplifies target sequences using very low amounts of DNA, and its sensitivity in detecting interspecies cross-contamination is similar to that of iso-enzyme analysis [37].

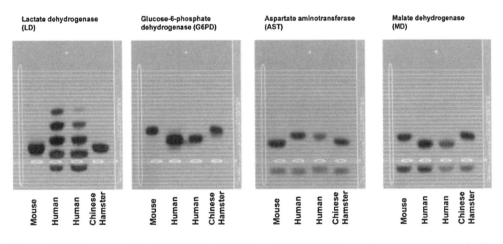

Figure 9.6 Isoenzyme Analysis (Authentikit) Species Identification - Isoenzyme analysis looks at a range of enzymes which are present in nearly all species but demonstrate heterogeneity between species. These isoenzymes have similar substrate specificity but different molecular structures which affects their electrophoretic mobility. Each species produces a specific mobility pattern.

The second method is DNA barcoding, which is becoming more prevalent as a protocol for species determination. This method is based on the use of the mitochondrial DNA gene, cytochrome c oxidase I (COI) as a genetic marker for species identification. A 648-base-pair region of this gene is used to generate a 'barcode'. Cell line DNA is extracted and sequenced, and the sequence data are entered onto the BOLD database of COI sequences for comparison and identification of a species. As of 2009, the database of COI sequences included at least 620 000 specimens from over 58 000 species of animals, larger than databases available for any other gene [37].

9.2.6.2 Cytogenetic analysis

Cytogenetic analysis is used to establish the common chromosome complement or karyotype of a species or cell line [38, 39]. Using this technique it is possible to detect both changes in cell cultures and the occurrence of cross-contamination between cell lines. The methodology has therefore been important in the quality control of cell lines used to produce biologicals, but is no longer considered necessary in certain cases, particularly where the final product does not contain cells but rather is a highly purified molecule [40]. The technique is complicated and labour intensive, and requires a high degree of training and experience. There are many different methods described for staining and banding chromosomes; these include Geimsa (G) banding, G11 banding, Quinacrine banding, chromosome painting and spectral karyotyping.

9.2.6.3 DNA fingerprinting

The discovery of hypervariable regions of repetitive DNA within the genomes of many organisms led to the development of DNA fingerprinting by Jeffreys and colleagues in 1985 [41], which can specifically identify individuals. A range of probes and methods have been developed to exploit the repetitive DNA sequences found throughout the animal kingdom [42]. Multi-locus probes identify many loci in a given genome. Using the multi-locus probes developed by Jeffreys (33.15 and 33.6), the chance of finding two unrelated human individuals with identical fingerprints is reported to be less than 5×10^{-19}, indicating the technique's resolving power. Multi-locus probes can also detect cross-contamination from a wide range of species because, under the correct hybridization conditions, they are able to cross-hybridize with a wide range of repetitive DNA sequences occurring in many species. Multi-locus fingerprinting can simultaneously identify intra- and inter-species cross-contaminations.

The Southern blot procedure used involves the digestion of extracted genomic DNA with the restriction enzyme *Hinf*I followed by separation of the fragments by agarose gel electrophoresis. After depurination *in situ*, the DNA is blotted onto a nylon membrane and hybridized with labelled multi-locus probes 33.6 or 33.15. The membranes are then exposed to X-ray film producing a characteristic fingerprint, as shown in Figure 9.7. Currently, chemiluminescent methods for labelling and detecting these probes are used in preference to radio-labelling. At ECACC this method is used for non-human cell line authentication [43, 45].

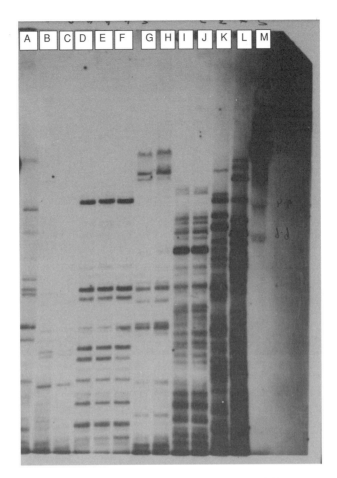

Figure 9.7 DNA fingerprint of MCBs and WCBs of non-human cell lines. DNA profiles showing the fidelity between banks. Lane A, J774A.1; Lane B, CRFK WCB; Lane C, CRFK MCB; Lane D, 23CLN WCB 1; Lane E, 23CLN WCB2; Lane F, 23CLN MCB; Lane G, NFS-SC1 WCB; Lane H, NFS-SC1 MCB; Lane I, Vero AC-free WCB; Lane J, Vero AC-free MCB; Lane K, K562 (control); Lane L, HeLa S3 (control); Lane M, Marker II. Uses multi-locus Jeffreys' probes 33.15 along with Southern blotting technology. The probes cross-hybridize with most common species but this profile requires visual interpretation, and comparison with other profiles can be subjective.

9.2.6.4 *Short tandem repeat profiling*

STR-PCR microsatellite genotype profiling is a quick and powerful tool for the verification of human cell lines only. ECACC has recently transferred to the Applied Biosystems AmpF*ℓ*STR® Identifiler® Plus PCR Amplification Kit, a semi-automated system for DNA profiling that amplifies 15 highly informative STR loci plus the gender marker amelogenin. These afford, on average, a 1 in 5×10^{12} probability of identity, producing a unique profile that can be compared with those held on ECACC's database (Figures 9.8 and 9.9).

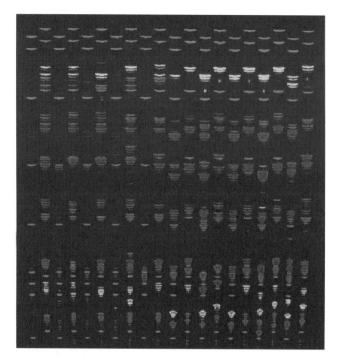

Figure 9.8 STR-PCR profile of a human cell line microsatellite DNA, as generated by the Applied Biosystems AmpFSTR® Identifiler® Plus PCR Amplification Kit. The colour coded banding pattern translates into digital code (see Figure 9.9), which is stored on a DNA profile database. See Plate 4 for the colour figure.

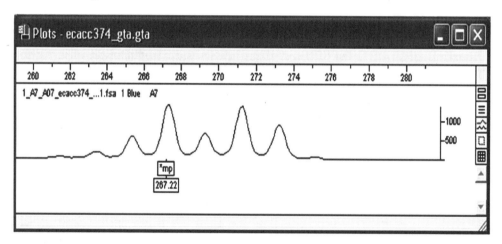

Figure 9.9 Electropherogram of STR-PCR of a human cell line translated into digital code from the fluorescent gel image. Occasionally allele calling and analysis of STR electropherograms can reveal multiple peaks on an allele, as shown here. While this can be caused by over-amplification of the DNA signal it can also be indicative of more than one sample type being present in the sample well.

The primers contained within the AmpFℓSTR® Identifiler® Plus set (D8S1179, D21S11, D7S820, CSF1P0, D3S1358, TH01, D13S317, D16S539, D2S1338, D19S433, vWA, TPOX, D18S51, Amelogenin, D5S818, FGA) allow genotype profiles to be compared with cell line profiles generated from other commercially available sets, such as AmpFℓSTR® SGM plus (Applied Biosystems) or Powerplex® (Promega), used by other culture collections. Cell line profiles are stored in the ECACC cell line profile database, which facilitates the comparison of profiles of all cell lines to determine any expected or unwanted profile matches (Table 9.5).

Table 9.5 STR-PCR DNA profiles for HeLa and derivative cell lines.

Cell Line Name	D11S902	D12S83	D16S404	D1S234	D2S165	D3S1292	D4S424	D5S436	D7S516	D8S260	%homology with HeLa
INT407*	9,3	6,6	1,1	9,3	10,2	9,9	10,2	5,7	2,2	10,8	90
WISH*	9,3	6,6	1,1	9,3	10,2	9,9	10,10	5,7	2,2	10,8	90
KB*	9,3	6,6	1,1	9,3	10,2	9,9	10,10	5,7	2,2	10,8	90
HEP2C	3,3	6,6	1,1	9,3	10,2	9,9	10,2	5,7	2,2	10,8	90
CHANG LIVER*	9,3	6,6	1,1	9,3	10,2	9,9	10,10	5,5	2,2	10,8	90
GIRARDI HEART*	9,3	6,6	1,1	9,3	10,2	9,9	10,2	5,7	2,2	10,8	100
MRC-5	12,13	2,7	5,5	5,6	15,3	6,10	3,6	3,9	2,3	3,3	0
293	2,3	3,3	1,1	5,5	6,6	2,2	6,6	7,7	2,3	3,5	20
THP-1	1,1	2,5	4,4	1,1	13,13	11,11	1,3	2,3	8,5	5,6	0
CACO2	9,9	2,3	1,8	3,3	1,2	2,2	**2,2**	11,11	2,2	3,7	30
HeLa	9,3	6,6	1,1	9,3	10,2	9,9	10,2	5,7	2,2	10,10	100
HeLa S3	9,3	6,6	1,1	9,3	10,2	9,9	10,2	5,7	2,2	10,10	100

The primer set used in this analysis (D11S902; D12S83; D16S404, D1S234, D2S165, D3S1292, D4S424, D5S436, D7S516, D8S260) were a set of allele primers designed in-house by ECACC and used between 1999-2003, and were replaced with the AmpFLSTR® Identifiler® Plus set.

*These cell lines, although given the names stated, are now known in fact to be HeLa.

9.2.7 Regulatory aspects

From the development of the first virus vaccines using cells in culture by Salk in 1954, the production of biologicals for human and veterinary use has been regulated by safety regulations issued by regulatory agencies. It is increasingly important that scientists engaged in the development of cell lines are fully aware of the regulations and guidelines from the outset if an eventual product from the cells is to gain regulatory approval. Several international guidelines describe requirements in the use of cell lines for the production of biological products. Many of these international agencies and their guidelines work cohesively in promoting a risk-based approach to the regulation of cell culture processes used in the preparation of bio-therapeutics, including the FDA, EMEA, World Health Organization (WHO) and International Conference on Harmonisation of Technical Requirements for Registration of Pharmaceuticals for Human Use (ICH) [43]. The practice of documenting the procedures, cultures and reagents used is important for promoting high-quality scientific work, as well as for the use of cell substrates in bio-therapeutics manufacturing [40, 43, 45]. Such documentation also provides traceability. Not only does one need to document the qualification of cell banks, it is also important to establish the stability of critical culture characteristics upon passage to ensure the reliable performance of cells derived from the cell banks.

The inconsistency of primary cultures has led to the development of human diploid cell lines, for example WI-38 and MRC5 [46, 47], which are fully characterized to a rigid standard that dictates the range of population doublings within which these cells are stable enough to provide a substrate for vaccine production, for example MRC5 between population doublings 30 and 40.

The present list of requirements (Table 9.6) involves not only the quality control of master and working cell banks, but also important details on early characterization data and the derivation (provenance) of the cell line and implications for late in-process testing. Guidelines established by the US FDA Center for Biologics Evaluation & Research in their *Points to consider* 1993 [48, 49], the CEC Committee for Proprietary Medicinal Products [49, 50] and the WHO [51], provide guidance on regulators' minimum requirements for qualification of cell lines or derived products to be used in the preparation of a pharmaceutical product. Cell-derived biological products and related tests are closely

Table 9.6 Some culture information required for regulatory approval of a cell culture-derived therapeutic. (This table is by no means exhaustive, and workers should refer to the current guidelines of the appropriate regulatory authority/authorities for fuller details.)

Provenance of cell line (tissue origin and history)
Manufacturing batch records and storage information of master and working cell banks
Culture requirements, including components and their sources
Growth characteristics, including validated limits on culture period
Sampling and testing procedures
Production, storage and test facilities
Quality control tests
    Karyology[a]
    Iso-enzyme analysis
    DNA fingerprinting/STR profiling
    Microbial testing (bacteria, fungi, mycoplasma and viruses)
    Retrovirus status
    Tests for contaminating host cell protein and DNA in final product
    Purification procedures/validation data
    Characterization of product

---

[a] May not be required in all cases.

evaluated by regulatory agencies to develop a full profile of the cell line and its product. Virus testing is considered to be of particular importance with exhaustive initial testing, clearance studies and in-process analysis [52].

Stem cell therapies and research are heavily regulated areas that are developing rapidly. In the UK regulators (Human Tissue Authority, Human Fertilization and Embryology Authority, and Medicines and Healthcare products Regulatory Agency (MHRA)) have jointly developed a Route Map for Stem Cell Research and Manufacture [53] as a reference tool for those who wish to develop a programme of stem cell research and manufacture, ultimately leading to clinical application.

9.2.8 Summary

The quality control of any cell line should be established from the outset of a research project or the development of a production process. Failure to do so could be expensive in terms of money and scientific reputation. Many research projects can lead to the commercialization of a product (diagnostic reagent/therapeutic use); therefore, it is important that at the outset of a project all necessary quality control measures are used to satisfy regulatory agencies and prepare the product for market.

It is our view that scientific journals and research funding agencies should continue to take a more active role in ensuring that basic quality control is performed [54]. This could be achieved by insisting that cell lines used in the research have been tested for microbial contamination and have had their identity verified prior to the publication and/or further funding of the scientific research.

9.3 Troubleshooting

Good cell culture and banking practices provide the best basis for success in the tissue culture laboratory. Cell lines should not be maintained in continuous culture for a number of reasons:

- risk of microbial contamination

- loss of characteristics of interest (e.g. surface antigen or antibody expression)

- genetic drift particularly in cells known to have an unstable karyotype (e.g. CHO, BHK 21)

- loss of a cell line due to senescence (reaching the limit of its finite life span – see Chapter 4), for example human diploid cells such as MRC-5

- risk of cross-contamination with other cell lines

- increased consumables and staff costs.

These potential problems can be eliminated by reliance upon a robust tiered banking system where MCBs are established and qualified for the production of working cell banks as described in Chapter 6.

Implementation of a tiered banking system ensures that

- material is of a consistent quality
- experiments are performed using cultures in the same range of passage numbers
- cells are only in culture when required
- original characteristics of the cell line are retained.

9.3.1 Quality control considerations

The main areas of quality control that are of concern for tissue culture laboratories are

- the quality of the incoming reagents and materials
- the provenance and integrity of the cell lines
- the avoidance of microbial contamination.

9.3.1.1 Reagents and materials

Reagents and materials are a potential source of contamination. In recent years, scrutiny has been placed on bovine serum that has been the source of BVDV. Porcine trypsin is also a potential source of *M. hyorhinis* and porcine parvovirus. Good-quality reagents and media are now widely available from reputable manufacturers of tissue culture reagents and supplements.

Manufacturers are aware of the requirements to screen for mycoplasma and BVDV and generally supply Certificates of Analysis with their products, complete with product numbers, names and lot numbers, which combined with the tissue culture laboratory's records form the basis of your bank history. For critical reagents such as FBS it is advisable to ensure that thorough in-house testing is performed to assess the level of batch to batch variation in your own work

9.3.1.2 Provenance and integrity of cell lines

Evidence of the provenance and integrity of incoming cell lines must be obtained in order to ensure that the cell lines received into the tissue culture laboratory are fit for purpose. Cell lines obtained from a reputable supplier will have confirmed identities and be certified clear of known contaminants. These are assurances that should be sought from any cell lines supplier (complete with Certificates of Analysis).

Where cell lines have been received by a less formal route it is the responsibility of the recipient to ensure that characterization of the cell line is performed under quarantine conditions before it is released for cell banking and general use.

9.3.1.3 Microbial contamination

Microbial contamination can be largely avoided by rigorous application of the principles described in Chapter 4, along with the careful selection, use and maintenance of equipment as outlined in Chapter 1. A schedule of routine testing of cultures within the laboratory should be implemented as a screening programme to monitor the status of work in the tissue culture laboratory.

9.3.2 Environmental monitoring

The employment of an environmental monitoring programme is advisable to monitor the condition of the tissue culture laboratory and its support areas. Critical areas such as media preparation rooms, Class II MSCs and incubators can be monitored using TSA settle plates and contact plates and to determine the cleanliness and suitability of the 'clean' areas for performance of aseptic tissue culture. Plates should be exposed in specific monitoring locations for a period of 2 (incubators/hot rooms) to 4 h (ambient and cold rooms). Further details of environmental monitoring methods are given in Chapter 1, *Protocol 1.3*. Comparison of samples taken from benches, under vents and from other surfaces against the critical locations such as the MSC(s) allows any trends with respect to the ingress of contamination to the laboratory to be determined.

Specification for acceptable microbial counts, alert and alarm levels should be agreed, and an agreed procedure devised for action to be taken in the event of contamination being found in any critical areas.

9.3.3 Good practice in the cell culture laboratory

- Use the relevant personal protective equipment at all times (laboratory coat or gown, gloves and eye protection when required (the last of these is not usually necessary when working in an MSC)) in order to protect yourself from hazards associated with the work.

- Class II MSCs offer protection both to you (from particles generated during your work) and also your cultures (from particles in the laboratory environment or generated by you).

- Overshoes, and disposable hats to cover your hair, can be used to reduce the level of particulates you generate in the laboratory.

- Use visors and thermal protective gloves and apron where handling LN (see Chapter 4, Section 4.2.8.1).

- Keep work surfaces free of clutter.

- Correctly label all vessels, bottles and other containers with contents, date and any other relevant information

- Only handle one cell line at a time and ensure any reagents or paperwork (including labels) are cleared away after each line.

- Work surfaces should be cleaned with 70% v/v isopropyl alcohol (or similar surface disinfectant – see Chapter 2, Section 2.2.4.2) between manipulations.

- Retain separate bottles of media for each cell line, and aliquot reagents from larger stock bottles before use.

- Quality control media and reagents before use. Batches should be used sequentially as far as possible.

- Examine cultures and media daily for any signs of contamination or degradation.

- Cardboard packaging should be prohibited form the tissue culture laboratory.

- Screen cells for mycoplasma on a regular basis

- Ensure that the facilities (HVAC) and equipment are cleaned, maintained and/or repaired at regular intervals.

- Antibiotics should not be used continuously in the tissue culture laboratory (see Chapter 4, Section 4.1.2.1.*i* Antibiotics).

- Remove all waste from the laboratory as soon as possible.

- The number of operators within the laboratory at any one time should be minimized.

- Handle untested cell lines in a quarantine laboratory away from the main tissue culture laboratory.

- Subculture cell cultures when they reach 70–80% confluence and do not allow the cultures to become confluent.

- Laboratories, equipment and facilities should be regularly cleaned down to minimize the level of microbes in the working environment.

References

★ 1. Human Tissue Authority Codes of Practice (2009) Human Tissue Authority consultation on revised Codes of Practice prepared under the Human Tissue Act 2004: A rationale. Approved by UK Parliament in July 2009 and have been brought into force via Directions 002/2009. http://www.hta.gov.uk/policiesandcodesofpractice/codesofpractice.cfm (Accessed September 2010). – *Recommended reading for guidance on human tissue procurement, processing, storage and cell line establishment.*

2. Commission Directive 2004/23/EC of the European Parliament and of the Council of 31 March 2004, on setting standards of quality and safety for the donation, procurement, testing, processing, preservation, storage and distribution of human tissues and cells. http://eurlex.europa.eu/LexUriServ/site/en/oj/2004/l_102/l_10220040407en00480058.pdf (Accessed September 2010).

3. Commission Directive 2006/17/EC of 8 February 2006 implementing Directive 2004/23/EC of the European Parliament and of the Council as regards certain technical requirements for the donation, procurement and testing of human tissues and cells. http://eurlex.europa.eu/LexUriServ/site/en/oj/2006/l_038/l_03820060209en00400052.pdf (Accessed September 2010).

4. Commission directive 2006/86/EC of 24 October 2006 implementing Directive 2004/23/EC of the European Parliament and of the Council as regards traceability requirements, notification of serious adverse reactions and events and certain technical requirements for the coding, processing, preservation, storage and distribution of human tissues and cells. http://eurlex.europa.eu/LexUriServ/site/en/oj/2006/l_294/l_29420061025en00320050.pdf (Accessed September 2010).

★★★ 5. Health & Safety Executive (2005) Biological agents: managing the risks in laboratories and healthcare premises. http://www.hse.gov.uk/biosafety/biologagents.pdf (Accessed September 2010). – *Guidance on handling of biological samples including containment, use and disposal of cell cultures.*

★★★ 6. Health & Safety Executive (2005) The Genetically Modified Organisms (Contained Use) (Amendment) Regulations 2005. http://www.hse.gov.uk/biosafety/gmo/ guidance/amendingregs.pdf (Accessed September 2010). – *Guidance on handling of recombinant cell lines including containment, use and disposal.*

★ 7. Hay, R.J. (1988) The seed stock concept and quality control for cell lines. *Anal. Biochem.*, **171**, 225–237. – *Major publication describing basic concepts of cell banking.*

★ 8. Nelson-Rees, W.A., Daniels, D.W. and Flandermeyer, R.R. (1981) Cross-contamination of cells in culture. *Science*, **212**, 446–452. – *Major publication addressing the issue of cross contaminated and mis-identified cell lines.*

9. Code of Federal Regulations (1986) Animals and animal products. 9. *Detection of mycoplasma contamination.* Office of the Federal Register National Archives and Records Administration, Washington DC, Ch. 1. Part. 113.28. p. 379.

10. Council of Europe (COE) (2007) *European Pharmacopoeia*, 6th edn. Part 2. *Bacterial tests.* European Directorate for the Quality of Medicines (EDQM), Strassbourg.

★ 11. Robinson, L.B., Wichelhausen, R.H. and Roizman, B. (1956) Contamination of human cell cultures by pleuropneumonia-like organisms. *Science, NY*, **124**:1147–1148. – *First publication identifying mycoplasma-infected cell cultures.*

12. McGarrity, G.J. and Kotani, H. (1986) Detection of cell culture mycoplasmas by a genetic probe. *Exp. Cell Res.*, **163**, 273–278.

13. Doyle, A., Hay, R., Ohno, T. and Sugiwara, H. (1990) Resource centres, in *Living Resources for Biotechnology – Animal Cells* (eds A. Doyle, R. Hay and B.E. Kirsop), Cambridge University Press: Cambridge, UK, pp. 5–15.

★ 14. Cowan, S.T. and Steel, K.J. (1979) *Manual for Identification of Medical Bacteria.* Cambridge University Press, Cambridge, UK. – *Reference publication on mycoplasma identification.*

15. Harasawa, R., Uemori, T., Asada, K. and Kato, I. (1992) *Acholeplasma laidlawii* has tRNA Genes in the 16S-23S spacer of the rRNA operon. *J. Bacteriol.*, **174**, 8163–8165.

16. Mowles, J.M. (1990) Mycoplasma detection, in *Methods in Molecular Biology*, vol. **5** (eds J. W. Pollard and J. M. Walker). The Humana Press, Clifton, NJ, pp. 65–74.

17. Macsween, K.F. and Crawford, D.H. (2003) Epstein-Barr virus – recent advances. *Lancet Infect. Dis.*, **3**, 131–140.

18. Duberg, A., Janzon, R., Bäck, E., *et al.* (2008) Surveillance and outbreak reports: the epidemiology of hepatitis C virus infection in Sweden. *Eurosurveillance*, **13**, 1–5.

19. Zuckerman, A.J. and Harrison, T.J. (1990) Hepatitis B virus and hepatitis D virus, in *Principles and Practice of Clinical Virology*, 2nd edn (eds A.J. Zuckerman, J.E. Banatavala and J.R. Pattison), John Wiley & Sons, Chichester, UK, pp. 153–172.

20. Biggar, R.J., Schmidt, T.J. and Woodall, J.P. (1977) Lymphocytic choriomeningitis in laboratory personnel exposed to hamsters inadvertently infected with LCM virus *J. Am. Vet. Med. Assoc.*, **171**, 829–832.

21. Garnick, R.L. (1996) Experience with viral contamination in cell culture. *Dev. Biol. Stand.*, **88**, 49–56.

22. Festen, R. (2007) Understanding animal sera: considerations for use in the production of biological therapeutics, in *Medicines from Animal Cell Culture* (eds G. N. Stacey and J. M. Davis), John Wiley & Sons, Chichester, UK, pp. 45–58.

23. Croghan, D.L., Matchett, A. and Koski, T.A. (1973) Isolation of porcine parvovirus from commercial trypsin. *Appl. Microbiol.*, **26**, 431–433.

24. Hay, R.J. (1991) Operator-induced contamination in cell culture systems. *Dev. Biol. Stand.*, **75**, 193–204.

25. Lloyd, G., Bowen, E.T.W., Jones, N. and Pendry, A. (1984) HFRS outbreak associated with laboratory rats in UK. *Lancet*, **1** (8387), 1175–1176.

26. CBER (1993) *Points to Consider in the Characterization of Cell Lines Used to Produce Biologicals*. US FDA, Rockville, MD. http://www.fda.gov/downloads/BiologicsBloodVaccines/SafetyAvailability/UCM162863.pdf (Accessed September 2010).

27. Peebles, P.T. (1975) An *in vitro* focus-induction assay for xenotropic murine leukemia virus, feline leukemia virus C, and the feline–primate viruses RD-114/CCC/M-7. *Virology*, **67**, 288–291.

28. Rowe, W.P., Pugh, W.E. and Hartley, J.W. (1990) Plaque assay techniques for murine leukemia viruses. *Virology*, **42**, 1136–1139.

29. Hertig, C., Pauli, U., Zanoni, R. and Peterhans, E. (1991) Detection of bovine viral diarrhea (BVD) virus using the polymerase chain reaction. *Vet. Microbiol.*, **26**, 65–76.

30. Chen, D., Nims, R., Dusing, S., *et al.* (2008) Root cause investigation of a viral contamination incident occurred during master cell bank (MCB) testing and characterization – A case study. *Biologicals*, **36**, 393–402.

31. Raines, A., Story, B., and Priola, S.A. (2004) Susceptibility of common fibroblast cell lines to transmissible spongiform encephalopathy agents. *J. Infect. Dis.*, **189**, 431–439.

32. Dawson, M. (1993) Bovine spongiform encephalopathy (BSE) and fetal bovine serum: an outline of hazard assessment, in *Cell and Tissue Culture: Laboratory Procedures* (eds A. Doyle, J. B. Griffiths and D. G. Newell), John Wiley & Sons, Chichester, UK, pp. 7B:7.1–7B:7.5.

33. CPMP and CVMP (2004) Note for guidance on minimizing the risk of transmitting animal spongiform encephalopathy agents via human and veterinary medicinal products. Official Journal of the European Union, 2004/C24/03. http://www.ema.europa.eu/docs/en_GB/document_library/Scientific_guideline/2009/09/WC500003698.pdf (Accessed September 2010).

★★★ 34. Capes-Davis A, Theodosopoulos G, Atkin I, *et al.* (2010) Check your cultures! A list of cross-contaminated or misidentified cell lines. *Int. J. Cancer*, **127**, 1–8. – *Recent* *publication providing comprehensive list of cell lines that have been cross-contaminated, and mis-identified cell lines.*

★ 35. Nardone RM (2007) Eradication of cross-contaminated cell lines: a call for action. *Cell Biol. Toxicol.*, **23**, 367–372.– *Recent publication proposing that all cell lines described in scientific papers and proposed for use in sponsored research be authenticated prior to publication or use.*

36. Buehring, G.C., Eby, E.A. and Eby, M.J. (2004) Cell line cross-contamination: how aware are mammalian cell culturists of the problem and how to monitor it? *In Vitro Cell Dev. Biol. Anim.*, **40**, 211–215.

37. Losi, C.G., Ferrari, S., Sossi, E., Villa, R. and Ferrari, M. (2008) An alternative method to isoenzyme profile for cell line identification and interspecies cross-contaminations: cytochrome b PCR-RFLP analysis. *In Vitro Cell Dev. Biol. Anim.*, **44**, 321–329.

38. Rooney, D.E. and Czepulkowski, B.H. (1986) *Human Cytogenetics: A Practical Approach.* IRL Press, Oxford, UK.

39. Macgregor, H. and Varley, J. (1988) *Working with Animal Chromosomes*, 2nd edn, John Wiley & Sons, Ltd, Chichester, UK.

40. Centers for Biological Evaluation and Research (CBER) and Drug Evaluation and Research (CDER) and International Conference on Harmonisation Guidance for Industry (ICH) (CBER/CDER/ICH), (2001) Q7A *Good Manufacturing Practice Guidance for Active Pharmaceutical Ingredients.* US FDA, Rockville, MD, USA. www.fda.gov/downloads/RegulatoryInformation/Guidances/ucm129098.pdf (Accessed September 2010).

★ 41. Jeffreys, A.J., Wilson, V. and Thein, S.L. (1985) Hypervariable 'minisatellite' regions in human DNA. *Nature*, **314**, 67–73. – *Original paper identifying hypervariable DNA in human cells.*

★ 42. Burke, T., Dolt, G., Jeffreys, A.J. and Wolff, R. (1991) *DNA Fingerprinting: Approaches and Applications*, Birkhäuser Verlag, Basel. – *Application of hypervariable DNA analysis to paternity testing and population studies.*

43. International Conference on Harmonisation (ICH). *Q5D Guidance on Quality of Biotechnological/Biological Products: Derivation and Characterization of Cell Substrates Used for Production of Biotechnological/Biological Products*; http://www.ich.org/LOB/media/MEDIA429.pdf (Accessed September 2010).

★ 44. Stacey, G.N., Bolton, B.J. and Doyle, A. (1992) DNA fingerprinting transforms the art of cell authentication. *Nature*, **357**, 261–262. – *Application of hypervariable DNA analysis to cell banking.*

45. Medicines and Healthcare products Regulatory Agency (2007) *Rules and Guidance for Pharmaceutical Manufacturers and Distributors 2007 – the 'Orange Guide'*, Pharmaceutical Press, London. http://www.mhra.gov.uk/Publications/Regulatoryguidance/Medicines/Othermedicinesregulatoryguidance (Accessed September 2010).

46. Hayflick, L., Plotkin, S.A., Norton, T.W. and Koprowski, H. (1962) Preparation of poliovirus vaccines in a human fetal diploid cell strain. *Am. J. Hyg.*, **75**, 240–258.

47. Wood, D.J. and Minor, P. (1990) Use of human diploid cells in vaccine production. *Biologicals*, **18**, 143–146.

48. Food and Drug Administration, Office of Biologics Research and Review (1993) *Points to consider in the manufacture and testing of monoclonal antibody products for human use.*

U.S. Department of Health and Human Services, Food and Drug Administration, Center for Biologics Evaluation and Research, February 28, 1997.

49. CEC Committee for Proprietary Medicinal Products (1988) *Trends Biotechnol.*, **6**, G5.

50. CEC Committee for Proprietary Medicinal Products (1988) *Trends Biotechnol*, **6**, G1.

51. World Health Organization (1987) *Acceptability of Cell Substrates for Production of Biologicals* Technical report series, **747**. WHO, Geneva, Switzerland.

52. International Conference on Harmonisation (ICH) (1999) *Viral safety evaluation of biotechnology products derived from cell lines of human or animal origin, Q5A (R1), Step 4*. International Conference on Harmonization of Technical Requirements for Registration of Pharmaceuticals for Human Use, Brussels, Belgium. http://www.ich.org/LOB/media/MEDIA425.pdf (Accessed September 2010).

53. Interim UK Regulatory Route Map for Stem Cell Research and Manufacture: Version 12.03.2009 http://www.advisorybodies.doh.gov.uk/genetics/gtac/IntcrimUKSCroutemap 120309.pdf (Accessed September 2010).

54. Mowles, J.M. and Doyle, A. (1990) Cell culture standards – time for a rethink? *Cytotechnologist*, **3**, 107–108.

55. Bolton, B.J., Packer, P. and Doyle, A. (2002) The quality control of cell lines and the prevention, detection and cure of contamination, in *Basic Cell Culture: A Practical Approach*, 2nd edn (ed. J.M. Davis), Oxford University Press, Oxford, UK, pp. 295–323.

Websites

ECACC: www.hpacultures.org.uk

ATCC: www.atcc.org

DSMZ: www.dsmz.de

Riken: www.riken.go.jp/engn/

Lonza: www.lonza.com

Invitrogen: www.invitrogen.com

Cell Bank Australia: www.cellbankaustralia.com

10

Systems for Cell Culture Scale-up

Jennifer Halsall[1] and John M. Davis[2]

[1]Eden Biodesign, Liverpool, UK
[2]School of Life Sciences, University of Hertfordshire, Hatfield, Hertfordshire, UK

10.1 Introduction

Traditionally, the small numbers of cells that could be grown in plates, flasks or Petri dishes were sufficient for research purposes. However, as more and more applications are devised for cells and cell-derived products, the demand for large numbers of cultured animal cells is increasing, exemplified by the 20 000-l bioreactors now used for this purpose by some companies.

The goal of scale-up is to satisfy projected market demand and make the product available at an affordable price by realizing economies of scale. One strategy for scale-up is to replace a large number of small batches with a limited number of large batches, resulting in a reduction in the number of procedures, the labour required, and the amount of quality control and other testing. This approach to scale-up involves an increase in risk. For example failure, for whatever reason, of one small batch among many may be tolerable, but the loss of a huge single batch could be catastrophic to the producer, and possibly also (in the case of biological medicines) to the patient population. Thus successful scale-up demands reliability, in terms of the process and its control, services and the hardware used. In determining the best route for scale-up, a balance must be struck between the size of the unit process, and consequently the number of batches required per year, while also ensuring the most efficient use of staff time and existing facilities.

While the various scale-up systems available can be classified in a variety of ways, the most useful for the individual seeking to scale up an existing culture is to categorize them by the nature of the cells for which they are useful. For example, systems suitable

Animal Cell Culture: Essential Methods, First Edition. Edited by John M. Davis.
© 2011 John Wiley & Sons, Ltd. Published 2011 by John Wiley & Sons, Ltd.

for attached or anchorage-dependent cells rely on increasing the surface area available for cell attachment, whereas systems suitable for cells that grow in suspension largely involve volumetric scale-up.

It should be noted that in some cases the same cells may attach to a substrate or grow in suspension, depending on the conditions under which they are cultured. For example, certain strains of CHO cells may attach to a surface if grown in serum-containing medium, yet grow in suspension in protein-free medium. This underlines the importance of ascertaining the essential properties of cells and defining the medium of choice as soon as possible during process development. In reality it is unlikely that this will have been completed prior to some degree of scale-up, but all changes become more difficult and more expensive the later they are introduced during development.

Although a vast array of cell culture systems exist, because of the constraints of space only the most commonly used laboratory systems for cell culture scale-up, their advantages, limitations and applications will be examined here, and some key methods described. For broader coverage of the topic, see reference 1. It is acknowledged that certain systems not covered here are used widely for certain applications. For example, hollow-fibre systems are used extensively for the culture of monoclonal antibody-producing cell lines in the diagnostics industry, and protocols for their use have been published elsewhere [2, 3]. Similarly, other culture systems will have their own specialist literature, and in the absence of coverage here readers should consult this.

10.2 Methods and approaches

10.2.1 Adherent cells

10.2.1.1 Roller bottles

The use of rotating bottles for large-scale cell culture was first described in 1933 [4], since which time roller systems have been used for culturing a great many different types of attached cell. Clean glass bottles, for example 2.5-l Winchester bottles, were used for many years, and now purpose-designed reusable glass bottles and disposable plastic vessels are available. The inner surface of the bottle or vessel is used as a cylindrical growth surface. Cells are introduced in a limited volume of medium, and the vessel is then placed in a horizontal position on or in an apparatus that will rotate it slowly around an axis parallel to that of the cylindrical growth surface. The cells attach to the vessel wall and are cyclically immersed in and then removed from the bulk of the medium as the vessel rotates. When not immersed, there is only a thin film of medium covering the cells, thus permitting efficient oxygen transfer.

Protocol 10.1 describes a method for the cultivation of HEK293 cells in roller bottles. For a given cell line, the optimum rotation rate must be determined and will depend at least in part on the attachment efficiency of the cells. Often a slow speed, for example 0.1–0.5 rpm is employed to encourage the cells to adhere to the vessel surface, and then once attachment is complete the rotation rate may be increased to 1.0 rpm or higher. Roller bottles with different surface coatings are available in order to aid cell attachment, for example CellBIND® from Corning. Some workers using cells that attach efficiently have recommended a different approach, using 0.2–0.4 rpm during the attachment phase, then 0.08–0.16 rpm once attachment is complete [5]. If the rotation rate is too slow, the cells may dry out before they are reimmersed in the bulk of the medium. Systematic studies have permitted the mixing patterns in roller bottles to be characterized and modelled mathematically [6,7]. From these studies and others [8] it is clear that for best results, optimization of rotation rates should be performed with the vessel/cell/medium combination actually in use. Periodic reversal of the direction of rotation may be useful in facilitating cell adhesion to the vessel walls [6]. In addition, the introduction of a vertical rocking motion can improve mixing within the bulk of the medium [7].

In order to avoid oxygen limitation in sealed vessels, it has been recommended that the gas-volume: medium ratio should not be less than $5:1$ [5]. Alternatively, vented roller bottles are available. Another approach to improving oxygen transfer has been to attach an external recirculation loop to a roller bottle [9] and although highly effective this greatly increases the complexity of the system, negating one of the roller bottle's greatest assets: its simplicity.

A single flat-surfaced roller bottle will usually have a surface area of between 490 and $1800\,cm^2$. This area may be increased by using a design with a ribbed culture surface, which can increase the surface area 2.0–2.5-fold without increasing the external dimensions of the vessel. A further increase in surface area has been achieved without increasing the dimensions of the vessel by adding concentric cylinders [10] or plastic spiral films [11], but such vessels do not seem to be widely available. Although longer vessels can be used, further scale-up is usually achieved simply by increasing the number of units used (Figure 10.1). This makes scale-up from the laboratory to initial production

Figure 10.1 Scale-up of roller bottle culture is achieved by increasing the number of roller bottles (Photograph courtesy of Eden Biodesign).

scale and beyond relatively simple, as there is no change in the unit process. However, large numbers of units have to be handled and incubated, so economies of scale can only be realized through automation of the various steps involved in roller bottle handling. The largest-scale industrial systems are almost completely automated and have a capacity of tens of thousands of roller vessels. Such industrial systems have been in use in the vaccine industry since at least the mid-1960s [12, 13]. The emphasis now seems to be either on the replacement of human operatives with robots in order to reduce labour costs further and decrease the potential for contamination [14] or to convert these processes so that they can be performed in bioreactors.

A number of modern biological medicines have been produced in large-scale roller bottle facilities, including a number of vaccines as well as recombinant erythropoietin (by Amgen, Johnson & Johnson, and Janssen). However, many of these have now been converted to bioreactor processes, either with or without microcarriers (see Sections 10.2.1.3 and 10.2.2.4).

PROTOCOL 10.1   Cultivation of HEK293 cells in roller bottles

Equipment and reagents

- Roller apparatus and 37 °C incubator

- 10% CO_2 supply

- Sterile plugged Pasteur pipettes

- 850-cm^2 roller bottles

- Mid-log-phase HEK293 cells

- Pre-warmed, complete culture medium (DMEM + 2 mM L-glutamine + 10% FBS)

- Dulbecco's phosphate-buffered saline without Ca^{2+} and Mg^{2+} (CMF-DPBS)

- Trypsin-EDTA.

Method

1 Seed mid-log-phase HEK293 cells in 850-cm^2 roller bottles at 1.5×10^4 viable cells/cm^2 in fresh complete culture medium to a working volume no greater than 300 ml[a].

2 Gas the cultures using 5% CO_2 (introduced via a sterile plugged Pasteur pipette) for 30–60 s per bottle and secure the cap.

3 Incubate at 37 °C, 0.1 rpm for 16–24 h.

4 16–24 h after seeding, increase the roller speed to 0.5 rpm and continue incubation until the monolayer is 80–90% confluent[b].

To recover the cells:

5 Aspirate and discard the culture medium.

(continued)

6 Wash the cells gently twice, using 40–50 ml CMF-DPBS per 850-cm^2 roller bottle and discard^c.

7 Add 15 ml of trypsin per 850-cm^2 roller bottle and incubate at 37 °C, 0.5 rpm for 5–10 min^d.

8 Add 35 ml of complete medium per roller bottle to fully resuspend the cells.

9 Centrifuge the cells at 233 g for 5 min.

10 Discard the supernatant. Resuspend the cell pellet in fresh complete medium.

11 Sample the culture and determine the viable cell density (see Chapter 4, *Protocol 4.5*).

Notes

^{a}For every cell line the optimum seeding density and incubation conditions should be investigated.

^{b}For every cell line the optimum rotation rate should be identified.

^{c}The wash step may be achieved efficiently using the roller apparatus at 0.5 rpm for approximately 5–10 min per wash.

^{d}Extensive exposure to trypsin (e.g. beyond 15 min) may damage the cells.

10.2.1.2 Stacked plates

Stacked-plate culture systems contain a number of flat culture surfaces stacked in parallel, one above the other within a single unit, in order to increase the culture surface area that can be manipulated in one operation. Unlike roller bottles, they require no continuous agitation and can be thought of for most purposes as large culture flasks, with a similar culture environment.

Initial scale-up is achieved by increasing the number of surfaces within a unit and in current commercially available systems there may be up to 40 within a single unit (Figure 10.2). Media and other solutions are added through an access port and distributed between the layers by tilting the unit. Manufacturers of stacked-plate systems offer comprehensive guidance for successful handling of the units. Standard cell culture seeding densities and enzymatic dissociation methods may be applied as a starting point for stacked-plate culture, with later optimization of the protocols.

The requirement to move the whole unit, complete with its contents, into different orientations means that the largest unit that can realistically be handled by an unaided operator is a 10-stack (10 parallel culture surfaces). Both manual and electrically operated rigs are available to facilitate handling of a 40-stack. Beyond this, scale-up is accomplished by the use of multiple units. Similarly to roller bottles, further economies of scale are difficult to realize with stacked-plate systems other than by the use of automation. For example, electrically operated rigs can be used to handle multiple 40-stacks simultaneously (Figure 10.3).

One limitation of stacked-plate systems is that there is no means of direct access to the culture surface, and removal of cells by scraping, for example, is not possible. Microscopic observation of multi-layer stacks of more than two layers is not generally possible, and often representative single layer units are employed in parallel in order to monitor culture progression.

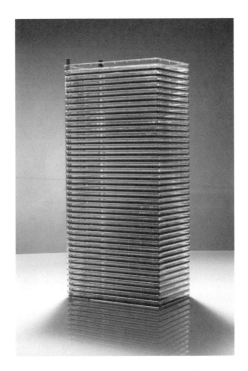

Figure 10.2   A 40-layer cell factory (Photograph courtesy of Nunc).

Figure 10.3   Four 40-layer cell factories on an electric handling rig (Photograph courtesy of Nunc).

A CO_2-enriched atmosphere may be created by manually gassing the unit using an external CO_2 supply (although very great care must be exercised in order to avoid over-pressurizing the vessel with consequent rupture of the relatively fragile internal seals) and then plug-sealing the vessel. However, until recently, it was difficult to maintain such an atmosphere in an effective dynamic manner in order to ensure adequate buffering with CO_2/HCO_3^--buffered media as used with many mammalian cells. This problem has now been overcome by Nunc with the introduction of 'active gassed' cell factories that are engineered with a gas distribution system. A CO_2/air mixture is pumped constantly or intermittently at a low flow rate (typically 20 ml per square cm of culture surface per hour, easily attained by using an aquarium-type pump) to this distribution system, which ensures even distribution both within and between layers, resulting in efficient buffering and gas exchange.

Stacked-plate culture systems have been available for over 30 years, and have been used with a wide range of cell lines [15–25] and at the industrial scale for vaccine manufacture [26].

10.2.1.3 Microcarriers

Microcarrier beads were conceived as a way of improving the volumetric efficiency (surface area per unit volume) of culture systems for attached cells. Using these tiny beads, surface areas in excess of $30 \, cm^2$ per cm^3 of culture medium are easily attainable for use in simple batch culture [27], approximately 10 times more than can be achieved in T-flask or stacked-plate culture. Higher values can be attained using more intensive culture modes, for example fed-batch or perfusion processes. Additionally, if one compares surface area with total culture unit volume, a real measure of the amount of incubator/laboratory space required for scale-up, then microcarriers in a spinner flask or stirred-tank bioreactor will attain around $10 \, cm^2/cm^3$, whereas a 225-cm^2 T-flask will give approximately $0.2 \, cm^2/cm^3$ and a 10-stack cell factory around $0.55 \, cm^2/cm^3$.

Protocol 10.2 outlines the essential steps that must be taken in order to successfully optimize a microcarrier culture, including important considerations for the initial choice of microcarrier. A wide range of microcarriers is now commercially available (e.g. from GE Healthcare, Nunc, Percell Biolytica AB, Sigma-Aldrich and Solohill Engineering), made from a variety of materials including glass, dextran, cellulose, plastic and gelatine. Microcarrier manufacturers offer comprehensive support and protocols for the successful use of microcarriers. *Protocol 10.3* details preliminary conditions for establishing microcarrier cultures, including methodology for seeding and harvesting cells from microcarriers. The initial time and associated costs involved in establishing appropriate protocols for effective microcarrier culture are considered to be the main limitation, particularly when compared with roller bottle and stacked-plate systems that are relatively quick and simple to deploy.

Another advantage of using microcarrier beads is that the culture of attached cells can be carried out in the same type of equipment that is employed for the homogeneous stirred culture of cells growing in suspension [28], usually with slight modifications to help keep the relatively rapidly sedimenting microcarriers suspended. This has numerous benefits in terms of volumetric rather than linear scale-up, environmental control, direct sampling, and observation of the cells, which is difficult in some other systems for culturing attached cells, such as stacked-plate systems. These benefits have led to the widespread use of microcarrier culture, particularly in the vaccine industry, where this technology was first commercialized nearly 30 years ago [29, 30].

Macroporous microcarriers have also been utilized in both packed-bed [31] and fluidized-bed systems. Formerly, a fluidized-bed system using weighted collagen carriers was available from Verax Corp., and more recently a similar unit (the Cytopilot) was marketed by GE Healthcare for use with their Cytoline carriers. However, manufacture of both of these systems has been discontinued, despite the fact that they were capable of volumetric efficiency beyond that attainable in a homogeneous suspension system [32]. Microcarriers have also been successfully utilized for the culture and cryopreservation of human embryonic stem cells [33].

PROTOCOL 10.2 Strategy for successful utilization of microcarriers

Method

1 Select the most appropriate type of microcarrier taking into account the considerations in Table 10.1.

2 Use manufacturer protocols and support to optimize cell adherence and dissociation methodologies at small scale.

3 Optimize cultivation conditions; in particular, the agitation speed must be developed in order to keep the microcarriers in suspension and also to minimize shear.

4 The optimized process may then be appropriately scaled up volumetrically.

Table 10.1 Considerations for selection of the most appropriate microcarrier.

Factor	Considerations
Composition	Microcarrier beads must not be toxic to the cells. This was a problem during the early years of microcarrier use, but has since been largely overcome [34]. Similarly, the monomeric material from which they are made, and any other substances such as surface coatings that are liable to leach from the beads, must not be inhibitory to cell growth
Coatings	A variety of surface coatings are available to encourage the attachment of particular cell types, while others are made from materials that can be enzymatically digested in order to release the cells with minimum cellular damage

(continued)

Table 10.1 (*continued*)

Factor	Considerations
Porosity	Most microcarrier beads are designed such that cells will grow on their surface. However, some are specifically designed to be macroporous so that the cells actually grow within the body of the bead. This may be advantageous in terms of protecting the cells from bead-bead impacts and high liquid shear environments that can be damaging to cells [35], although these mechanisms may only become significant at high agitation rates [36, 37]. Effective removal of the cells from such macroporous beads by trypsinization may be difficult
Diameter	Each microcarrier bead must have the ability to carry several hundred cells. A large number of beads/cm^3 is required, both to ensure that the suspension is homogeneous and to obtain the required volumetric efficiency. Given these constraints an optimum balance is often obtained with bead diameters in the 150–230 μm range. Bead size distribution should be minimal to ensure an even distribution of cells between the beads and to avoid preferential colonization of smaller or larger beads [38].
Density	The beads must be dense enough not to float, but not so dense that they are difficult to keep in suspension. Values between 1.02 and 1.04 g/cm^3 are most frequently employed (except in fluidized-bed applications)
Charge	Originally DEAE-Sephadex A-50 beads were used [24], but the density of positive charges on these was too high for optimum cell attachment and growth. Reducing the charge density overcame the problem [39, 40] and nowadays negatively charged and amphoteric surfaces are also used
Shape	Most microcarrier beads are spherical although this shape is not essential. Cylindrical cellulose carriers can be used [28] and Nunc currently markets flat, hexagonal polystyrene carriers
Transparency	A highly transparent bead material aids microscopic examination of the attached cells

PROTOCOL 10.3 Establishment of microcarrier cultures

Equipment and reagents

- Microcarriers

- Stirred-tank bioreactor or spinner flask

- Mid-log-phase cells

- Pre-warmed complete culture medium

(*continued overleaf*)

- Dulbecco's phosphate buffered saline without Ca^{2+} and Mg^{2+} (CMF-DPBS)
- Cell dissociation agent (e.g. Trypsin-EDTA)

Method

1 For every litre of the required culture volume, suspend 20 g of microcarriers in 400 ml of de-ionized water or CMF-DPBS (equivalent to 20 ml per gram of microcarriers).

2 Once the microcarriers have settled, remove the rinse liquid and replace it with the same volume of fresh CMF-DPBS.

3 Sterilize the microcarriers by autoclaving at 121 °C for 30 min using a cycle suitable for use with liquids (see Chapter 2, Section 2.2.1.1).

4 Allow the liquid to cool, and the microcarriers to settle to the bottom of the container. Aseptically aspirate and discard the supernatant liquid.

5 Suspend the microcarriers in c. 90% of the required final volume of culture medium[a]. Add this suspension to the culture vessel and allow to equilibrate (for example, for 1 h at 37 °C).

6 Using routine methodology for the cell line, prepare sufficient cell inoculum to achieve a population density of approximately $3–5 \times 10^4$ viable cells per cm^2 of microcarrier surface area[a,b].

7 Add the cell inoculum to the equilibrated microcarriers.

8 Agitate the culture to facilitate homogeneous cell attachment[c,d].

9 Add pre-warmed culture medium to achieve the desired final working volume[a].

10 Once cells are attached, maintain the culture according to cell growth requirements[e].

To harvest the cells:

11 Allow the microcarriers to settle (or harvest by centrifugation at 200 g, 5 min) and remove the culture medium.

12 Rinse the microcarriers using 50 ml of CMF-DPBS per gram of microcarriers[f]

13 Discard the CMF-DPBS rinse and allow the microcarriers to settle (or harvest by centrifugation at 200 g, 5 min).

14 Dissociate the cells from the microcarriers using 5–10 ml of trypsin-EDTA per gram of microcarriers, and incubate at 37 °C, 5–10 min.

15 Agitate the culture to prepare a homogeneous single cell suspension.

16 Remove the microcarriers by filtration using an appropriate pore size to separate the cells from the microcarriers, for example 40–70 μm.

17 Clarify the cells by centrifugation at 200 g, 5 min.

18 For continued culture, resuspend the cells in pre-warmed complete culture medium at the desired seeding density.

Notes

[a] These three volumes (in steps 5, 6 and 9) must be adjusted relative to one another to achieve the final required culture volume.

(continued)

^{b}The seeding density must be optimized for each cell line and microcarrier type.

^{c}Continuous agitation at 40–70 rpm may be appropriate for cells that adhere with high efficiency. For cells that do not adhere efficiently, an intermittent agitation strategy may be more effective with agitation for 1 min at 30–60 rpm followed by no agitation for 10–30 min.

^{d}During cell attachment the microcarriers may settle out, restricting the availability of oxygen to the cells. Such oxygen depletion may inhibit cell attachment and growth and can be avoided by optimizing the agitation strategy for the cell attachment step.

^{e}The agitation speed may be increased at this stage after the cells have fully adhered to the microcarriers, in line with cell growth requirements.

^{f}The rinse step may be achieved using gentle agitation, for example stirring at 40 rpm for 10–15 min. Vigorous mixing should be avoided so as not to dislodge cells from the microcarriers.

10.2.2 Suspension cells

Although most of the systems described below were initially developed for use with cells growing in suspension, all of them can be adapted for use with attached cells growing on microcarrier beads.

10.2.2.1 Roller bottles

Roller bottles, previously discussed for use with adherent cells in Section 10.2.1.1, offer similar advantages for the cultivation of suspension cells, in terms of simplicity and linear scale-up. Typically, greater rotation rates are employed than for adherent cells in order to maintain a homogeneous suspension and avoid oxygen limitation.

Protocol 10.4 outlines a method for the cultivation of Sp2/0-Ag14 suspension cells in roller bottle culture. Since suspension cells are amenable to efficient volumetric scale-up, bioreactor processes (see Section 10.2.2.4) with controlled environments are often favoured over roller bottle culture in order to meet larger scale demands, for example above 10 l.

PROTOCOL 10.4   Roller bottle culture of a suspension cell line (Sp2/0-Ag14)

Equipment and reagents

- Roller apparatus and 37 °C incubator

- Supply of CO_2 in air, with the percentage of CO_2 matched to the medium in use (see Chapter 4, Section 4.1.2.1)

- Sterile plugged Pasteur pipettes

(continued overleaf)

- 850-cm^2 roller bottles
- Mid-log-phase Sp2/0-Ag14 cells
- Pre-warmed, complete culture medium.

Method

1 Seed mid-log-phase Sp2/0-Ag14 cells into an 850-cm^2 roller bottle, diluting them with fresh culture medium to give a working volume no greater than 350 ml and a cell population density of 1×10^5 viable cells/ml[a]

2 Gas each bottle using the CO_2/air mixture (introduced via a sterile plugged Pasteur pipette) for 30–60 s per bottle.

3 Incubate at 37 °C, 5.5 rpm for 2–4 days[b].

4 Gently swirl the roller bottle to ensure a homogeneous suspension.

5 Sample the culture and determine the viable cell density (see Chapter 4, *Protocol 4.5*).

6 Repeat steps 1–3 to passage the culture.

Notes

[a] For every cell line the optimum seeding density and incubation conditions should be investigated.

[b] For every cell line the optimum agitation rate should be identified.

10.2.2.2 Shake flasks

Erlenmeyer-style flasks have been used for the culture of mammalian cells since the 1950s. These flasks are secured to an orbital shaker apparatus, which mixes the contents of the flask and keeps the cells in suspension. This culture method is particularly useful for small to moderate volumes of cells having high oxygen requirements, such as insect cells. For efficient gas exchange and mixing, a maximum working volume one-fifth the total capacity of the flask is recommended, for example a 50-ml working volume in a 250-ml flask. Both reusable and disposable flasks are available in sizes ranging from 50 ml to 6 l. Baffles can be added to aid mixing (although these are predominantly used for microbial cultures), and vented caps can be used to increase gas exchange. *Protocol 10.5* outlines the cultivation of an insect cell line in Erlenmeyer culture.

Scale-up is achieved linearly, by the use of multiple flasks. For very large scale, this approach is unsuitable, as it is labour-intensive and requires extensive agitated incubation capacity. Some automation for shake flask culture, such as the Sonata (The Automation Partnership), has been developed. Medium-scale requirements may be met by the use of spinner flasks (Section 10.2.2.3) and to meet large-scale requirements, bioreactor processes with carefully controlled environments offer a volumetric scale-up solution for suspension cells (Sections 10.2.2.4 and 10.2.2.5).

PROTOCOL 10.5   Cultivation of Sf9 cells in Erlenmeyer culture

Equipment and reagents

• Shaking incubator

• Sterile 250-ml vented Erlenmeyer flask

• Mid-log-phase Sf9 cells

• Pre-warmed culture medium.

Method

1 Seed mid-log-phase Sf9 cells into a 250-ml vented Erlenmeyer flask, diluting to a final population density of 2×10^5 viable cells per ml in a 50-ml working volume[a].

2 Incubate the culture at 27 °C, 120–140 rpm, 3–4 days[b].

3 Sample the culture for determination of the viable cell population density and percentage viability (see Chapter 4, *Protocol 4.5*).

4 Repeat steps 1 and 2 to passage the culture.

Notes

[a] For every cell line the optimum seeding density and incubation conditions should be investigated.

[b] The quoted agitation rate is for a shaking platform with a 1-inch (2.5-cm) orbit. If an incubator with a different sized orbit is used, the optimum agitation rate may be different. For every cell line, the agitation rate should be optimized.

10.2.2.3 Spinner flasks

Spinner flasks are cylindrical, agitated vessels, varying in capacity from around 100 ml to 36 l, and represent an intermediate level of scale-up between flasks and bioreactors. They are far cheaper to purchase and run than a small bioreactor of equivalent size. Typically, spinner flasks need fewer utilities than bioreactors. For example, bioreactors may require three-phase electricity, steam, cooling water, compressed air, O_2, CO_2 and N_2 supplies, as well as provision for liquid drainage and the venting of gases, and these can be expensive to install and maintain, whereas a spinner flask may be incubated simply inside a CO_2 incubator on a magnetic stirrer plate. Typical agitation rates for spinner flasks are in the region of 60–120 rpm.

Different formats are available with different methods of agitation, different stirrer configurations, different aspect-ratio vessels, different cap types and different degrees and ease of access to the gas and liquid phases. In general, the access options increase with the size of the vessel. Examples of two types are shown in Figure 10.4.

Spinner flasks are limited as scaled-down models of bioreactors for process development, because far less instrumentation and control is available than on a bioreactor.

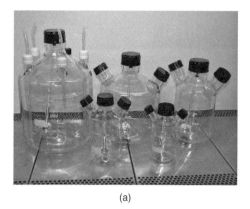 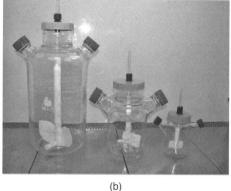

(a)   (b)

Figure 10.4 Two different types of spinner flasks. Both types are operated with the use of a magnetic stirrer base beneath the vessel. (a) Techne spinner flasks. Note the ball-ended stirrer bar fixed by a flexible mount to the cap. The ball end stirs the culture by travelling in an orbit around the central raised area of the flask bottom. (b) Bellco spinner flasks intended for use with microcarriers. The stirrer bar rotates around a central axle, and in this case carries broad vanes to help keep microcarrier beads (which sediment more rapidly than individual cells) in suspension.

Spinner flasks cannot usually accommodate *in situ* probes, and are not supplied with feedback controllers. Above the 20-l scale, manual handling of spinner flasks becomes difficult and bioreactor culture may be preferable at this scale.

Spinner flasks may be useful in the early stages of cell inoculum growth for large-scale systems; for the small-scale production of material, either during the 'proof of principle' or early clinical trial stage of development of a biological medicine; or for the production of diagnostics.

10.2.2.4 Bioreactors (fermenters)

For industrial-scale production of animal cells and their secreted products, by far the commonest method of culture is submerged culture in stirred-tank or, less commonly, airlift bioreactors (Figures 10.5 and 10.6 respectively). This technology has been in use in the brewing and other industries for a great many years and the principles and engineering involved are well understood [41–46]. Thus early attempts to use bioreactors for large-scale mammalian cell culture employed the technology and designs that had been developed for microbial fermentation. It soon became apparent that, although many of the issues and concerns remained the same (e.g. pH control, mixing (including gas/liquid mixing), oxygen mass transfer) the special characteristics of mammalian cells required that corresponding adaptations be made to bioreactor hardware and control strategies [47]. For example, mammalian cells are more sensitive to shear and therefore marine impellers rotating at reduced speeds replaced the rapidly rotating Rushton impellers used for microbial fermentation. In order to improve oxygen transfer at the headspace, and thus reduce the requirement for sparging and the consequent potential for cell damage caused by bubble disengagement, a high diameter to height ratio is preferred. Additionally, since mammalian cells typically have lower oxygen demands than microbial cells, the amount

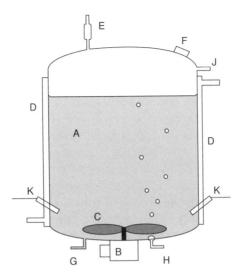

Figure 10.5 Simplified diagram of a stirred-tank fermenter for use with mammalian cells. A, cell suspension; B, motor drive to impeller; C, marine impeller; D, jacket for water/steam; E, off gases port with condenser; F, viewing port; G, liquid addition port; H, gases-in line to sparger; J, gases-in line to headspace; K, probes (pH, DO, etc.).

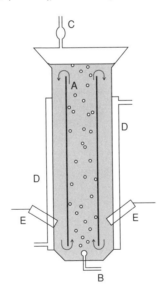

Figure 10.6 Simplified diagram of an airlift fermenter for use with mammalian cells. A, cell suspension; B, gas inlet; C, off gases port with condenser; D, jacket for water/steam; E, probes (pH, DO, etc.). Bubbles of an air/CO_2 mixture, introduced at the bottom centre of the bioreactor, rise up through the central draft tube. This gives the cell suspension within the draft tube a lower pseudo-density than that outside the draft tube. Thus this column of liquid/gas rises. The bubbles disengage at the headspace, and the liquid without bubbles descends outside the draft tube.

of sparging may be limited and bubble sizes can be restricted by using a sintered sparge head.

The change from spinner flask to bioreactor will generally occur when the required culture volumes are 10 l or above. Useful guides to purchasing bioreactors are available [48, 49]. Up to the 15-l scale, glass vessels with sterilization by autoclaving are still an option. However, manual handling issues, and the fragility and poor thermal conductivity of glass rapidly become an issue. Consequently, larger bioreactors are typically made of stainless steel, and are sterilized *in situ*. *Protocol 10.6* outlines the key steps and some starting conditions for the preparation of a stirred-tank bioreactor culture for mammalian cells.

More recently, disposable stirred-tank bioreactors have become available, such as the Hyclone single use bioreactor (SUB). Although consumables for disposable bioreactor systems may seem expensive, the initial capital outlay is often much less than for traditional bioreactors and maintenance requirements are reduced. In addition, costly sterilization and cleaning procedures (along with the expensive and time-consuming validation thereof) are avoided with sterile disposable systems.

As stirred-tank bioreactors are so widely used, there is a wealth of process data available for many different cell lines and product types. The consistent process control achieved using bioreactors allows the development of robust processes that are readily reproduced. In addition, bioreactor processes may be intensified by the use of fed-batch or continuous/perfusion culture modes [50].

Scale-up of all bioreactor vessels, particularly stirred tanks, cannot be proportional. For example, doubling a vessel's dimensions while retaining the same three-dimensional shape will increase its volume eightfold, but the air/liquid interface area of the headspace will only increase fourfold, which therefore decreases the role that headspace gas exchange can play in oxygenation of the culture. Similarly, as a stirred tank increases in diameter it will usually be necessary to increase the diameter of the impeller in order to ensure good mixing. Yet if the impeller rotation rate is kept the same, the shear at the tips will increase. Many other interacting factors will also vary to different extents. Thus for scale-up, a good understanding of the cells' physical and metabolic requirements (e.g. shear sensitivity, oxygen requirements) as well as the characteristics of the medium (e.g. density, viscosity, foaming properties) becomes essential in order to define the specification of the bioreactor. Currently, the largest stirred-tank fermenters in use for animal cell culture have volumes of 20 000 l, whereas the largest airlifts are 5000 l.

PROTOCOL 10.6　Batch cultivation in a 2-l working volume stirred-tank bioreactor

Equipment and reagents

- Stirred-tank bioreactor, controller and supporting utilities (e.g. air, O_2, CO_2, N_2)
- Sterile auxiliaries including addition, sample and harvest bottles, c-flex tubing, appropriate connectors and clamps
- Mid-log-phase cells

(continued)

- Sterile culture medium

- Base solution, filter sterilized (e.g. 7.5% sodium bicarbonate)

- Sterile antifoam

Method

1 Prepare the bioreactor head plate with appropriate inlet and outlet lines (cells, media, base, antifoam, gas inlet to overlay, gas inlet to sparge, gas outlet, sample/harvest line).

2 Carry out maintenance on the oxygen sensor according to manufacturer's instructions and fit into head plate.

3 Maintain and calibrate the pH sensor according to the manufacturer's instructions and fit into head plate.

4 Sterilize the bioreactor and associated bottles[a].

5 Inoculate the vessel with 1.5 l of complete culture medium and allow to equilibrate to operating conditions:[b]

   - temperature at 37 °C controlled by heated jacket

   - DO at 40–50% saturation, controlled by an air overlay and O_2 supplied (as required) to the sparger

   - pH at 7.0–7.3 controlled by CO_2 and base addition

   - agitation at 70–90 rpm controlled by stirrer motor with a marine impeller.

6 Once equilibrated, perform fresh DO and pH calibrations according to manufacturers' instructions. If necessary (i.e. if significant adjustment of the calibration of either probe was required), allow the medium to re-equilibrate to the desired operating conditions.

7 Inoculate the vessel at the desired seeding density with mid-log-phase cells[c].

8 Antifoam may be added as required (a few drops at a time) to prevent excessive foam formation[d].

9 Sample the bioreactor regularly to monitor cell growth and viability, and to assess product titre. Samples may also be used for metabolite analyses in order to better understand and optimize the process.

10 Harvest the bioreactor contents at a suitable harvest point, which may be determined over a time-course or based on cell viability.

Notes

[a] During sterilization the gas outlet line must not be clamped off.

[b] Set points must be optimized for each cell line/process.

[c] The maximum working volume of the bioreactor and the volumes of any additions to be made, including base, must be taken into account.

[d] It is recommended to limit the use of antifoam as it may be difficult to remove during downstream processing.

10.2.2.5 Wave-type bioreactors

Bags suitable for use as vessels for cell culture have been available for over 20 years [51]. Initial applications generally used static, gas-permeable bags in CO_2 incubators, and culture volumes were consequently limited to a few litres at most by the need for oxygen to diffuse through the bag walls. A far more suitable system for use in scale-up was first commercialized by Wave Biotech in 1998, and similar systems are now also available from Sartorius-Stedim and Applikon. These systems basically consist of a sterile, single-use, non-gas-permeable pillow-shaped bag that is mounted on a rocking apparatus (see Figure 10.7). The bag is inflated with a CO_2/air mixture suitable for the medium to be used, then the medium is introduced into the bag (and warmed, if necessary) after which cells are added. The bag retains a substantial headspace of CO_2/air, typically equal in volume to that of the medium. Mixing is achieved and gas exchange facilitated by the reciprocating rocking motion of the motorized platform on which the bag is mounted. Shear levels are claimed to be very low, as the cells mostly move with the bulk of the liquid, but the rocking must be regulated in order to avoid excessive foaming. The bags come fitted with a fill tube, harvest tube, inlet filter, exhaust filter, sampling port, constant pressure relief valve and ports for *in situ* pH and dissolved oxygen probes. These systems

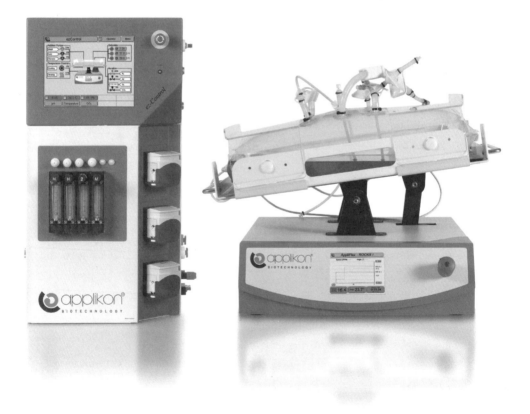

Figure 10.7 Wave-type bag system (Photograph courtesy of Applikon.).

have been used with animal cells both in suspension [52, 53] and on microcarriers [54], as well as with plant and prokaryotic cells.

With retention of bag geometry and adjustment of the tilting parameters, linear scale-up is claimed from 100 ml up to 500 l [55, 56]. Although the makers claim that there is no intrinsic limit to scale-up, other factors such as mechanical and safety problems associated with a large reciprocating mass could become a problem, as could temperature control which is still achieved by heat transfer through the bag wall.

The main advantages of the system, other than its simple principle and low-shear environment, are largely due to the disposable nature of the bag. Thus there is no cleaning, sterilization (except possibly of probes) or associated validation to be performed by the user, nor is there any need to replace seals or undertake many of the other time-consuming maintenance procedures that are necessary for bioreactors. *Protocol 10.7* outlines the key steps in the use of a rocking bag bioreactor system, using the Appliflex bioreactor (Applikon) as an example system and some suggested operating conditions for mammalian cells.

Rocking bag bioreactor systems are now widely used in facilities licensed for the production of human therapeutics. In at least some cases the system is used to grow cell inocula for seeding larger, more conventional culture systems rather than for the final full-scale manufacture of product. Nevertheless, up to a certain scale the system has attractive features. Manufacturers of these rocking bag bioreactors have developed a range of options for pH, DO and other sensors. These include reuseable options where the end user inserts their chosen sensors into the pre-sterilized bioreactor; and pre-sterilized sensors that form part of the disposable bioreactor itself.

PROTOCOL 10.7 Cultivation of mammalian cells in the Appliflex bioreactor

Equipment and reagents

- Appliflex rocking bag bioreactor, controller and supporting utilities (e.g. air, O_2, CO_2, N_2)
- Sterile auxiliaries including addition, sample and harvest vessels, c-flex tubing, appropriate connectors and clamps
- Sterile pH and DO sensors in the compatible probe holders
- Mid-log-phase cells
- Sterile culture medium
- Base solution, filter sterilized (e.g. 7.5% sodium bicarbonate).

Method

1 Make all appropriate aseptic connections (including the fitting of sensors) to the pre-sterilized bioreactor bag.

(continued overleaf)

2 Load the bioreactor bag onto the rocking platform, set the back pressure relief valve to 1 l/min and connect the bioreactor to the controller.

3 Inflate the bag with air[a].

4 Inoculate with medium and allow to equilibrate to operating set points:[b]

   • temperature at 37 °C controlled by a heated mat underneath the bioreactor bag

   • DO at 40–50% controlled by O_2 supplied (as required)

   • pH at 7.0–7.3 controlled by CO_2 and base addition

   • agitation at 8° and 10 rpm, controlled by the rocking platform.

5 Once equilibrated, perform fresh DO and pH calibrations according to manufacturers' instructions. If necessary (i.e. if significant adjustment of the calibration of either probe was required), allow the medium to re-equilibrate to the desired operating conditions.

6 Inoculate the vessel at the desired seeding density with mid-log-phase cells[c].

7 Sample the bioreactor regularly to monitor cell growth and viability and to assess product titre. Samples may also be used for metabolite analyses in order to better understand and optimize the process.

8 Harvest the bioreactor contents at a suitable harvest point, which may be determined over a time-course or based on cell viability.

Notes

[a] The back pressure relief valve prevents over-inflation of the bag.

[b] The operating set points and gassing configuration must be optimized for each cell line/process.

[c] The maximum working volume of the bioreactor and the volumes of any additions to be made, including base, must be taken into account.

10.3 Troubleshooting

• **Oxygen demand.** All animal cells require oxygen, but the poor solubility of oxygen in water (around 0.2 mmol/l at 37 °C) can present a significant barrier to the supply of adequate quantities to cells in large-scale culture. At small scale, supplying the oxygen demands of cells in culture is rarely a problem, and the rate of diffusion of oxygen through the few millimetres of medium separating the vessel headspace from the cells is usually sufficient to satisfy the cells' requirements. However, as the culture volume and the concentration of cells increases, supplying adequate amounts of oxygen to the growing cells can become the major problem posed by scale-up. Initially this can be addressed by stirring (or rocking) the culture (but beware of problems caused by shear [see below]), and then by introducing a flow of gas through the culture headspace, and subsequently by increasing the rate of headspace gas flow. Care must be taken with the choice of gas used for this purpose, particularly with CO_2/HCO_3^--buffered media, as insufficient CO_2 in the mixture will increase the pH of the medium. Evaporation of water from the medium may also become a problem and may necessitate the use of a condenser on the off-gases line. Eventually, sparging (bubbling of gas through the medium) may be required. Again, careful control of the CO_2 concentration may be required, and bubble damage (see below) could become a problem.

- **Nutrient and waste product gradients.** Cells in culture deplete nutrients from, and secrete waste products into, the surrounding medium. In a static (i.e. non-stirred) culture, a gradient of nutrients and waste products will tend to form around the cell, with the cell sited at the point with the lowest nutrient and highest waste product concentrations. Only diffusion (and possibly a little convective mixing) will dissipate these gradients. Considering the cell culture as a whole, the higher the local cell concentration, the steeper these gradients will tend to be. Clearly, for large-scale culture this is inefficient, and mixing mechanisms must be employed. However, these very mechanisms may introduce further problems, including shear and bubbles. Biochemical analysis of cultures during scale-up can assist in identifying suitable culture conditions and developing effective feeding regimes that minimize nutrient depletion.

- **Shear.** Because they lack a cell wall, animal cells are far more susceptible to the damaging effects of shear than are prokaryotes. However, most mixing mechanisms induce shear. Balancing the damaging effects of shear against the requirements for the mass transport of oxygen to the cells and the minimization of chemical gradients can be a major challenge in scale-up optimization. Although animal sera contain substances (e.g. albumin) with natural shear-protectant properties, there has been an increasing move to serum-free and protein-free cell culture processes. This has been driven by the increasing reluctance of regulatory authorities to accept processes involving serum, as well as the benefits for subsequent purification of cell-derived products. Thus shear has become a particular issue in the development of modern large-scale mammalian cell culture processes, affecting agitator design and speed as well as vessel design. Pluronic F68™ is one chemical that can be added to such cultures as a shear protectant (see also below).

- **Bubble damage.** The introduction of bubbles into a cell suspension can aid both oxygen transport and mixing. The disadvantage is that cells can be damaged upon disengagement of the bubbles at the vessel's liquid/head-space interface. This effect can be controlled by the addition of surface-active agents such as Pluronic F68™ to the medium [57]. The effect of using alternative sparge pipes that produce bubbles of different sizes may also be investigated in relation to bubble damage. However, the efficiency of gas transfer and mixing will also be affected and must be carefully considered.

- **Logistics.** Small culture vessels and their contents can be transported and manipulated by hand, moved from one environment to another (e.g. from an autoclave to a safety cabinet to an incubator), and warmed up or cooled down quickly. As scale increases, transportation and manual manipulation become increasingly difficult, and eventually impossible, at which stage everything must instead be brought to the culture system, and environmental changes effected *in situ*. Thus, for example, steam-in-place sterilization takes over from autoclaving, and other solutions have to be found where tilting of a vessel was required at small scale. Similarly, bringing medium up to the correct incubation temperature may only take minutes at small scale, but in a large fermenter may take many hours. These factors have very important impacts on both the design and operational costs (including down-time) of large-scale systems.

References

1. Davis, J.M. (2007) Systems for cell culture scale-up, in *Medicines from Animal Cell Culture* (eds G.N. Stacey and J.M. Davis), John Wiley & Sons, Ltd, Chichester, UK, pp. 145–171.

2. Hanak, J.A.J. and Davis, J.M. (1995) Hollow fibre bioreactors : The Endotronics Acusyst-Jr. and Maximizer 500, in *Cell and Tissue Culture: Laboratory Procedures*, Section 28D:3 (eds A. Doyle, J. B. Griffiths and D. G. Newell), John Wiley & Sons, Ltd, Chichester, UK.

3. Davis, J.M. (2007) Hollow-Fiber Cell Culture, in *Animal Cell Biotechnology: Methods and Protocols*, 2nd edn (ed. R. Poertner) The Humana Press Inc., Totowa, NJ, pp. 337–352.

★ 4. Gey G (1933) An improved technic for massive tissue culture. Am. J. Cancer, **17**, 752–756 – *First description of the use of roller bottles for mammalian cell culture.*

5. Griffiths, J.B. (1995) Roller bottle culture, in *Cell and Tissue Culture Laboratory Procedures*, Section 28A1 (eds A. Doyle, J. B. Griffiths and D. G. Newell), John Wiley & Sons, Ltd, Chichester, UK.

6. Muzzio, F.J., Unger, D.R., Liu M. *et al.* (1999) Computational and experimental investigation of flow and particle settling in a roller bottle bioreactor. *Biotechnol. Bioeng.*, **63**, 185–196.

7. Unger, D.R., Muzzio, F.J., Aunins, J.G. and Singhvi, R. (2000) Computational and experimental investigation of flow and fluid mixing in the roller bottle bioreactor. *Biotechnol. Bioeng.*, **70**, 117–130.

8. Tsao, E.I., Bohn, M.A., Ostead, D.R. and Munster, M.J. (1992) Optimization of a roller bottle process for the production of recombinant erythropoietin. *Ann. N.Y. Acad. Sci.*, **665**, 127–136.

9. Berson, R.E., Pieczynski, W.J., Svihla, C.K. and Manley, T.R. (2002) Enhanced mixing and mass transfer in a recirculation loop results in high cell densities in a roller bottle reactor. *Biotechnol. Prog.*, **18**, 72–77.

10. Knight, E. (1977) Multisurface Glass Roller Bottle for Growth of Animal Cells in Culture. *Appl. Environ. Microbiol.*, **33**, 666–669.

11. Griffiths, J.B. (2001) Scale-up of suspension and anchorage-dependent animal cells. *Mol. Biotechnol.*, **17**, 223–238.

12. Nardelli, L. and Panina, G.F. (1976) 10-years experience with a 28,800 roller bottle plant for FMD vaccine production. *Develop. Biol. Standard.*, **37**, 133–138.

13. Panina, G.F. (1985) Monolayer growth systems: multiple processes, in *Animal Cell Biotechnology*, vol. 1 (eds R.E. Spier and J.B. Griffiths), Academic Press, London, UK, pp. 211–242.

14. Kunitake, R., Suzuki, A., Ichihashi, H. *et al.* (1997) Fully-automated roller bottle handling system for large scale culture of mammalian cells. *J. Biotechnol.* **52**, 289–294.

15. Siegl, G., deChastonay, J. and Kronauer, G. (1984) Propagation and assay of hepatitis A virus in vitro. *J. Virol. Meth.*, **9**, 53–67.

16. Goetz, R., Kolbeck, R., Lottspeich, F. and Barde, Y.A. (1992) Production and characterization of recombinant mouse neurotrophin-3. *Eur. J. Biochem.* **204**, 745–749.

17. Bishop, N.E., Hugo, D.L., Borovec, S.V. and Anderson, D.A. (1994) Rapid and efficient purification of hepatitis A virus from cell culture. *J. Virol. Meth.*, **47**, 203–216.

18. Dickinson, L.A. and Kohwi-Shigematsu, T. (1995) Nucleolin is a matrix attachment region DNA-binding protein that specifically recognizes a region with high base-unpairing potential. *Mol. Cell. Biol.* **15**, 456–465.

19. Litjens, T., Bielicki, J., Anson, D.S. *et al.* (1997) Expression, purification and characterization of recombinant caprine N-acetylglucosamine-6-sulphatase. *Biochem. J.*, **327**, 89–94.

20. Nelson, K., Bielicki, J. and Anson, D.S. (1997) Immortalization and characterization of a cell line exhibiting a severe multiple sulphatase deficiency phenotype. *Biochem. J.*, **326** 125–130.

21. Ohtaki, T., Ogi, K., Masuda, Y. *et al.* (1998) Expression, purification, and reconstitution of receptor for pituitary adenylate cyclase-activating polypeptide. Large-scale purification of a functionally active G protein-coupled receptor produced in Sf9 insect cells. *J. Biol. Chem.* **273**, 15464–15473.

22. Lee S-Y, Kim, S.H., Kim, V.N. *et al.* (1999) Heterologous gene expression in avian cells: potential as a producer of recombinant proteins. *J. Biomed. Sci.*, **6**, 8–17.

23. Lay, A.J., Jiang, X.M., Kisker, O. *et al.* (2000) Phosphoglycerate kinase acts in tumour angiogenesis as a disulphide reductase. *Nature* **408**, 869–873.

24. Suehiro, K., Mizuguchi, J., Nishiyama, K. *et al.* (2000) Fibrinogen binds to integrin alpha(5)beta(1) via the carboxyl-terminal RGD site of the alpha-chain. *J. Biochem. (Tokyo)*, **128**, 705–710.

25. Loewen, N., Bahler, C., Teo, W.L. *et al.* (2002) Preservation of aqueous outflow facility after second-generation FIV vector-mediated expression of marker genes in anterior segments of human eyes. *Inv. Ophth. Vis. Sci.*, **43**, 3686–3690.

26. Hagen, A.J., Aboud, R.A., DePhillips, P.A. *et al.* (1996) Use of nuclease enzyme in the purification of VAQTA, a hepatitis A vaccine. *Biotechnol. Appl. Biochem.*, **23**, 209–215.

★ 27. van Wezel, A.L. (1967) Growth of cell-strains and primary cells on micro-carriers in homogeneous culture. *Nature*, **216**, 64–65. – *First description of the use of microcarrier beads for mammalian cell culture.*

★★ 28. Reuveny, S. (1990) Microcarrier culture systems, in *Large-scale Mammalian Cell Culture Technology* (ed. E.S. Lubiniecki), Marcel Dekker, New York, USA, pp. 271–341. – *A useful evaluation of anchorage-dependent cell propagation systems.*

29. Meignier, B., Mougeot, H. and Favre, H. (1980) Foot and mouth disease virus production on microcarrier-grown cells. *Dev. Biol. Stand.*, **46**, 249–256.

30. Montagnon, B.J., Fanget, B. and Vincent-Falquet, J.C. (1984) Industrial-scale production of inactivated poliovirus vaccine prepared by culture of Vero cells on microcarrier. *Rev. Infect. Dis.*, **6** (Suppl. 2), S341–S344.

★★ 31. Looby, D. and Griffiths, J.B. (1988) *Cytotechnology* **1**, 339–346. – *An excellent review of the use of animal cells and appropriate culture systems.*

32. GE Healthcare, (2010) http://www.gelifesciences.com/aptrix/upp01077.nsf/Content/DiscontinuedProducts?opendocument&itemid=11002914&newitemid=&text=VSFCYTO PILOT MINI REACTOR. (Accessed November 2010).

33. Nie, Y., BergendahI, V., Hei, D.J. *et al.* (2009) Scalable culture and cryopreservation of human embryonic stem cells on microcarriers. *Biotechnol Prog.*, **25**, 20–31.

34. Giard, D.J., Thilly, W.G., Wang, D.I.C. and Levine, D.W. (1977) Virus production with a newly developed microcarrier system. *Appl. Env. Microbiol.*, **34**, 668–672.

★★★35. Cherry, R.S. and Papoutsakis, E.T. (1988) Physical mechanisms of cell damage in microcarrier cell culture bioreactors. Biotechnol. Bioeng., **32**, 1001–1014. – *A useful discussion of the mechanisms of cell damage during microcarrier culture.*

★★★36. Croughan, M.S., Hamel, J-F. and Wang, D.I.C. (1987) Hydrodynamic effects on animal cells grown in microcarrier cultures. Biotechnol. Bioeng., **29**, 130–141. – *A detailed explanation of the hydrodynamic effects exerted on animal cells grown in microcarrier cultures.*

37. Croughan, M.S., Hamel J-F. and Wang, D.I.C. (1988) Effects of microcarrier concentration in animal cell culture. *Biotechnol. Bioeng.*, **32**, 975–982.

38. van Wezel, A.L. (1985) Monolayer growth systems: homogeneous unit processes, in *Animal Cell Biotechnology*, vol. 1 (eds R. E. Spier and J. B. Griffiths), Academic Press, London, UK, pp. 265–282.

39. Levine, D.W., Wang, D.I.C. and Thilly, W.G. (1977) Optimizing parameters for growth of anchorage-dependent mammalian cells in microcarrier culture, in *Cell Culture and its Application* (eds R. T. Acton and J. D. Lynn), Academic Press, New York, NY, pp. 191–216.

40. Levine, D.W., Wang, D.I.C. and Thilly, W.G. (1979) Optimization of growth surface parameters in microcarrier cell culture. *Biotechnol. Bioeng.*, **21**, 821–845.

41. Bailey, J.E. and Ollis, D. (1986) *Biochemical Engineering Fundamentals*, 2nd edn, McGraw-Hill, Columbus, OH.

42. van't Riet, K. and Tramper, J. (1991) *Basic Bioreactor Design*, Marcel Dekker, Abingdon, UK.

43. Doran, P. (1995) *Bioprocess Engineering Principles*, Elsevier, Amsterdam, the Netherlands.

44. Asenjo, J. and Merchuk, J. (1994) *Bioreactor System Design*, Marcel Dekker, Abingdon, UK.

45. Stanbury, P.F., Whitaker, A. and Hall, S. (1999) *Principles of Fermentation Technology*, 2nd edn, Elsevier, London, UK.

46. Nielsen, J., Villadsen, J. and Liden, G. (2002) *Bioreaction Engineering Principles*, 2nd edn, Springer, New York, NY.

47. Hasler, L., Butz, R. and Graf, P. (2006) *Transparency, Form and Function. Fermenter Manufacturing – Art*, Three Point Publishing House, CH-8636 Wald, Switzerland.

★★ 48. Cino, J. and Frey, S. (1996) 20 Tips for purchasing research fermentors and bioreactors: a practical guide for researchers, Part 1. *BioPharm*, **9**, 52–56. – *Along with reference 49, a useful guide to purchasing bioreactors.*

49. Cino, J. and Frey, S. (1997) 20 tips for purchasing research fermentors and bioreactors: a practical guide for researchers, Part 2. *BioPharm*, **10**, 42–48.

50. Lim, A.C., Washbrook, J., Titchener-Hooker, N.J. and Farid, S.S. (2006) A computer-aided approach to compare the production economics of fed-batch and perfusion culture under uncertainty. *Biotechnol. Bioeng.*, **93**, 687–697.

51. Schoof, D.D., Gramolini, B.A., Davidson, D.L. *et al.* (1988) Adoptive immunotherapy of human cancer using low-dose recombinant interleukin 2 and lymphokine-activated killer cells. *Cancer Res.*, **48**, 5007–5010.

★ 52. Singh, V. (1999) Disposable bioreactor for cell culture using wave-induced agitation. *Cytotechnology*, **30**, 140–158. – *Description of the development of a disposable bioreactor for cell culture that employed wave-induced agitation.*

★★ 53. Fries, S., Glazomitsky, K., Woods, A. *et al.* (2005) Evaluation of disposable bioreactors: rapid production of recombinant proteins by several animal cell lines. *BioProcess Int*, October, 36–44. – *A review of suitable bioreactor types for animal cells.*

54. Namdev, P. and Lio, P. (2000) Assessing a disposable bioreactor for attachment-dependent cell cultures. *BioPharm*, **13**, 44–50.

55. Pierce, L.N. and Shabram, P.W. (2004) Scalability of a disposable bioreactor from 12–500 L. *BioProc. J.*, **3**, 1–6.

56. WAVE Bioreactor Systems. www.wavebiotech.com (Accessed September 2010).

57. Handa-Corrigan, A. (1990) Oxygenating animal cell cultures: the remaining problems, in *Animal Cell Biotechnology*, vol. 4 (eds R.E. Spier and J.B. Griffiths), Academic Press, pp. 123–132.

11

Good Laboratory Practice in the Cell Culture Laboratory

Barbara Orton

Quality Assurance Department, Bio-Products Laboratory, Elstree, Hertfordshire, UK

11.1 Introduction

11.1.1 What if ...

You joined a new cell culture laboratory recently and one morning the manager calls you into his office to tell you that a senior colleague has been suddenly taken seriously ill. As you are less busy with experimental work than other staff he asks you to take over. The missing scientist was working on some very important projects (though the manager is not clear precisely how many cultures might need to be checked and taken over) and you are told that if you just look in the laboratory book you will easily be able to carry on the cell work he was doing. The manager's confidence in you is encouraging so you hurry off to find out what needs to be done.

The book is quickly located (clearly well used on an almost daily basis for several months) and you turn to the last pages. However, you find some of the write-ups not very easy to follow (did he use personal abbreviations?) and several pages have additions squeezed into the margins, which appear to be further relevant details. It is not the way *you* would have recorded the experiments, but you manage to identify two different cultures in one of the incubators that are part of the missing colleague's work.

The first one is a standard culture you have worked with before so you are familiar with the usual medium. Just to be sure, you check the laboratory book and find the record of the most recent passage, which is marked with an asterisk for which you can find no

Animal Cell Culture: Essential Methods, First Edition. Edited by John M. Davis.
© 2011 John Wiley & Sons, Ltd. Published 2011 by John Wiley & Sons, Ltd.

explanation. The rest of the team are also mystified by this, but point you to the refrigerator your colleague used, where you find just one bottle of the medium you expect. You check the expiry date and prepare to passage the cells. Then one of the others remarks that the refrigerator where you found the medium was making a strange noise last week and he thought it might have been running rather warm. He advises against using the medium, so you start turning back the pages in the laboratory book to try to find the recipe. After some searching you find an undated record of preparing the medium, which followed the standard recipe, but from the adjacent pages the preparation was about 2 months ago, and the expiry date on the label does not match your expectation for storage of this medium. You wonder about this, but are able to follow the recipe. The sterilization method was not recorded, but you are aware that only 0.2-μm filtration can be used and so you prepare the medium and perform the passage, writing down in your own way what you have done.

The second culture is of cells completely unknown to you. The culture flask is labelled with a code and date, and referring to the laboratory book you find the entry on that date says only that the culture was passaged 'as previously'. You resign yourself to searching earlier pages to find more details, such as the split ratio and medium, but after reading entries for the previous month you still have not found the information. In consequence you do not feel sufficiently knowledgeable to work with the culture in question and you have to ask for assistance, explaining that your previous experience has not included this culture and without further written details you will need someone to explain what requires doing.

In a laboratory operating to formal Good Laboratory Practice (GLP) the above scenario would be rather different. First, the manager would know from the 'master schedule' exactly which studies the missing scientist was involved in as 'Study Director'. This document would also uniquely identify the studies so you would then refer to the 'study protocols' (or plans) for further information. A GLP study protocol must describe in detail all the required methods and materials for the experiments (possibly with reference to 'standard operating procedures'). Only individuals with appropriate recorded training should perform any activities, in order to ensure that work is done correctly. The records completed during the study must all be identified to the study and clearly state exactly what was done, the methods and materials used, all measurements and observations. All this paperwork must be stored safely and be available for future reference. These GLP study records would enable any individual reasonably familiar with the type of work to understand precisely how the experiment was performed and thus to repeat it using the same methods, materials and equipment, and reproduce the results.

11.1.2 Chapter aim

The aim of this chapter is to describe the historical background that led to the development of GLP regulations, in order to provide the context for the requirements, and indicate some practical methods of complying in a cell culture laboratory. Some differences between the US Food and Drug Administration (FDA) and UK requirements will also be discussed.

11.2 Background to GLP

In the 1970s investigation by the authorities at some laboratories performing non-clinical animal safety studies uncovered serious issues. These laboratories had been conducting experiments to provide data for companies hoping to manufacture new drugs and submitting applications for licenses to the FDA.

The findings included:

* poor procedures

* inadequate records of the experiments

* staff not qualified or sufficiently experienced to undertake the tasks they were performing

* fraudulent practices

* instances of fabricated data

* exclusion of non-favourable results

* significant inconsistencies between laboratory records and the reports submitted to the FDA

* reports of tests that had not been conducted.

As a result of the problems found at these laboratories the FDA published Part 58 of the Code of Federal Regulations (CFR) in 1978 [1] to regulate such safety studies. This has since revised in some sections.

Following the first US GLP regulations, the Organization for Economic Cooperation and Development (OECD) group of countries published an agreed set of GLP Principles (1981), which was later revised in 1997 and published in1998 [2]. Other countries such as Japan have also issued their own variations on GLP, although many have adopted versions very close to the OECD 1998 document. The European Directive [3], and consequently the UK GLP Regulations [4], are consistent with the OECD 1998 Principles.

Countries with GLP in place operate national programmes to inspect laboratories, often on 2-yearly cycles for routine monitoring. In the UK the GLP inspectorate, part of the Medicines and Healthcare products Regulatory Agency (MHRA), is responsible for performing such inspections and the Guide to the GLP Regulations [5] explains how the GLP Regulations are applied and gives information about the inspections performed to confirm compliance.

Because the OECD countries (including Europe, USA, Canada, Australia and Japan) all agreed to the 1998 GLP Principles they also agree to accept data from studies performed in other member countries under the Mutual Acceptance of Data (MAD) agreement. This is intended to reduce the need for repeat testing or inspections in foreign countries, while allowing the regulatory authorities to have confidence in foreign laboratories' results. To support the MAD scheme, joint visits are performed by the different national GLP inspectors in order to harmonize inspection procedures and standards. Membership of the OECD MAD group is now growing with the addition of new countries.

Most national regulations are based on the OECD GLP Principles [2], which set out the underlying principles. Thus laboratories can implement different systems to meet the requirements. For example, one important GLP principle is that all personnel should be adequately trained for the work they are expected to perform. This means a laboratory needs to develop a system to record training, in order to be able to demonstrate that personnel have been adequately trained. It might be that a laboratory wishes to perform and document training against individual procedures. Alternatively, training could be provided for activities that depend on a number of procedures. Significantly different training systems could be developed and still be compliant with the same GLP regulations, and it would be for the national GLP inspectors to assess each laboratory.

It is the regulatory authorities responsible for approving or licensing new drugs or chemicals who determine which experiments need to be performed, and sometimes the particular methods to be used. When they are satisfied that the work meets the GLP standards they can use the data to consider if the product should be approved or licensed.

11.3 General GLP principles

11.3.1 GLP objectives

The main objectives of GLP are to ensure the integrity of data, and that it is of adequate quality to permit the assessment by regulatory authorities of the hazards associated with chemicals, pharmaceuticals or other products. GLP is concerned with the organizational processes and systems supporting the laboratory work, and that the experiments can be reconstructed from archived materials.

The objectives of GLP should also be the objectives of all good laboratories, and therefore even when work is not intended for submission as safety data it would be well worth following the GLP principles.

11.3.2 Responsibilities

GLP assigns many responsibilities to various roles within the laboratory – management, Study Director, study personnel, Quality Assurance (QA) and archive staff. Figure 11.1 shows an organogram for a GLP laboratory.

11.3.2.1 Management

Management has overall responsibility for the laboratory and its compliance with Good Laboratory Practice, including:

- Establishing the major GLP systems such as:
  - training
  - control of procedures
  - test item identification and handling.

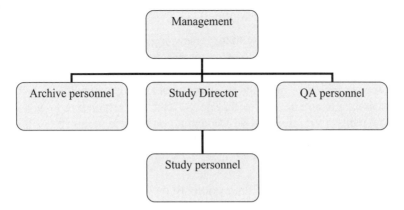

Figure 11.1 Staff organogram for a GLP laboratory. The responsibilities of each category of staff are described in the main text.

- Providing the necessary resources:
  - facilities (e.g. controlled temperature storage, archive)
  - equipment (adequately validated)
  - supplies
  - available staff.
- Supporting the QA unit (see Section 11.3.9):
  - ensuring that audits are appropriately closed after any necessary actions are completed.

In larger organizations many of these responsibilities may be delegated to different levels of management.

One management responsibility is to ensure a master schedule is maintained. This needs to include key information about all the work of the laboratory, indexed by study. It will therefore indicate for each study the unique study identification, Study Director, test item, test system, study type, sponsor and current status. One way to implement a master schedule is to use a simple database or spreadsheet, indicating the study status by the dates for the key stages of protocol authorization, experiment start and end, study report issue and transfer to archive. Additional details may also be useful to the laboratory, for example other staff involved, intermediate stages of lengthy studies or use of key equipment (which might preclude other studies starting), and interim report issue.

Management must formally appoint the other key staff in the laboratory, that is each Study Director, the archivist and QA personnel. All these roles must be independent of each other to ensure there is no conflict of interest, and everyone should understand their roles thoroughly so that there are no gaps in responsibilities.

11.3.2.2 Study Director and personnel

The Study Director is appointed by management to oversee the GLP study and ensure its compliance. This is a key role that requires both in-depth scientific knowledge of the work of the study, and also a good understanding of GLP. Specific responsibilities of the Study Director include issuing the study protocol and any amendments, as well as the final report, assessing any deviations that may occur, and responding to any findings in study-related audits, (see reference 6). Depending on the way a laboratory is operating, the Study Director may or may not be performing practical work, but he or she must be available to staff for any queries relating to the study, and at all times be aware of the study's progress.

Study personnel are responsible for working to the study protocol and any applicable standard operating procedures (SOPs), for making adequate records of all required data and for promptly reporting to the Study Director any problems that may occur.

11.3.2.3 Quality Assurance personnel

The QA staff must perform adequate monitoring to assure that the studies are in compliance with the GLP requirements (see Section 11.3.9). QA should report to both management and all Study Directors within the laboratory any findings from audits, in case there are similar situations in other studies. It is also one of QA's responsibilities to ensure that any adverse findings are addressed in a timely fashion by the appropriate personnel.

In order to be aware of the ongoing activities of the laboratory, QA will need access to the master schedule and all study plans and relevant SOPs. Guidance is given on QA responsibilities in an OECD consensus document entitled *Quality Assurance and GLP* [7].

11.3.2.4 Archive personnel

Archive staff, usually a main archivist and at least one deputy, are appointed by management and maintain the archive. It is their responsibility to keep the archived material indexed to allow ready retrieval, and ensure all movements into and out of the archive are properly recorded (see Section 11.3.8). Where data are archived electronically, the IT department may also be involved in ensuring that records remain available for the required retention period (see Section 11.3.8).

11.3.3 Training

All personnel need to have records to show they have appropriate training, qualifications and experience to undertake their roles within the laboratory. There will be some job-specific training, but also training in GLP for all staff so that they understand their responsibilities and the requirements of GLP. Each member of staff should have a job description in their training file, as well as records to show training in procedures or activities they perform in the laboratory. Both procedural and GLP training records will need to be updated periodically as methods may change, and it is also expected that staff will receive regular 'refresher' training in GLP.

11.3.4 Procedures

It is one of the basic principles of GLP that routine operations should be described in SOPs. Management is responsible for:

• ensuring that appropriate and technically valid SOPs are established and followed

• approving all SOPs and revisions.

Therefore, SOPs will be needed for many laboratory operations and methods, and it is logical first to document how the SOPs will be issued and controlled, and the responsibility for these tasks. Some kind of coding system will be needed to reference the SOPs. A laboratory may choose to code documents to differentiate between, for example, practical methods, equipment operation and administrative procedures (such as the preparation of study protocols) within the SOP reference number. Version numbers or dates must be indicated to keep track of changes – it is usual to use numbers and define an effective date to allow for a period of staff training before introducing the new procedure or change. The level of authorization needs to be decided and it could be that different managers sign the different types of procedures. There is commonly a QA signature on SOPs, although this is not a specified requirement of the GLP Regulations, which do however clearly state that SOPs must be current. The requirement for maintaining the currency of documents does mean a system for review is necessary; this may allow 're-authorization' if no changes are needed to the SOP. Any superseded versions of SOPs must be retained in the archive in order that they remain available for study reconstruction in the future.

Providing the SOPs are authorized, reviewed, readily available where needed and recalled when superseded, it is possible to either issue controlled paper copies or allow users to view and use the electronic versions. An electronic SOP system does have some advantages, and even with such a system printing by users can be permitted providing there is a mechanism to reduce the likelihood of keeping an old printed copy, for example including an automatic date of printing with the statement 'valid only on date of printing'.

11.3.5 Records

In a GLP laboratory raw data are defined as the original worksheets, notes, records in laboratory books, recorded data from automated instruments, and so on, and must all be retained. In certain circumstances, for example if an instrument printout is of a size or shape such that it does not allow easy archiving, verified copies may be substituted for the originals. In FDA-regulated work, electronic records may be used if the requirements of 21CFR Part 11 [8] are satisfied. Part 11 sets out the conditions for the FDA to consider electronic records to be as reliable as paper records, including system security and protection, secure electronic audit trails and electronic signatures. Guidance for application is given in Guidance for Industry Part 11 [9].

All raw data, whether specific to one study or generic (such as equipment maintenance records), should be attributable to a member of staff. Experimental data recorded manually must be clearly legible and recorded promptly and accurately, on a designated form or book, and the operator must sign, or initial, and date the entries. For data collected automatically, for example temperature monitoring, the person responsible for setting up the instrument should be known. Many automated instruments now include log-in systems to show such traceability, but otherwise the printout should be signed or initialled.

Anyone can accidentally write down an incorrect number, but if this does happen, the change must be done so that the previous value remains legible; the change must then be explained, initialled and dated. Similarly, in an electronic system there should be an audit trail of any changes with a record of the reason and the operator who modified the value.

A laboratory can choose whether to use notebooks or forms for records completed manually; both have advantages and could be used in different situations.

For routine records where certain details should always be included, it is often worth developing a standard form that can serve to prompt for all the required information. One example of such a form could be for reagent preparation, to ensure that

- the quantities, expiry date and batch numbers of all component chemicals,

- the expiry date and storage conditions of the reagent once prepared,

- the equipment number of any balances used, and

- the values of any final pH and conductivity measurements

are all recorded.

During a study a specially designed form may also be used, in order to include instructions for particular methods as well the records of results, and additional space for any other observations made during the course of the work.

On the other hand, it may be more appropriate to use a bound logbook for equipment maintenance where, for example, the date and service details are recorded. Such a log would then remain usable for several years and the equipment history could be easily followed.

11.3.6 Test item and test system

The test item in a GLP study is the substance whose safety is under investigation. There need to be procedures in place to ensure that its integrity is maintained (and hence that the study results are valid), for example by storing at an appropriate temperature, and ensuring that the containers are clearly labelled and handled to avoid cross-contamination. In order to be able to show the test item is properly accounted for, records of the quantities received into the laboratory and subsequently used need to be maintained. It should be possible to reconcile the test item quantities both during and on completion of the experiment. A simple form to record the amount logged in, and subsequent usage would be sufficient.

The test system of a study is the biological (or sometimes physical, chemical or combination) system used in a study to investigate the safety of the test item. Biological systems should be assessed on receipt into the laboratory before use in the study, and proper procedures for handling established. All test systems must be handled appropriately to ensure that the validity of study results is maintained.

11.3.7 Facilities and equipment

The management of the laboratory needs to ensure that appropriate facilities and equipment are available for the studies to be performed. This will include adequate laboratory areas, designed for handling hazardous biological agents if applicable, with sufficient storage to segregate materials as necessary, and to maintain any reagents and test items that need to be frozen or refrigerated. Equipment of adequate design and capacity for the work must be provided. For automated equipment, if the work is not regulated by the FDA, the amplification of the GLP Principles given in the OECD consensus document for computerized systems [10] should be considered.

Suitable autoclaves (see Chapter 2, Section 2.2.1.1) should be installed and demonstrated to be able to sterilize the various types of loads. Each cycle, for sterilizing or treating waste, needs to be tested so that the autoclave can be shown to be 'fit for purpose'. Once installed a schedule of routine maintenance and appropriate calibration should be established, for example 6-monthly service by a contractor, to include calibration of the gauges, in addition to weekly leak rate tests and daily warm-up/test cycles. For normal use operational instructions need to be documented in a SOP, either in detail or by reference to a manufacturer's manual. All maintenance and operational records must be filed, and eventually archived.

Where safety cabinets are used (see Chapter 1, Section 1.2.3.1, and Chapter 2, Section 2.2.5.5) these need to be shown to perform adequately. Tests would include regular, perhaps monthly, measurement of the airflow in different parts of the cabinet and records that these meet the defined requirements. The frequency and complete instructions for the airflow tests, together with all other necessary checks, would be described in one or more SOPs for the use and maintenance of cabinets. These procedures would need to describe also the appropriate cleaning and decontamination measures necessary to ensure there is no cross-contamination between cell cultures. The required maintenance, often by external engineers, should also be documented and records of all these services kept.

The period between any routine calibration or verification of equipment must be decided by management, but must be documented in the relevant SOP. Although all personnel are responsible for confirming before use that equipment has been checked, it is usual to visibly label the items with the most recent and next due calibration date. If an item of equipment, such as a balance, has a greater capacity than is routinely used, then calibration may be performed over the normal range, and then the item labelled to state the calibration range, with an instruction not to use outside this range. This can be justified providing the label is prominent so that accidental use outside the calibrated range is unlikely.

11.3.8 Archives

One of the fundamental requirements of GLP is that all the records needed in order to be able to 'reconstruct' a study must be kept for the period defined by the appropriate regulatory authority. For data that will be submitted to the FDA the 21 CFR Part 58 [1] states that records must be retained for the shortest of the following periods:

- 2 years following approval of a submission by the FDA (except for studies supporting Investigational New Drug (IND) or Investigational Device Exemption (IDE) applications)

- 5 years following submission of the data in support of an IND or IDE application

- otherwise, 2 years after completion of the study (e.g. if the data is not submitted to the FDA).

However, in the UK, the MHRA has not clearly stated the required period for GLP safety data for pharmaceutical products. Many laboratories use a period of 10 years from completion of the study. Other countries may require retention for different periods, therefore the relevant regulatory authority should be asked to confirm the necessary period.

The materials which need to be archived are all those which would be necessary to show in detail how the study was performed, that is all the study-specific records such as the study protocol, all raw data and the final study report, but also the supporting documentation. This includes records to show that all personnel had been appropriately trained and all equipment adequately maintained and calibrated at the time of the work. In addition, the SOPs that were effective at that time must be available; hence all superseded versions must be kept. The records of audits are specifically mentioned in the OECD Principles and the UK GLP Regulations among the items that need to be archived.

In the US, FDA GLP states that a sample of each batch of test item shall be retained for studies of longer than 4 weeks' duration. The OECD Principles require retention of test item samples unless the study is 'short-term', where short-term studies are defined as being of short duration and utilizing widely used routine techniques. It would be sensible, if unsure whether it is required to retain the test item, to check with the appropriate regulatory authority.

The GLP requirements for the operation of the archive are intended to ensure the materials are under the control of archive staff and a complete chain of custody is maintained. Thus at the end of the study (or possibly at suitable points within a longer experimental period), the Study Director is responsible for ensuring that all the study materials are transferred to the archive. There needs to be a detailed record of what is handed over to the care of the archive staff, and both parties should sign this. Any non-study-specific records, such as equipment maintenance, controlled temperature storage or QA audit files, also need to be archived at appropriate intervals, perhaps annually.

Once documents are in the custody of archive staff, access to them should be carefully controlled. Normally no other personnel would be allowed into an archive unaccompanied, and if materials are removed for any reason this must be recorded; they should be returned within a reasonable period, and thoroughly checked by archive staff on return.

The required records may be retained electronically instead of as hard copy if desired, providing that similar controls are in place, and measures taken to ensure the records remain readable for the required period. This may mean transferring files to a different medium, with appropriate checks on the transfer process. Such an archive may facilitate read access, for example if scanned images are stored on a network drive; however, adequate procedures for back-up and restoration will be needed.

The archive facility needs to have suitable precautions in place to protect the contents from deterioration, for example by flood, fire or pests. Although strict environmental control measures are not specifically required for GLP archives in the UK, suitable conditions are documented in the standard BS5454:2000 Recommendations for the Storage and Exhibition of Archival Documents [11]. Guidance on interpretation of GLP archiving requirements is available from the MHRA [12] and the OECD [13].

11.3.9 QA and audits

The GLP Regulations require an audit programme to be operated within the laboratory. The audits must be performed by staff independent of the area or study being audited, to ensure they are unbiased. The objective of such audits is to confirm that the work complies with GLP standards; however, they can also lead to quality improvements and better systems.

The main difference in GLP audits between the FDA and UK regulations is that 21 CFR Part 58 states QA need to 'inspect each non-clinical study', that is audit during the experimental stage. There can be some short GLP studies where this is not very practical and therefore the wording of the UK regulations (following OECD) allows any combination of facilities, process and study audits to be used, providing the audit programme is 'sufficient to ensure the compliance of all studies'.

Facilities audits would be those of a laboratory's systems that support the studies, for example archives, SOP control, training, equipment calibration and maintenance. Process audits could be performed on routine laboratory operations that are not study-specific, although they may not be applicable in some facilities. However, 'process' audits could also be used if many similar short studies are conducted, auditing one example study periodically and thereby using the audit as applicable to similar studies performed within, say, 3 months. It is the expectation of the UK GLP inspectors that facilities audits are performed once and process audits four times per year.

Under OECD GLP, all study protocols must be reviewed by QA to ensure they are GLP compliant, and for both the FDA and MHRA all final study reports must be audited (see Section 11.4.3). All the activities of QA personnel need to be described in SOPs, and there should also be a mechanism for independent confirmation that QA is performing audits appropriately (see reference 14).

11.4 Study performance

The general flow of a regulatory GLP study is illustrated in Figure 11.2.

11.4.1 Planning and initiation

Once the need for a study is identified, management should appoint a Study Director and ensure that any necessary resources are made available. The Study Director should check

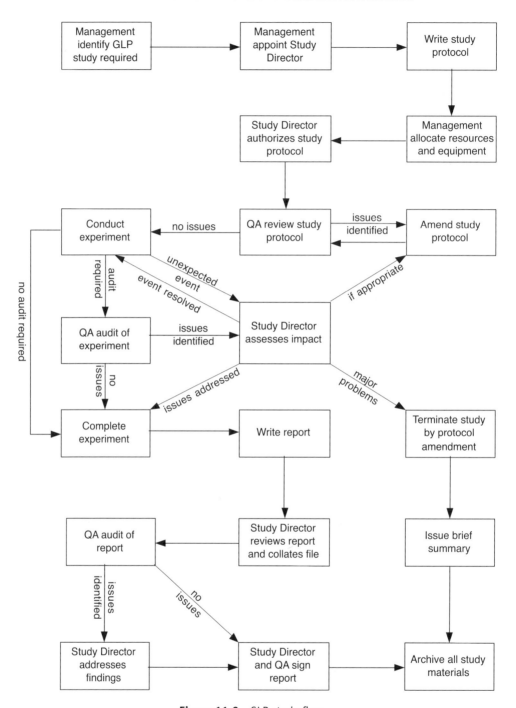

Figure 11.2 GLP study flow.

that he/she understands the objective of the study, in order to design the experimental work appropriately.

It is the Study Director's decision how to define the study. A study can be short or long, completed in a few days or in progress for a year, but whatever the protocol defines as the study must then be written up in one study report. There can be situations when smaller or larger studies are more appropriate.

Both scientific and GLP aspects of the study should be considered in the preparation of the study protocol. Before the protocol is authorized the Study Director should ensure that the appropriate equipment, supplies and staff will be available at the required times to achieve the study objective and complete the work in a timely fashion. The Study Director should also inform QA of the intended study. This may be achieved by circulating a draft protocol, and in any case it is a GLP requirement in the UK that QA verify that the study protocol is GLP-compliant. It is not necessary that QA sign the protocol, although this would be one means to show that the QA review had been done. The only authorizing signature required in the UK is that of the Study Director, but in the USA the FDA GLP regulations do require the date the sponsor approved the protocol. However, many laboratories do include a management signature on the protocol, and this would show that management were aware of the study and therefore agreed to provide the necessary resources.

The protocol needs to include the study objective, identification of the test and any reference item, the sponsor and test facility, Study Director, proposed experimental start and completion dates and details of the methods, including data to be recorded and any statistical techniques. The instructions in the study protocol, together with any referenced SOPs, must comprise sufficient information for staff to carry out the experimental work. Once signed and dated by the Study Director, this becomes the formal plan and should be distributed to all appropriate personnel.

11.4.2 Study conduct

During the experimental stage of a study not everything may go as expected. Whatever problems do occur, personnel should record what actually happened and inform the Study Director promptly. The Study Director must then decide whether the unexpected event has any impact, which may require a 'note to file' to record this judgement. Annotations can be made directly on the study records where the incident was noted, but it can be clearer to separate such comments and explanations using notes to file. Providing that the Study Director has commented appropriately, the location is a matter of choice.

It may be that a future stage of the study needs to be changed, in which case the Study Director should prepare a protocol amendment. This amendment needs to describe and justify the proposed change, and be authorized by dated signature of the Study Director.

If not all the intended experiment can be performed, then this should also be documented and explained by the Study Director. On occasion it may be necessary to terminate the study before completion, which would necessitate a short report and the retention of all records as normal.

11.4.3 Study report

When the experimental work has been completed, then a final study report needs to be prepared. It is usually easiest if those who performed the work write up their activities, although the Study Director remains responsible for the report. Once he or she has collated all the files, and reviewed and revised any draft report prepared by others, the report must be submitted for QA audit. This is an independent check that the report accurately reflects the study's raw data and the GLP compliance of the work. However, it is not for QA to perform a complete verification check of all results in the report; it is the Study Director who signs a statement accepting responsibility for the report. If any issues are found with the report, then once they have been satisfactorily addressed by the Study Director, QA will prepare a statement for the report. This lists all relevant audits with their dates and the dates of reporting the audit findings.

The final study report will describe the methods, results and conclusions of the work. In addition, the GLP Regulations require the report to include information on the location of the archived material, the dates of study initiation (the date the Study Director signs the study protocol), study experiment start and finish, and the address of the test facility. The Study Director must also sign a formal GLP compliance statement for the study, and if there were major deviations from the GLP principles this statement would need to describe these.

11.5 Good Manufacturing Practice

Good Manufacturing Practice (GMP) is another quality standard required by the FDA and MHRA, for the commercial-scale manufacture of medicines. The purpose of GMP is to ensure that the medicine is safe, and there are documented GMP regulations in the USA [15], and UK [16]. The fundamental requirements of GMP and GLP are similar in such areas as facilities, equipment, records, trained personnel and internal audits, but the specifics are tailored to either manufacturing or laboratory work respectively.

Although some laboratories may perform tests to GMP, this may be a matter of convenience due to other work on the site. Each standard is appropriate and mandatory in different situations. For work that constitutes a laboratory safety study as defined in the relevant GLP regulations [1, 4], the regulatory authorities may not accept data that were not generated and reported in compliance with GLP.

11.6 Summary

Formal GLP is a quality system for laboratories performing safety studies. The underlying purposes of Good Laboratory Practice are to ensure the integrity and validity of the data generated and to allow reconstruction of the study from the archived materials. Where the results are submitted to regulatory authorities they are used to evaluate the risks associated with the test items and form part of the basis on which the authorities may authorize the licensing of new products.

There are inevitably additional costs entailed in meeting all the requirements of the GLP Regulations, and unless a laboratory is required to fully comply with these regulations because regulatory safety studies are performed, the management of a cell culture laboratory may decide that it is not necessary to set up all the systems detailed above. However, the main principles of GLP can be usefully applied in all cell culture laboratories to ensure the experimental data is reliable and study reports accurate.

References

★★★ 1. FDA (1978) and revisions. *Good Laboratory Practice for Nonclinical Laboratory Studies*, US Food and Drug Administration, Silver Spring, Code of Federal Regulations Title 21, Part 58 – *The legal requirements for studies performed in the USA*.

2. OECD (1998) *OECD Principles of Good Laboratory Practice (as revised in 1997)*, OECD, Paris, OECD Series on Principles of GLP and Compliance Monitoring Number 1.

3. Official Journal of the European Union (2004), *Directive 2004/10/EC of the European Parliament and of the Council on the Harmonisation of Laws, Regulations and Administrative Provisions Relating to the Application of the Principles of Good Laboratory Practice and the Verification of Their Applications for Tests on Chemical Substances*, Office for Official Publications of the European Communities, Luxembourg.

★★★ 4. Department of Health, (1999) *The Good Laboratory Practice Regulations 1999*, HMSO, London, Statutory Instrument 1999 No. 3106 – *The standard against which the UK inspectors will audit*.

★★ 5. The UK Good Laboratory Practice Monitoring Authority (2000) *Guide to UK GLP Regulations 1999*, MHRA, London, UK. – *A useful guidance document for the UK*.

★★ 6. OECD (1999) *The Role and Responsibilities of the Study Director in GLP Studies*, OECD, Paris, OECD Series on Principles of GLP and Compliance Monitoring Number 8 – *A helpful description of this crucial role*.

★★ 7. OECD (1999) *Quality Assurance and GLP*, OECD, Paris, OECD Series on Principles of GLP and Compliance Monitoring Number 4 – *Valuable guidance for QA personnel*.

★★ 8. FDA (1997) *Electronic Records; Electronic Signatures; Final Rule*, US Food and Drug Administration, Silver Spring, Code of Federal Regulations Title 21 Part 11 – *Requirements for the use of electronic records in the USA*.

★★ 9. FDA (2003) *Guidance for Industry Part 11, Electronic Records; Electronic Signatures – Scope and Application*, US Food and Drug Administration, Silver Spring – *Further guidance on the interpretation of Part 11 by FDA inspectors*.

10. OECD (1995) *The Application of the Principles of GLP to Computerised Systems*, OECD, Paris, OECD Series on Principles of Good Laboratory Practice and Compliance Monitoring Number 10.

11. BSI (2000) *BS5454:2000 Recommendations for the Storage and Exhibition of Archival Documents*, BSI, London.

12. MHRA (2006) *Good Laboratory Practice Guidance on Archiving*, MHRA, London.

13. OECD (2007) *Establishment and Control of Archives that Operate in Compliance with the Principles of GLP*, OECD, Paris, OECD Series on Principles of Good Laboratory Practice and Compliance Monitoring Number 15.

14. MHRA (2007) *GLPMA Expectations for Audit of the Quality Assurance Programme*, MHRA, London.

15. FDA (1978) and revisions. *Current Good Manufacturing Practice for Finished Pharmaceuticals*, US Food and Drug Administration, Silver Spring, Code of Federal Regulations Title 21, Part 211.

16. MHRA (2007) *Rules and Guidance for Pharmaceutical Manufacturers and Distributors 2007*, Pharmaceutical Press, London.

Websites

MHRA inspections: http://www.mhra.gov.uk/Howweregulate/Medicines/Inspectionandstandards/index.

FDA CFR search site: http://www.accessdata.fda.gov/scripts/cdrh/cfdocs/cfcfr/cfrsearch.cfm.

OECD: http://www.oecd.org/ehs.

Index

The following page number fomat key is used: *italic* refers to figures, **bold** refers to tables, underlined refers to protocols.

Animal Cell Culture: Essential Methods, First Edition. Edited by John M. Davis.
© 2011 John Wiley & Sons, Ltd. Published 2011 by John Wiley & Sons, Ltd.

344 INDEX